D1796916

Drip Irrigation for Agriculture

Initially associated with hi-tech irrigated agriculture, drip irrigation is now being used by a much wider range of farmers in emerging and developing countries. This book documents the enthusiasm, spread and use of drip irrigation systems by smallholders but also some disappointments and disillusion faced in the global South. It explores and explains under which conditions it works, for whom and with what effects. The book deals with drip irrigation 'behind the scenes', showcasing what largely remain 'untold stories'.

Most research on drip irrigation use plot-level studies to demonstrate the technology's ability to save water or improve efficiencies and use a narrow and rather prescriptive engineering or economic language. They tend to be grounded in a firm belief in the technology and focus on the identification of ways to improve or better realize its potential. The technology also figures prominently in poverty alleviation or agricultural modernization narratives, figuring as a tool to help smallholders become more innovative, entrepreneurial and business minded. Instead of focusing on its potential, this book looks at drip irrigation-in-use, making sense of what it does from the perspectives of the farmers who use it, and of the development workers and agencies, policymakers, private companies, local craftsmen, engineers, extension agents or researchers who engage with it for a diversity of reasons and to realize a multiplicity of objectives. While anchored in a sound engineering understanding of the design and operating principles of the technology, the book extends the analysis beyond engineering and hydraulics to understand drip irrigation as a sociotechnical phenomenon that not only changes the way water is supplied to crops but also transforms agricultural farming systems and even how society is organized. The book provides field evidence from a diversity of interdisciplinary case studies in sub-Saharan Africa, the Mediterranean, Latin America and South Asia, thus revealing some of the untold stories of drip irrigation.

Jean-Philippe Venot is a researcher at the French National Research Institute for Sustainable Development (IRD/UMR G-EAU) and is affiliated to the Water Resources Management group of Wageningen University, the Netherlands. He is currently based at the Royal University of Agriculture in Phnom Penh, Cambodia.

Marcel Kuper is a senior irrigation scientist at the International Agricultural Centre for Research and Development (CIRAD/UMR G-EAU), France, and a visiting professor at the Agricultural and Veterinary Sciences Institute Hassan II in Rabat, Morocco.

Margreet Zwarteveen is professor of Water Governance at IHE Delft Institute for Water Education, and at the Governance and Inclusive Development Group of the University of Amsterdam, both in the Netherlands.

Earthscan Studies in Water Resource Management

For more information and to view forthcoming titles in this series, please visit the Routledge website: www.routledge.com/books/series/ECWRM/

Drip Irrigation for Agriculture

Untold Stories of Efficiency, Innovation and Development

Edited by Jean-Philippe Venot,
Marcel Kuper and Margreet Zwarteveen

LONDON AND NEW YORK

from Routledge

First published 2017
by Routledge
2 Park Square, Milton Park, Abingdon, Oxon OX14 4RN

and by Routledge
711 Third Avenue, New York, NY 10017

Routledge is an imprint of the Taylor & Francis Group, an informa business

British Library Cataloguing-in-Publication Data
A catalogue record for this book is available from the British Library

Library of Congress Cataloguing-in-Publication Data
A catalog record for this book has been requested

ISBN: 978-1-138-68707-3 (hbk)
ISBN: 978-1-315-53714-6 (ebk)

Typeset in Bembo
by Apex CoVantage, LLC

MIX
Paper from
responsible sources
FSC
www.fsc.org   FSC® C013056

Printed and bound in Great Britain by
TJ International Ltd, Padstow, Cornwall

Contents

Modernization and agrarian change

Poverty and development

Figures

Plates

Tables

Contributors

Fatah Ameur is a PhD student at the Agronomy and Veterinary Sciences Institute Hassan II in Rabat (Morocco) and at AgroParisTech (France). He is an agricultural engineer and he has obtained a master's degree in irrigation and water control. Currently, he conducts research about agrarian change in relation with the access to groundwater in North Africa.

Maya Benouniche has a PhD in Agronomy from the *Institut Agronomique et Vétérinaire Hassan 2 (IAV)*, Rabat, Morocco and the University of Montpellier 2, France. In her PhD she shows how drip irrigation can be understood as a sociotechnical innovation whereby different actors engage with the technology so as to give sense and meaning to it. Following her PhD, she conducted a post-doctoral at IAV and the University of Wageningen to study drip irrigation dynamics in the Saïss Region of Morocco.

Harm Boesveld is a Lecturer and Researcher for more than 15 years at the Water Resources Management Group of Wageningen University, the Netherlands. He is trained in tropical land and water management and worked for 10 years as agricultural consultant in fruit production prior to his work at Wageningen University. His work at Wageningen University focuses around irrigation methods with specialisation in sprinkler and drip irrigation, (regulated) deficit irrigation, crop water requirements, irrigation scheduling and low-cost water technologies for smallholder farmers.

Vera Borsboom concluded her thesis on the discourse of low-cost water technologies in Zambia as part of her MSc programme in International Development Studies at Wageningen University in 2012. She currently works as Programme Manager Social Entrepreneurs Exchange & Development at the Euclid Network.

Lisa Bossenbroek obtained her PhD in rural sociology at the University of Wageningen (the Netherlands). As part of her research she studied the role of young people in agrarian dynamics and how agricultural labor relations alter along gender lines. Currently, she works as a post-doc at the École de Gouvernance et d'Economie in Rabat, Morocco. Her research focuses on the interaction of agrarian change and gender relations and subjectivities in Morocco and India.

Sami Bouarfa is a Senior Irrigation and Drainage Scientist at IRSTEA, the French Research Institute of Science and Technology for Environment and Agriculture. He has 20 years of experience in irrigation and drainage issues in France, North-Africa, China and Uzbekistan. His research focuses on the environmental impacts of irrigation. He is presently the President of the Technical Committee of AFEID, the French National Committee of the International Commission on Irrigation and Drainage (ICID). He is also Deputy Head of the 'Water Department' at IRSTEA.

Eduardo Chía is a researcher at the French National Institute of Agronomic Research (INRA) in the Department of Science for Action and Development. He holds a PhD and a habilitation in Economics and Management Sciences. His research focuses on rural development and the adaptation of farming and aquaculture enterprises to the new production requirements - Sustainable Development, decentralization, climate change. He uses action-research methods to conduct research and produce actionable knowledge and support the actors in the production of sociotechnical and organizational innovations. He has extensive experience as a researcher and expert in international organizations such as FAO, IUCN, ECLAC, on rural development issues, governance and territorial innovations in Europe, Africa, the Caribbean and Latin America.

Charlotte de Fraiture is Professor of Land and Water Development at UNESCO-IHE, Centre for Water Education based in Delft, the Netherlands. Her research interests include sustainable use of water for agriculture, irrigation and ecosystem services, and water for food security.

Mostafa Errahj is Assistant Professor and Researcher at the Development Engineering Department of the National College of Agriculture of Meknes (Morocco). He holds an MSc Degree in Extension Sciences for Agronomy and undertakes a PhD Thesis with the University of Gent (Belgium). His research focuses on differentiation processes at play within family farms; collective action and coordination, and processes toward the autonomy of farmer organizations.

Manuel Escobar is a Sociologist with a MSc in Local Development. He is a Consultant at the Institute of Public Policy at the Universidad Católica del Norte, Coquimbo. Chile.

Jean-Marc Faurès is Senior Officer at the UN Food and Agriculture Organization (FAO) and contributes to the organisation's strategic programme on sustainable agriculture since its creation in 2014. Prior to that he worked for more than 20 years as a water resources management expert with FAO, and was involved in multiple projects dealing with water policy, irrigation management and modernisation, pro-poor water technologies, as well as global perspective studies on water and agriculture.

Marta García-Mollá is an Associate Professor at the Department of Economy and Social Sciences of the Universitat Politècnica de València and member of theValencian Center for Irrigation Studies (CVER). She has worked on research projects in the area of water policy and planning, related to the economics of water in agriculture and the water costs in Mediterranean irrigation systems.

Ali Hammani is a Professor and Head of the Department of Water, Environment and Infrastructure of the Agronomy and Veterinary Sciences Institute Hassan II in Rabat (Morocco). He has coordinated several research project related to irrigation and water management. He has supervised more than 80 MSc theses and he is supervising and co-supervising 5 PhD Theses. He has about 60 publications in peer-reviewed journals and congresses with reviewed proceedings.

Tarik Hartani is a Senior Water Scientist with 20 years of professional experience in the Mediterranean region. He has obtained a PhD in agricultural drainage at the Ecole Centrale of Paris (France). He is presently a Professor at the Tipaza University Center, an Associate Senior Researcher at the National School of Agronomy (Algeria) and, a member of the editorial board of the international journal 'Cahiers Agricultures' (France). He supervises, and co-supervises several PhDs in the field of irrigation management, and has been the local coordinator of several international research projects.

Paul Hebinck is a Rural Development Sociologist specialized in agrarian transformation processes in Africa with an emphasis on land reform, small-scale farming and rural livelihoods. He carries out various longitudinal research projects in West Kenya, Zimbabwe, South Africa and Namibia on the social-material dimensions of livelihoods.

Daniela Henriquez Encamilla is a Sociologist with a diploma in Social Research. She is presently a Researcher at the Institute of Public Policy at the Universidad Católica del Norte, Coquimbo, Chile.

Jaime Hoogesteger is Assistant Professor at the Water Resources Management Group, Wageningen University, the Netherlands. He has done extensive research in Ecuador and Mexico on issues related to water governance, participation and social movements from a political ecology perspective. At present his research focuses on understanding the relations between water governance, agro-export chains and rural livelihoods in Guanajuato, Mexico.

Famke Ingen-Housz concluded her thesis on gendered patterns of treadle pump use in Zambia as part of her MSc programme in International Land and Water Management at Wageningen University in 2009. Currently, she is a Policy Advisor at the Waterboard 'Hoogheemraadschap van Delfland', the Netherlands.

Andrew A. Keller is President of Keller-Bliesner Engineering and an Adjunct Professor at Utah State University. A licensed civil engineer with a PhD in Irrigation Engineering, Andy specializes in agricultural development, water resources planning, and the design and management of irrigation. Andy has extensive US and international experience working with farmers, irrigation

districts, and government and nongovernment organizations. He is passion-
ate about wealth enabling opportunities for small-scale farmers worldwide.

Dinesh Kumar, PhD, has more than 20 years of professional experience in
the water and agriculture sector, undertaking research, consulting, action
research, and training. He is also an Associate Editor for *Water Policy* and an
Editorial Board Member for *International Journal of Water Resources Development.* He is now Executive Director, Institute for Resource Analysis and
Policy, Hyderabad, India.

Marcel Kuper is a Senior Irrigation Scientist with more than 25 years of professional experience in Asia, Africa and the Mediterranean region. He has
obtained a PhD in water resources management at the University of Wageningen (the Netherlands), and a habilitation to supervise PhD students at
the University of Montpellier II (France). He is presently a senior researcher
at CIRAD, and a Visiting Professor at the Agricultural and Veterinary Institute Hassan II in Rabat (Morocco). He supervises, and co-supervises, several
PhDs in the field of irrigation management, and has been a project leader of
several international research projects.

Khalil Laib is a PhD student at the National High School of Agronomy of
Algiers, Algeria. He is an assistant lecturer at the University Center of Tipaza,
Algeria. He has more than eight years experience with institutional and
engineering projects in the field of agriculture. His work focuses on innova-
tion and adaptive practices of using irrigation water use in the arid areas of
North Africa.

Bruce Lankford is Professor of Water and Irrigation Policy with more than
30 years experience in agriculture, irrigation and water resources management, and has recently been a co-Director of the University of East Anglia's
Water Security Research Centre. His research covers irrigation infrastructure
and management in sub-Saharan Africa; the use of serious games in natural
resource management; resource use efficiency and the paracommons; food
and water security metrics; river basin management; inter-sectoral water
allocation; water and ecosystem services, and conceptualizing and calculating the water-energy-food nexus.

Caroline Lejars holds a PhD in Economics and Management Science from
AgroParisTech (Paris Institute of Technology for life, food and environmen-
tal science). She has professional experience in French overseas territories
(Reunion Island, New Caledonia), in Southern Africa and in North Africa.
Her field of expertise encompasses analysis of agricultural and agri-food
sectors, supply and value chain management, and decision support systems.
She is presently Senior Researcher at CIRAD-(UMR G-EAU) and Associ-
ate Professor at the Agricultural and Veterinary Institute Hassan II in Rabat
(Morocco). Her most recent work analyzes the vulnerability and adaptive
capacity of North Africa's agricultural groundwater economies.

Janwillem Liebrand works as a Researcher at the Water Resources Management group of Wageningen University, the Netherlands. He is also affiliated with the Department of Human Geography and Planning, International Development Studies, Utrecht University and a guest lecturer at the UNESCO-IHE Institute for Water Education, Delft. Recently, he completed a PhD thesis on masculinities among irrigation engineers and water professionals in Nepal. He currently works on a critical analysis of farmer-led irrigation development in Mozambique, investigating its potential and processes of formalization. He is currently also involved in the Hindu Kush Himalayan Monitoring and Assessment Programme (HIMAP).

Anaïs Marshall is a Geographer. She defended her doctoral thesis in 2009 at the University of Paris 1 Panthéon Sorbonne. Her doctoral work focused on the social and environmental impacts of the development of agro-industry on the coastal desert of Peru. She then expanded her research areas and works currently on local impacts and rural dynamics engendered by the process of globalization in several Peruvian and Argentine oasis (mainly Mendoza). Since 2011, she is a lecturer at the University of Paris 13 North where she teaches geography to Bachelor and Master students.

Doug Merrey holds a PhD in social anthropology. He has spent his career working within interdisciplinary teams carrying out applied research on water management for agriculture, and water resources management, in developing countries. He has lived and worked for about 36 years in India, Pakistan, Sri Lanka, Egypt and South Africa. He spent over 20 years at IWMI and held various positions including Deputy Director General for Programs, and was the founding Director for Africa. He has numerous publications. Currently residing in Florida, US, Doug now works as an independent consultant.

Bruno Molle is a Senior Researcher from IRSTEA, Montpellier, France, managing the PReSTI, a research and technology platform that deals with irrigation technology and practices to make a better use of irrigation water. His recent work focuses mainly on issues related with the performance of irrigation (equipment and practices) when reusing waste water.

Peter Mollinga is Professor of Development Studies at SOAS University of London, UK. His research focuses on the relationship between water and development, with a geographical focus in South Asia and Central Asia. His research is informed by and contributes to debates on the cultural political economy of water, comparative water politics, boundary work and interdisciplinarity. He supervises several PhDs in related fields.

Mohamed Naouri is a PhD candidate with particular interests in water management and innovation systems. Prior to enrolling in his thesis, he worked as Irrigation Engineer then Project Manager at the Algerian Company of Rural Engineering. He holds a master's degree in irrigation and water control from the Agricultural and Veterinary Institute Hassan II in Rabat (Morocco) and a magister degree in Innovative Irrigation Systems from the National High School of Agronomy of Algiers (Algeria).

Mar Ortega–Reig is a Researcher at the Institute of Water and Environmental Engineering (IIAMA) of the Universitat Politècnica de València. She has worked in the field of agricultural water management, focusing on the efficiency of water use in agriculture, water governance, drought adaptation and collective irrigation management in the Valencia Region.

Jean Payen has more than 35 years experience on irrigation and sustainable natural resources management. During his carrier, he worked for UN organizations and multilateral development banks. He is now a senior independent consultant and has notably been involved in six micro-irrigation promotion projects over the last 10 years in Africa, Asia and Central America.

Nynke Post Uiterweer is a sociotechnical irrigation engineer with a keen interest in the human aspects of irrigation practices and the use and development of technologies. Her work has focused on Latin America, Southern Africa and Spain. Currently she works as policy adviser education at Wageningen University and Research.

Brent Rowell is currently an Extension Specialist for international and sustainable agriculture in the Department of Horticulture at the University of Kentucky. He received his PhD degree in Horticulture from Cornell University in 1984. In addition to 12 years as Extension Vegetable Specialist and Professor at the University of Kentucky, he lived and worked in developing countries for 15 years promoting small farm vegetable production through extension and research in Bangladesh, Cambodia, Thailand, and Myanmar. He recently returned from a six-year assignment in Myanmar to introduce low-cost drip irrigation as a means of raising small farm incomes.

Carles Sanchis–Ibor is a Researcher at the Valencian Center for Irrigation Studies (CVER) of the Universitat Politècnica de València and Lecturer at the Department of Geography of the Universitat de Valencia. He has worked on the irrigation systems of the Valencia Region for 20 years, analyzing the institutions, policies, water management and the historical evolution of these socio-ecological systems.

Henk van den Belt is Assistant Professor in Philosophy at Wageningen University, the Netherlands. He is interested in the philosophy of science, technology and innovation, with a special focus on modern life sciences like biotechnology and synthetic biology.

Saskia van der Kooij holds a PhD degree in Water Management from the Wageningen University in the Netherlands. Fascinated by the interactions between society, technology and the environment, her interdisciplinary research centres around the introduction of drip irrigation in farmer managed irrigation systems. She has teaching experience in courses on irrigation and water management and she has working experience in the Netherlands, Morocco, Spain and Benin.

Jan Douwe van der Ploeg held the Chair of Transition Processes at Wageningen University until 2016. Prior to that, he held the chair for Rural Sociology. Jan Douwe is involved in several research initiatives as well as public and policy debates on the future of farming in Europe and beyond as illustrated by his recent book on the *New Peasantries*.

Claudio Vasquez is Engineer, Corporative Manager in CEAZA, Center for advanced studies in arid areas, Coquimbo Region, Chile.

Gert Jan Veldwisch works as an Assistant Professor in the field of Irrigation and Development. He is based at the Water Resources Management Group of Wageningen University. His research revolves around the organisation of (irrigated) agricultural production processes with a particular interest in farmer-led irrigation development and a geographic focus on sub-Sahara Africa.

Jean-Philippe Venot is Senior Researcher at the *Institut de Recherche pour le Développement* (IRD), and currently based in Phnom Penh, Cambodia. He is also affiliated to the Water Resources Management Group, Wageningen University. His most recent work draws from anthropology and development and STS to unravel the discursive framing and dynamics of drip irrigation promotion in sub-Saharan Africa and delta management in Southeast Asia.

Jeroen Vos obtained his PhD at Wageningen University with an investigation on the performance of the management of large irrigation systems by Water Users' Associations in the desert Coast of Peru. He is now Assistant Professor in Water Governance with the Water Resources Management group of Wageningen University. Prior to that, he lived for several years in Peru and Bolivia where he worked as an advisor to different development cooperation organizations on topics of water policy and water user's organizations. He now investigates – and teaches courses on – water governance, institutional reform in the water sector, and irrigation and development. His academic interest is the combination of technical analysis with institutional, social and economic concepts. Themes of interest include water stewardship certification and the debate on water use efficiencies.

Jonas Wanvoeke is Rural Sociologist with extensive professional experience in the field of agricultural research for development in West Africa. He recently completed his PhD with the Water Resource Management Group at Wageningen University in the Netherlands. His PhD focused on understanding the discrepancies between the dissemination and the use of low-cost drip irrigation technology in Burkina-Faso.

Robert Yoder lives in Vermont, US, and divides his time between international water management consulting and assisting his son operate a farm-to-market horticulture farm. He was Technical Director for iDE from 2003 through 2010 and resident in Ethiopia for four years during that period. Prior to that he was Senior Technical Advisor for the consulting firm ARD for 12 years. In the 1980s he joined IWMI as Resident Scientist in Nepal, managing research on farmer managed irrigation systems. He was engaged in hydropower development in Nepal for eight years in the 1960s and 70s. He received a PhD degree in Agriculture and Biological Engineering from Cornell University in 1986.

Margreet Zwarteveen is Professor of Water Governance at the IHE Delft Institute for Water Education and the University of Amsterdam. Trained as both an irrigation engineer and a social scientist, Margreet is interested in water allocation policies and practices, focusing on questions of (gender-) equity and justice. Her current research includes a project which explores how new investments in irrigation systems along the Nile in Ethiopia, Sudan and Egypt re-allocate water and water-related benefits, tracing what this means for and according to different groups of people, including experts. In another project, Margreet looks at the mobility of Dutch Delta experts and expertise to examine the production of evidence under conditions of uncertainty. In her work, Margreet favours an interdisciplinary approach, seeing water allocations as the outcome of interactions between nature, technologies and society.

Acknowledgments

The editors and the authors would like to thank the external reviewers who helped in improving the quality of the contributions through their thoughtful comments: Patricia Avila, James Ayars, Trevor Birkenholtz, Floriane Clément, Fernando Eguren, Mark Giordano, Meredith Giordano, Armando Guevara, Jan Hendriks, Andy Keller, Jacob Kijne, Laurens Klerkx, Freddie Lamm, Luciano Mateos, Lyla Mehta, Rodrigo Mena, François Molle, Peter Mollinga, Jean Payen, Chris Perry, David Rohrbach, Hilmy Sally, Jim Sumberg, Gerardo van Halsema, Pieter van der Zaag, Karen Villhoth, Flip Wester and Philip Woodhouse. We also thank the researchers, development practitioners, decision makers and farmers with whom we interacted over the last few years. Their ideas and experience of drip irrigation form the basis of this work. Several organizations made this work possible, notably Wageningen University, the CIRAD, the Institut de Recherche pour le Développement (IRD), and UNESCO-IHE. Finally, we acknowledge the financial support of the Netherlands Organisation for Scientific Research (NWO), through the funding of the 'Drip Irrigation Realities in Perspective' project (313–99–230).

Foreword

I was in Myanmar when I received an emailed invitation to write a foreword to this book. It was my sixth trip to Myanmar in two years as an investor and the irrigation engineering director in a for-profit social enterprise working with smallholder farmers to produce in international and regional markets. An important part of our strategy in Myanmar is the introduction of irrigation to raise high-value crops during the dry season.

Drip has been our primary irrigation solution in Myanmar. We use drip irrigation for many, mostly technical, reasons – its flexibility to fit nearly any shape field; its scalability; its adaptability to nearly any topography, soil and crop; its high-uniformity of application; its suitability for fertigation; its minimal run-off, etc. We also believe drip irrigation has the potential for greater productivity of land, water, labor and energy compared to other irrigation methods, although we do not have rigorous scientific proof of this. But we are under no illusion that drip irrigation saves water in the sense of making more water available downstream.

Part of our drip irrigation solution in Myanmar also includes the conversion of traditional paddy rice to drip irrigated rice so that the soil tilth needed for high-value crops exists in the dry season in former paddy fields. This also has the benefit of increasing canal command areas during the wet season when water is abundant but access is limited due to inadequate diversion and conveyance capacity.

I have been a developer and promotor of affordable drip irrigation solutions for smallholder farmers in Africa, South Asia, and Eastern Europe since the turn of the century. I have worked with governmental and non-governmental organizations in international development deploying drip irrigation. I have developed low-cost technological solutions to drip irrigation and written manuals and guides for those working with smallholders implementing drip irrigation. I have worked with major global drip manufactures designing suitable drip irrigation system solutions for small-plot farmers. My family has used drip irrigation for years in our vegetable and flower gardens. In short, I am a strong believer in the potential of drip irrigation to enhance agricultural production for small-scale farmers.

But despite seeing this potential for drip irrigation, and spending considerable effort over the past two decades promoting drip irrigation as a key wealth-enabling tool for smallholders, including my current engagement in Myanmar,

I have yet to see its true adoption by this targeted population. Smallholder farmers are not re-electing at scale to buy more drip irrigation and maintain and replace old drip systems without the continued intervention of implementers. Furthermore, though I have witnessed the poverty alleviation potential from adopting drip irrigation at the individual farmer level, I know not what the societal impacts are of the disruption.

While I still have hope for success in Myanmar (because we have shifted the emphasis of our approach away from enabling drip technology to engaging willing smallholder producers in year-round production supplying high-value markets), we all have much to learn before smallholders in the global South adopt drip irrigation at scale. And, that assumes drip irrigation is appropriate (socially, environmentally and technically) in the various contexts. That is where this book comes in and why it is so important.

This book, with chapters written by some of the leading researchers and thinkers in development, documents the eagerness with which drip irrigation is being promoted for poverty alleviation, preservation of water resources, food security, irrigation modernization and agricultural intensification. In so doing, it explores and describes the conditions where drip irrigation works, for whom, and to what effects.

I am an engineer, and while I believe I have a holistic approach and listen to the farmer customers I work with, I know I miss important contextual clues and that some of my long-held, professional views may hinder me from a broader understanding of the consequences of the transformation to drip irrigation. The approach taken in this book analyzes drip irrigation from 'a larger development and environmental governance perspective to understand the meanings and impacts of a shift to drip in terms of social equity and environmental sovereignty at different levels.' This is key, and for me, one of the most important contributions of this book, because it offers a way of viewing drip irrigation and its implications that otherwise may be missed by the well-meaning, but, perhaps, typical practitioner.

This book also challenges some of my long-held beliefs. For instance, it questions the supposition that smallholders seek to be entrepreneurs. Whereas International Development Enterprises (iDE), of which I am a director, contends that 'entrepreneurs are everywhere' and investing in them with the right business skills, training, loans, and products designed with them in mind, is the most efficient way to solve poverty.

Whether philosophically or informationally, reading this book will deepen the understanding of promotors, disruptors, implementers, practitioners, designers, manufacturers, policymakers, financers, donors and users of drip irrigation in the broader implications to society and the environment resulting from drip irrigation's widespread adoption by smallholders.

By Andy Keller

In the mid-1990s, I visited a research site in western Africa where irrigation technologies were tested and demonstrated. The research institute had installed a system consisting of a bucket hanging two meters above ground, distributing water through a perforated plastic pipe about 10 m long that was running along a row of salads. A worker was asked to fill the bucket, which he did by collecting water in the nearby river, and stepping on an unstable ladder to reach the bucket. Water flew nicely through the pipe to the soil, to the satisfaction of those present. While we were all busy watching this 'low cost drip system' that was supposed to revolutionise the way farmers performed irrigation in the region, I looked at the worker. He had gone back to his own home garden, a few meters farther, and was watering his salads with an old water can.

An anecdote? Maybe, but, to a certain extent, this situation is representative of the many inconsistencies that accompanied, and still accompany the discussions on the role of drip irrigation in addressing today's major challenges regarding water management in agriculture.

Irrigation has always played an important role in development programmes since the green revolution that was predicated on the rapid expansion of agricultural production through extensive use of improved seed varieties, irrigation and chemicals, in particular fertilisers. In many countries, irrigation has represented, and still represents to a certain extent, a large share (often more than 50%) of public investments (and lendings) in agriculture and rural development programmes. There is no doubt that irrigation has helped in boosting agricultural production and productivity, and helped keeping up with the rapidly growing demand for food and other agricultural products.

With time, irrigation investment programmes became increasingly sophisticated. It soon became clear that increasing production could not anymore be the only goal to which irrigation was supposed to contribute. In many places, water became scarce, and efforts turned toward increasing the efficiency in the use of water in agriculture. Production was increasing but poverty and food insecurity were – and are still – stagnating, with most of the poor struggling to make a living out of agriculture. Issues of waterlogging, salinization and degradation of water-based ecosystems added to the increasing complexity of the water-agriculture-development nexus. Finally, in many places, came the idea that the only solution for agriculture was to move toward a vibrant sector, well connected to the rest of the economy, associated with modern technologies.

Probably the most important finding for development operators in the agricultural sector is the acknowledgment that technology doesn't solve problems by itself, and that investment programmes must give much more consideration to the social and institutional dimension of development. This is not new, and efforts for more people-centred, participatory approaches have been around for quite some time. Yet, technology remains the preferred entry point for many. Technology can be seen, is tangible and concrete. Pictures can be taken and stories can be told about technologies.

To many, drip irrigation represents the ideal technological solution for today's agriculture. It is modern, and efficiently replaces old and supposedly wasteful, gravity-based irrigation systems. It allows for more precise application of water, combined with fertilisers, and therefore helps increasing yields and productivity. In so doing, it offers opportunities for farmers, including smallholder farmers, to increase their production and their income; helps solving the world's water scarcity problems; reduces waterlogging and environmental degradation, and contributes to food security and poverty reduction. How exactly drip irrigation achieves all these results is rarely explained, and few studies focus on the real impact of drip irrigation.

That drip irrigation has a place in today's agricultural transformation is undeniable. The recent spread of drip irrigation in many countries shows that there is a market and there is scope for increased conversion to pressured irrigation, including drip as part of a broader move toward modern agriculture. Increasingly, large shares of agricultural production are modernising, there is an increasing demand for better control of production factors like water and fertilisers, and for labour-saving technologies. This is valid primarily for horticultural crops, vegetables and fruit trees which are often considered 'high value' crops for which investments in technologies like drip irrigation show a high return on investment. As markets for these products increase rapidly with urbanisation and the progressive adoption of more diversified diets, the future is great for drip irrigation technologies.

This, however, has little to do with the rationale behind the compulsive frenzy with which the development community has been, and is still in many cases, focusing efforts toward adoption of drip irrigation by smallholder farmers in developing countries. Too often, the reasons stated to promote drip irrigation are unfounded, the expected outcomes are unrealistic and the approaches to widespread adoption of drip irrigation unrealistic, leading to poor performance of development projects, abandonment of installed drip infrastructure and unsustainable patterns of investments. There are many reasons for this, usually related to a poor understanding of the conditions in which drip irrigation can be successfully applied; technical design flaws; underestimation of the implications for poor and unskilled farmers to adopt (and maintain) relatively complex technologies; a disconnect between technology, crops and market opportunities; and little consideration for the financial implications of a conversion to drip irrigation for farmers.

It is high time, therefore, that more clarity be brought to the debate about drip irrigation, its specificities, and the role it can play in agricultural development strategies, and that is exactly what this book helps in doing. By scrutinizing local realities, and analysing the results of past and on-going initiatives aiming at promoting drip irrigation, the book offers a series of interesting findings that have the potential to help making future initiatives to promote drip irrigation more effective and more sustainable.

With great scientific rigor, and no preconceived ideas, the different sections and chapters offer a picture of what is going on in a large variety of regions and socio-economic situations. It offers the reader an opportunity to learn about the real potential of drip irrigation. Some initiatives failed, and these failures are analysed, not in an attempt to denigrate the potential impact of drip irrigation but to understand the root causes of these failures and contribute to making future such initiatives more successful.

The book reminds us of the necessity to place drip irrigation back in the broader framework of socio-economic and agricultural development, to acknowledge the variety of local agricultural situations, and the large spectrum of farming realities that have different needs and capacities. In so doing, its puts back technology where it should be, as an opportunity for innovation that needs to be driven by needs and demand and not, or in any case not only, from the supply side. It shows the real reasons why farmers adopt drip irrigation, which are always far from the ideals and goals that drive the rationale used by large parts of the development community. It also shows how farmers are able to take advantage of a new technology but must adapt it to their specific requirements, and discusses progresses and innovations that such transformations induce. Through case studies, the book also highlights the implications of the adoption of drip irrigation in terms of social equity, water allocation and water savings, and offers new, and much more nuanced insights and concrete implications for policymaking.

Of course, this is not the end of the story. The infatuation of which drip irrigation benefits will continue to play an important role in agriculture development programmes, but the hope is that future initiatives can be better informed, that their expectations can become more realistic, and that their adoption, where justified, can contribute effectively to an agriculture that is more sustainable, more inclusive and more productive, where farmers can take advantage of new market opportunities and make best use of available technologies. For this to happen we need better knowledge and evidence, to which this book contributes, but we also need that decision makers, designers and promoters of drip irrigation accept to look beyond the technology itself, and remember that technology, at the end of the day, should always be about helping people.

By Jean-Marc Faurès

The first time I was induced to think beyond 'drip irrigation is a water saving technology', the line I was taught when I studied irrigation engineering at Wageningen University, the Netherlands, was around 1990. This was through field research by a master's student, Peter Commandeur, in the export oriented irrigated horticulture region near Murcia in the south of Spain. He studied the adoption of drip irrigation. The main finding of his research, at least the one that stuck in my mind, was that farmers' choices of field irrigation technology were only partially shaped by water (saving) concerns, but predominantly by the (family and migrant wage) labour requirements of different field irrigation methods, in terms of quantity and timing, for different crops. This seemed to me to warrant the adoption of a labour process perspective for the study of farmer level irrigation practice. In Marxist labour process theory, technologies, notably mechanisation, not only serve to enhance productivity and the efficiency of input use, but also to manage, control and eventually discipline labour (relations) in the production process.

A political economy perspective on drip irrigation can be taken from the farm labour process to the regional level by understanding farm level choices as part of a process of agrarian change – in the southern Spain case, the development of intensive irrigated horticultural production for export to the (super) markets of Western Europe. This then immediately shows that regional processes are embedded in even broader process – in this case Spain's entry to the common European market, of which the globalisation is ongoing (see for instance, the present public debate around the TTIP and CETA trade agreements).

In the age of neo-liberal capitalism that we live in, the theme of drip irrigation and development takes on another political economy dimension in the expansion of drip irrigation, or micro irrigation, as a business sector in its own right. In India, the region of my own research interest, this includes the entry of the private sector into larger scale irrigation project construction. Companies, with state support, now construct, and run for a number of years, lift irrigation schemes of thousands of hectares with centrally controlled sophisticated drip irrigation water delivery at their core. In addition, agribusinesses (e.g. sugar factories) have developed renewed interest in shaping and controlling farm-level production through drip irrigation. There is also a burgeoning business of sprinkler and drip irrigation for farm-level installation, associated with agricultural intensification. And then there is a growing market for low-cost drip irrigation for (very) small farmers. Clearly, drip irrigation is part of the ongoing commodification of agricultural production, as several chapters in this book illustrate for other parts of the world.

The water-saving narrative that is so closely associated with drip irrigation is one of the expectations and imagined functions loaded onto it. Drip irrigation has been an element, and sometimes a centre piece, of modernisation narratives, and the advocacy for and propagation of productivist approaches to agricultural development, be these focused on agribusiness centred models of development, or on small farmers accumulating themselves out of poverty. More recently, the ecological sustainability narrative has been added to that suite. These narratives operate at global level, in the international development research, policy and business community, but perhaps more importantly at the national level, where the concrete political, economic and discursive framings are constructed for the deployment of drip technology. This book has a lot to contribute in that respect as it focuses to a large extent on ground level deployments, moving beyond criticisms and critiques of drip irrigation related narratives as such. In a cultural political economy perspective on irrigation, which in my reading this book helps to develop, even when its explicit sources of theoretical reference are different, it is not only the (variety of) semiotic content as such that needs to be looked at, but particularly the processes of selection and retention of certain contents and the obscuration and marginalisation of other world views, narratives and policy instrumentation. A concern about the concrete developmental effects and impacts of expanding drip irrigation use provides the critical edge and grounded nature of such analysis.

This book therefore makes a significant contribution to not only irrigation studies but to interdisciplinary development studies more generally.

By Peter Mollinga

Introduction

Panda or Hydra? The untold stories of drip irrigation

Marcel Kuper, Jean-Philippe Venot, Margreet Zwarteveen

'It works!'

Irrigated areas in the world are witnessing a transformation from open canal systems to more 'modern' irrigation methods such as drip irrigation that convey water through closed pipe systems. Initially associated with hi-tech irrigated agriculture, drip irrigation is now being used by a wide range of farmers including smallholders in Asia, North Africa and Latin America. Enthused about its potential to save water, improve productivities, and combat poverty, national governments, international development aid organizations, local and global private-sector actors, as well as farmers have enthusiastically embraced drip irrigation. The technology also assumes a prominent place in policy documents and global environmental and development discourses, often figuring as a panacea to help solve contemporary water challenges.

Drip irrigation has left the technosphere in that it is no longer merely a pre-occupation of engineers. It frequently appears in popular media and wide audience publications, where it is referred to as 'arguably the world's most valued innovation in agriculture'.[1] Its attractiveness and popularity rests on several widely circulating 'success stories' about the wonders it is capable of bringing about. These include, first of all, its ability to conquer the desert and make it blossom, stories that have their origin in the early days of drip irrigation in Israel and California. Stories of the success of drip irrigation also include those recounting its importance in helping to bring about a water-wise modernization of agriculture. These stories have it that drip irrigation can help improve the efficiency and productivity of large-scale production of fruits and vegetables, while achieving water savings and improvements in water use efficiency (the statement 'more crop per drop', popularized by the former secretary general of the United Nations Kofi Annan, aptly captures this promise). There are, thirdly, also success stories about the capacity of the technology to help turn small-scale farmers into true rural entrepreneurs, who – in the process – work their way out of poverty.

When taken together, drip irrigation can be likened to the 'Panda' – the much loved and powerful symbol of the WWF[2] – in its widespread appeal and popularity. Indeed, the perpetual renewal of the purposes for which the same technology – well not exactly the same, variations on a theme is perhaps a better expression – is proposed is reminiscent of the apparent eternal juvenility and neotenous features of Pandas. Just like the Panda, drip irrigation's popularity turns it into something that many are keen to be associated with, as this offers a share in the reflections of its glow of goodness, modernity and greenness.

There is, nevertheless, also a minor strand of literature that engages more critically with drip irrigation, asking (im)pertinent questions about what drip irrigation really does and for whom. Huffaker and Whittlesey (2000) for instance questioned early on the Oregon water conservation programme's promotion of efficient irrigation methods. They pointed out how the programme narrowly focused on on-farm diversions, while neglecting the return flows at the basin level to conclude that: 'the popular Oregon legislative model may be the least effective in conserving water'. Others – such as the study of Kulecho and Weatherhead (2005) in Kenya – pointed at the low adoption rates of drip irrigations systems by smallholders to call into question the success of the technology. Such observations even brought other authors to speak of 'the mostly failed uptake of drip technology in sub-Saharan Africa' (Garb and Friedlander, 2014).

Do these more critical stories and findings mean that drip irrigation is merely a paper tiger, promoted by an international community looking for yet another silver bullet type but ineffectual solution to the world's problems, without any connection to field realities? Certainly not. Its rapid spread around the world over the last decades[3] shows that drip irrigation is not just a slogan but is effectively used by millions of farmers in the field. However, it is not sure that farmers adopt drip irrigation for the reasons advocated by the international community or policymakers. Many rather use it to suit their particular conditions and help the multiple problems they face, for example, to save labour or time. Moreover, there are many farmers who make a more or less conscious choice not to use drip irrigation, men and women who decided against engaging with it.

The serious questions raised by some about the desirability and effectiveness of expensive state-sponsored or internationally supported drip irrigation (development) projects have done little to dampen the policy enthusiasm for this technology. We suggest that this may be linked to its plasticity:[4] the fact that it can be mobilized as a 'solution' to a range of ever-changing and conjunctural international policy issues including environmental conservation, food security, gender equity, poverty alleviation and agricultural modernization. In this capacity, and to use yet another metaphor, drip irrigation may be likened to a many-headed Hydra, which grows stronger when one of his heads is slain.[5] Instead of falling out of fashion or being abandoned altogether, the technology continues going strong when confronted with criticisms or problems of adoption. For instance, the serious problems encountered with the promotion of low-cost drip irrigation in Africa (e.g. Garb and Friedlander, 2014) are so far mainly leading to revisions in dissemination strategies or in moving attempts to further spread it to yet another country. The belief that the technology is promising remains powerfully intact, even when ideas about what it is supposed to achieve – helping the poorest, commercializing agriculture, introducing value chain approaches – change. This makes it challenging to have a meaningful conversation about drip irrigation: its context and direction tend to continuously move from water saving to productivity to poverty alleviation. It becomes even more difficult when this hydra – drip irrigation – has become everyone's favourite policy, planning and development tool (why attack a Panda?).

Indeed, the Hydra-Panda combination that characterizes drip irrigation makes it difficult to talk differently – less positively and more agnostically – about what it is supposed to do (see Chapter 2). We experienced this firsthand when telling our preliminary results about what drip irrigation is and does at water and irrigation events: either what we said was politely ignored to move on to the next presentation about the technology's potential and promise, or our stories were dismissed as unfounded attacks on a technology that had 'demonstrated' its value. While agreeing that drip irrigation does not always work as designed or hoped, and accepting our evidence about the lack of farmer interest in the technology in West Africa, a representative of one of the largest drip irrigation equipment manufacturing companies worldwide, for instance, concluded the public debate about drip irrigation at the Stockholm Water Week in 2014 by stating: 'I just want to add one more point: it [drip irrigation] works!' The same happened at the first World Irrigation Forum in Mardin, Turkey, in 2013: a presentation focusing on the need to study drip irrigation 'in context' was followed by a discussion in which a senior Indian researcher stated: 'it is interesting, but in India, it is not like that, drip irrigation works, it saves water'.

The aim of this book

So what is this book about? Most studies on drip irrigation use plot-level studies to support claims about the technology's potential to save water or improve efficiencies; they are grounded in a firm belief in the technology, focus on ways to improve, or better realize its potential, and are distinctly managerial in their identification with engineers or policymakers and their use of narrow and rather prescriptive engineering or economic languages. This book, in contrast, looks at drip irrigation-in-use, tracing and making sense of what drip irrigation does in and from the perspectives of the everyday lives of the people who use it. The book stresses that not only farmers 'use' drip irrigation; many other actors also invoke the technology in their words and actions, using it for a diversity of reasons and to realize a multiplicity of objectives. These include development workers, national policymakers, international organisations and private companies, but also local craftsmen, engineers, extension agents or researchers (like ourselves). Our grounded understanding of drip irrigation reveals that there are much more complex dynamics at play when drip irrigation is used or promoted than the meta policy-narratives of efficiency, productivity and development that dominate the literature suggest. The book stresses that what drip irrigation is depends on contexts; there are, therefore, many different drip irrigations, just as there are many possible (often untold) stories about how the introduction and use of drip irrigation relates to broader processes of environmental and social change.

This book documents the enthusiasm, spread and use of drip irrigation in Africa, South Asia, Latin America and Europe in an attempt to explore and explain under which conditions it works, for whom and with what effects. While anchored in a sound engineering understanding of the design and operating principles of the technology, the book extends its analysis beyond engineering and hydraulics to understand drip irrigation as a sociotechnical phenomenon that not just changes the way water is supplied to crops but also transforms agricultural production systems and even how society is organized – sometimes in unexpected ways.

This book thus aims to unearth some of the untold stories of drip irrigation in use. How farmers and local craftsmen tinkered with drip irrigation to adapt it to different users and uses that were never envisaged at the design stage. How development professionals remained excited about drip irrigation, despite disappointing adoption rates in the field. Or how peasants felt rejected by the discourse and materiality of drip irrigation, and thus rejected the technology itself.

This introductory chapter first presents the three main discourses and narratives that have come to be closely associated with drip irrigation worldwide – i.e. water savings and efficiency; modernization of agriculture; and poverty alleviation and development. These by-now classic drip tales are at the heart of the current enthusiasm for and popularity of drip irrigation as a 'green technology'. The introduction then clarifies how the diverse chapters engage with these classic tales, often challenging the straightforward linear causalities between the technology and its effects that the tales suggest. By telling 'the untold stories' of 'drip irrigation in use', the diverse contributions shed a different and often sobering light on drip's water saving potential, allow deconstructing the standardized vision of what drip irrigation is and can help achieving, or question the often implicit assumptions in drip irrigation promotion strategies about what smallholders are, do, need and aspire to be.

By asking people from diverse 'communities of practice' to contribute to the book, we collected a range of sometimes contrasting viewpoints on and experiences with the dynamics and prospects offered by drip irrigation. Rather than presenting the final truth about drip irrigation, one of the explicit objectives of the book precisely is to provide the reader with a wide range of ideas, framings and perspectives, thereby further underscoring that drip irrigation can and does mean and do different things to different people. The book is meant for a wide audience: irrigation and development professionals, students, scientists, public authorities, donors. While the majority of the book's case studies have been conducted as part of a coherent research project in which many of the contributors to the book were involved, we have also invited a number of contributions by renowned specialists that were not part of our project.

Some of the book's chapters critically engage with the question of why the meta-narratives that surround drip remain so powerful in spite of growing evidence of disappointing results. These analyses remind us of Don Quijote de la Mancha's fight against windmills: the meta-narratives are powerful precisely because they are such wonderful stories – stories that, moreover, resonate remarkably well with powerful desires to believe that it is possible to combine growth with environmental conservation (sometimes referred to as market environmentalism) and with resulting normative ideals about how water responsibilities, rights and powers can best be distributed and arranged. It is far from certain that the untold stories we recount here will be as powerful, notably because many of them do clearly not speak from a vantage point of power or dominance. Be that as it may, we nevertheless hope that the book will help in diversifying and plurifying the debates about water saving, agricultural development and poverty that drip irrigation has come to be part of, showing how drip irrigation dynamics are part and parcel of broader development and modernization processes that reconfigure relations between the state, civil society and the private sector in ways that are sometimes difficult to predict. In this sense, the book also shows that drip irrigation provides an intriguing entry-point for understanding wider processes of socio-environmental change.

The meta-narratives and classical tales

This books looks at three meta-narratives in particular – i.e. efficiency and water preservation, productivity and agricultural modernization, and poverty and development. These three narratives can unfold independently, but are often mutually constitutive. We understand narratives not only as making sense of (complex) objects, events and processes through storytelling, but also as a mechanism whereby 'sense making by some can affect the sense making and behavior of others' (Röling and Maarleveld, 1999). In other words, we use the term 'narrative' to 'express how perspectives, "Leitbilds", metaphors, stories, images, theories, slogans, and axioms are woven together, become widely shared and dominate behavior' (Röling and Maarleveld, 1999). We will look at the plot of each narrative, but also at their characters. Character believability is an essential property of narratives because 'the events that occur in the story are motivated by the beliefs, desires and the goals of the characters' (Riedl and Young, 2010). The heroes (and villains) of each meta-narrative are not exactly the same, even though the same large categories of actors appear in all of them: policymakers, donors, government agents, engineers from private companies and farmers.

The conservation narrative: drip irrigation saves water due to a high irrigation efficiency

Van der Kooij et al. (2013) unravelled the plot of the conservation narrative that different researchers employ when presenting results of studies to assess and improve drip irrigation efficiency: 1) they use a specific framing of the problem, i.e. the global water crisis; 2) they suggest that drip irrigation is a (contribution) to solving this problem; 3) they refer to other studies highlighting the potential of drip irrigation to improve water efficiencies, 4) they present their case study, 5) they describe the results obtained, showing under which conditions the highest efficiencies can be obtained, 6) they recommend 'best irrigation practices' to obtain these. The problem with this story line is first that in almost all cases the results are obtained on experimental fields, with strictly defined and monitored boundary conditions. The assumption is that these are somehow comparable or translatable to what farmers do on theirs, but this is an assumption that is seldom verified. Second, recommended improvements refer to increases in irrigation efficiency, and thus to a decrease of 'losses' or 'wastage'. However, there is much ambiguity about what can be considered 'losses', as water saving at the plot level – the privileged scale of the vast majority of studies – does not necessarily imply any water saving at the river basin level (Seckler, 1996). The authors, therefore, conclude that 'there is no conclusive scientific evidence to support a general belief in drip irrigation as a water saving device or as a tool to help solve the water crisis' (van der Kooij et al., 2013).

The central actors of this narrative, perhaps surprisingly, are not the farmers but the engineers and irrigation scientists themselves. Farmers rarely make their appearance, although they do figure in the background – as those wasting water with 'traditional' irrigation practices, or as those unwilling to adopt modern irrigation technology – to help bring the rightness of the drip irrigation system and the engineer into sharper relief.

The productivity narrative: drip irrigation enables a higher productivity and a modernization of agriculture

This 'more crop per drop' narrative is closely interwoven with the water use efficiency narrative, but authors developing this narrative also criticize the narrow technical and plot-level focus of the irrigation efficiency narrative (e.g. Perry, 1999). The water productivity narrative was perhaps most extensively presented in the International Water Management Institute (IWMI)'s 2003 book on water productivity in agriculture (Kijne et al., 2003). It goes more or less as follows: 1) water and food security are closely related, as water (and not land) is likely to be the most limiting factor in increasing production (this goes back of course to the previous narrative, building on a perceived global water crisis), 2) the solution to food security is thus to be found in improving water management, and more specifically in increasing water productivity, 3) increasing water productivity for agriculture means getting more production, value or benefit '*from each drop of water*', 4) producing more food with less water will ease water scarcity and leave more water for other uses, including ecosystem services.

Contrary to the previous narrative, in this narrative, farmers assume the centre stage. They are often mentioned in the book and in the associated literature: 'Improvements in crop production can only be made at farm level. They result from the deliberate actions of individual farmers increasing production or the value of output with the same volume of available water or maintaining or increasing productivity using less water' (Kijne et al., 2003: 9). However, while farmers appear in the plot, their characters remain flat. Sometimes farmers appear as being opposed to irrigation system managers or bureaucrats, who – so the story goes – do not necessarily have the drive to increase water productivity. A distinction is sometimes made between large and small farmers, or between commercial farmers and smallholders. Yet, irrespective of such refinements, farmers' characters in these narratives remain rather unconvincing and one-dimensional. They tend to be depicted as classical *homo economicus* whose behaviours are mainly guided by financial and economic calculations and rationales. Hence, farmers appear as being primarily interested in producing more crop per drop: 'It can be expected that, when water becomes a real economic good, farmers are more inclined to adopt water saving technologies' (p. 63). As some chapters of this edited volume show, the difficulties in transferring best practices from the laboratory to the field provide just one indication that farmers may have other motivations for doing what they do: despite '*impressive successes*' initially (in this case the efficient management of rainwater to improve crop productivity), farmers went back to their '*normal practices*' when researchers went back to check 15 years later (Kijne et al., 2003: 200).

The poverty and development narrative: drip irrigation support the emergence of a vibrant and entrepreneurial smallholder based agriculture

The third encompassing narrative relates to the poverty alleviation potential of drip irrigation. Articulated first by development practitioners and researchers who were joined later on by global drip irrigation equipment manufacturers looking to further their corporate social responsibility policy, this narrative acquired widespread resonance in the late 1990s–early 2000s. This story starts by announcing that an agricultural revolution supported by drip irrigation (i.e. modernization) is happening, but that smallholders in developing countries are missing out on it (see Hillel, 1988). It continues by referring to the first two narratives; stressing the potential of drip irrigation in terms of improving agricultural productivity and water use efficiency; and then continues by relating the former to increases in income – hence, poverty alleviation (through the sales of agricultural products). What the story stresses, then, is the existence of a new spectrum of drip irrigation systems 'that could form the backbone of a second green revolution, this one aimed specifically at poor farmers in sub-Saharan Africa, Asia, and Latin America' (Postel et al., 2001, p. 3).

The narrative has two main characters. The first is an artefact, 'the drip kit' which is presented by its promoters as affordable, small size, and infinitely expandable (Polak, 2008). As such the kits provide a 'solution' to two major reasons why drip irrigation has, according to the story, eluded smallholders in developing countries for decades: high initial investment costs and complexity of use (the kits come together with a simple user manual). That they are said to be expandable (as for Lego, it is said that kits can be combined to one to another) also signals that farmers using them are not meant to continue to be locked in a 'poverty trap', something that positively distinguishes drip kits from other 'small-scale' technical innovations. The second character is the smallholder farmer him- or herself. In the story, smallholders are said to be poor because 'they do not earn enough money' (Polak, 2008). On the face of it, this statement seems difficult to challenge. However, it is also a statement that overlooks the structural and political dimensions of development and poverty – which are, as a result, being black boxed and pushed aside. In the story, the smallholder appears as a (potentially) innovative, entrepreneurial and business minded individual, who – if given a simple enough 'product' and appropriate support – will work his/her way out of poverty by selling vegetables for the market. What made this story particularly attractive to a range of actors in recent years is the fact that it brought together (1) an image of a smallholder modern enough to be supported, (2) and an approach that focused on adopting market-based principles for the promotion of the technology (see Venot, 2016 for a critique).

How to unearth the untold stories of drip irrigation?

The literature that inspired us in our quest for unravelling the untold stories was diverse. The overall framework of the book builds on science and technology studies (STS), envisaging technology as part of co-evolutionary socio-technical networks that weave together heterogeneous materials – artefacts, people, institutions, cultural meanings and knowledge – into working wholes (see Chapter 2 for an elaboration). Within this integrative framework, the questions the book addresses include: How is drip irrigation materially and discursively constructed and used in different water and development domains; which and whose goals do such constructions serve; and what are the social and environmental impacts of drip irrigation use? These questions are important inputs for larger reflections about how drip irrigation and the socio-natural transformations that it is part of and contributes to re-configure possibilities for democratically organizing agriculture and water governance.

Other sources of inspiration include the water sciences, in particular the irrigation performance debate that has yielded many valuable insights on how to look at irrigation technology in use (e.g. Burt et al., 1997). Another feature of this literature that we build on is the pragmatic approach to drip irrigation that is apparent in some of the reference works on the topic, showing not only the advantages but also the disadvantages of the technology: 'Microirrigation, as with other irrigation methods, cannot be used for all agricultural crops, land situations, or user objectives' (Lamm et al., 2006: 1). However, this common-sense and practical attitude of some engineers is not often found in more scientific articles that report on drip irrigation, and even disappears further when drip irrigation 'travels' to other circles (donors, policymakers, etc.). Finally, we draw from a limited number of studies that confront the theoretical and practical advances on, for example, performance indicators, with farmers' practices and motivations, instead of just focusing on studies in experimental settings under controlled conditions (e.g. Benouniche et al., 2014).

Recent advances in innovation systems research, in particular those applied to the agriculture sector, also inform many contributions of this book. This literature challenges linear technology transfer models to draw attention to the fact that 'innovation results not just from variation (trying something new), but also from selection (finding things better than what is currently used) and incorporation into long, complex processes where different types of knowledge are creatively used by a diversity of actors (Spielman et al., 2009). These insights are helpful to decentre the technology in research, instead moving the attention to the interactions between technology and society, which brings us back to STS studies.

In line with the third meta-narrative of drip irrigation, several contributions of this book also draw from the field of the anthropology of development, and a substream of work that highlights the importance of the interpretation of events over the events themselves (see for instance, Li, 2007; Mosse, 2005). This literature is extremely useful to understand the persistence of a paradox: that of a persistent positive imagery of drip irrigation among development professionals and the near absence of drip irrigation use by smallholders in sub-Saharan Africa. The 'will to improve' (Li, 2007) and a widely held belief in technology as a driver of change that still characterizes development interventions are strong enough to see in past failures, reasons to continue promoting a technology that 'has proved itself'.

Last, some chapters in the book were inspired by feminist technoscience studies, in particular their insistence on the methodological importance of starting inquiries from the perspective of those events, phenomena or people that do not fit prevailing categorisations or standards, or those that are not included in the actor-networks studied. What about the peasant farmers in Chile, Peru or Mexico who choose not to irrigate their crops with drip irrigation? And what about the sisters, daughters and wives of the Moroccan men who established their future and modernity by installing drip irrigation on their plots?

From a more practical standpoint, we took the three meta-narratives as a starting point for structuring the book. We then proceeded in two directions. First, we used field/plot dynamics to find out how the narratives unfolded in different contexts, looking both at the discourses and the materiality of drip irrigation-in-use. For example, a study was undertaken in Morocco to assess whether at all and if so how much water was saved during the implementation of drip irrigation (see Chapter 5). Second, we engaged with the meta-narratives by assessing how they had been elaborated, by whom and for what, and by investigating how they are part of broader processes of development and change. For example, a study on the implementation of low-cost drip irrigation in Burkina Faso extended the usual scope of analysis (focusing on (dis)-adoption dynamics) to explain the paradox of the framing of drip irrigation projects as successes even though drip irrigation use remains anecdotal (see Chapter 13). Decentralizing the analysis as such by revealing broader societal

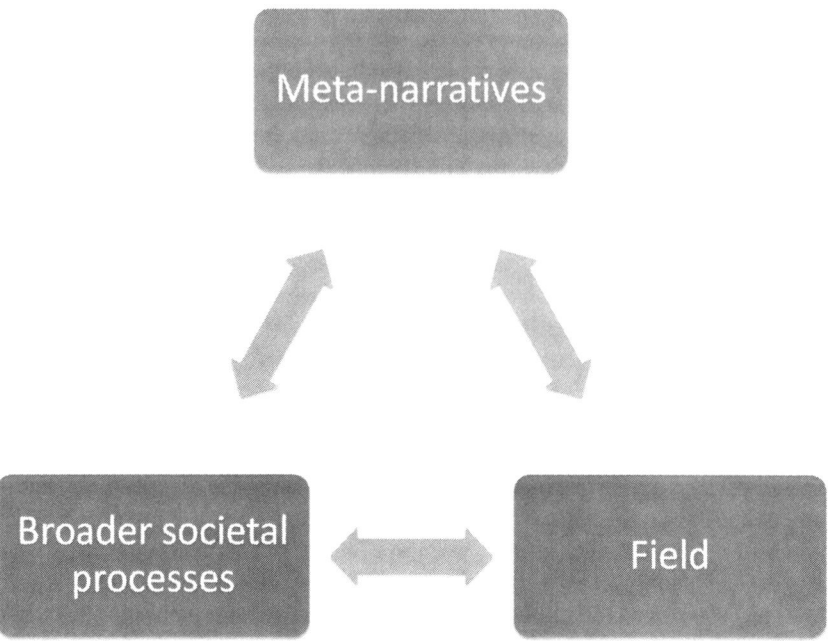

Figure 0.1 The research approach that inspired the contributions in this book.

processes and field realities allows, in turn, to critically analyse the making of the meta-narratives (see Figure 0.1).

The untold stories: contrasted views grounded on empirical evidence

As the debate on drip irrigation is dominated by overarching all-encompassing statements, there are many drip irrigation stories that remain untold. These include, for instance, the story about how millions of smallholders engage in practices of 'bricolage' to make hi-tech systems fit their specific circumstances (Chapter 15); stories about the fact that drip irrigation rarely travels out of development arenas in sub-Saharan Africa (Chapters 12 to 14); or again stories about how farmers adopt drip irrigation because it allows farming to become a white-collar profession rather than because it saves water (Chapter 6).

Before telling these stories, based on empirical evidence, the book starts with three overview chapters to provide a general picture of drip irrigation dynamics. Chapter 1 recounts how drip irrigation conquered the world, travelling to different continents, and how – broadly speaking – the meta-narratives unfold around the world. While doing so, the chapter highlights that drip irrigation progressively left the realm of engineers and experts to enter more global water policy and private business arenas. Chapter 2 provides a more theoretical background to the book by specifying how a practice-based understanding of drip irrigation enriches the discussion about the technology's performances in a range of contexts. Chapter 3 provides an engineering viewpoint showing that drip irrigation is not a standardized technological package, but is re-engineered by farmers, local craftsmen and engineers to fit specific bio-physical and socio-economic requirements.

Once the scene is set, the core of the book is then composed of three different sections, representing the three meta-narratives. The first section is about 'Efficiency and Water Saving' with three chapters on North Africa. Chapter 4 shows how the efficiency argument is often used to re-allocate water, which is 'yet-to-be-saved', to other uses, ultimately leading to an increase pressure on water resources. Chapter 5 zooms in on the paradox of continued promotion of drip irrigation based on claims of water saving that are rarely backed up with evidence. Illustrated by field evidence, this chapter shows that increased water productivity enhances the pressure on groundwater resources. Chapter 6 is also about groundwater, but looks at how drip irrigation in the Saïss Region in Morocco is instrumental in bringing about 'enclosures' of water, channelling it to some while making it increasingly difficult to access for others. The chapter thus shows that drip irrigation – at least in this specific context – forms part of and helps in bringing about larger transformations in water and land tenure relations.

The next section is about the water productivity meta-narrative, and more broadly about the modernization and agrarian change that drip irrigation forms part of and helps bring about. This section is composed of four chapters that narrate experiences with drip irrigation in South and Central America (Chile, Mexico and Peru), and North Africa. Chapter 7 shows how the state in Chile promoted drip irrigation for peasants to resolve a water crisis, which was in part created by promoting intensive export-oriented agriculture. It tells the story of how peasants refuse the technology, as the state (and its subsidy programme) as they say, does not 'accept our way of life'. Basically, official government discourses contrast and compare peasants to the virtual entrepreneurs they are to become according to the new neo-liberal doctrine. Chapter 8 shows how drip irrigation enabled large agribusiness companies to conquer the desert and produce luxury goods for export on the desert coast of Peru. The infrastructural works necessary to provide water to these companies were justified with the efficiency as well as the water productivity meta-narrative. The smallholders in the area also benefitted, to some extent, from this project, but the costs and benefits of drip irrigation development are unevenly distributed over the different stakeholders, while promised water savings are not realized. Chapter 9 shows how, in the state of Guanajuato in Mexico, drip irrigation subsidy programmes mainly benefitted large agro-industries, which were in a position to influence the formulation of such programmes. Drip irrigation in this context remains an elite technology, facilitating the accumulation of capital by these industries. Chapter 10 highlights that drip irrigation is the newest of a long suite of technological artefacts that have supported a dual vision of Moroccan agriculture over the last century. It provides two contrasting examples of technological change and rural development in Morocco. In the first case, the technology was implemented by the state in association with a large sugar-processing industry in a process that ignored the capacities of farmers to innovate and engage with modernization. Farmers were provided with a black-boxed technology and were made to confront the risks associated with the project, ultimately leading to its failure. In the second case, farmers took the initiative themselves, slowing down or accelerating the implementation of drip irrigation whenever required. By this continuous learning process, peasants succeeded to enhance their autonomy and resilience.

The section on 'Poverty and Development' is based on various case studies on the introduction of low-cost drip irrigation in Asia (India, Myanmar and Nepal) and in sub-Saharan Africa (Burkina Faso, Ethiopia and Zambia). Chapter 11 discusses the value and limitations of 'drip kits', showing that smallholders in India and Myanmar prefer by far having access to the different components than to be provided with prepackaged entire kits. The chapter also highlights the importance of significant and long-term research and development efforts, and notably the necessary technical and institutional redesign that is needed when introducing a technological artefact in a new context. Chapter 12 takes us inside Zambia's households, showing how drip irrigation tends to be taken up mostly by the more business-oriented entrepreneurial male farmers. This favours the intra-household bargaining positions of men and works as a mechanism that makes it more difficult for women to derive benefits from low-cost irrigation technologies. Chapter 13 takes place in Burkina Faso, showing the persistent enthusiasm of development actors for drip irrigation – who see and present it as an effective tool to alleviate poverty, improve food security and save water – even in the absence of sustained use of drip kits by farmers. The chapter then provides some explanation for why development actors continue promoting drip kits even when there is little empirical evidence to support the belief that, indeed, they deliver on these promises. Chapter 14 follows and reviews experiences with the donor-supported promotion of low-cost low-pressure drip irrigation kits in sub-Saharan Africa over the past 15 years, noting that no credible impact assessments of these programs have taken place. It investigates the 'vested-interests' hypothesis to explain the continued interest of donors in such technologies, and proposes positive advice to move forward, including the fact that researchers and development practitioners must get away from focusing on testing single technologies and instead should open up to multiple interested parties and a menu of technologies and practices.

After these three sections, closely linked to the three meta-narratives of drip irrigation, the next – and last – section of the book focuses on the alliances and networks in which drip irrigation innovation actually takes place. Chapter 15 looks at the process of 'bricolage', understood here as a creative process of learning and adaptation with smallholders co-steering the design process with engineers, thereby changing both the users' context and the drip irrigation technology itself. The chapter explains that the fact that drip irrigation lends itself to 'bricolage' helps explain its success as an innovation. Chapter 16 then focuses on farmer-led innovation of drip irrigation in Algeria's Sahara. It describes a true 'innovation factory' that involved thousands of farmers who gradually improved the technology for different local conditions. The factory produced technical innovations, but also local actors capable of innovating in new situations. Chapter 17 moves away from the farmers and presents the 'grey' support sector, providing farmers with equipment, advice and credit. Based on a case study in Morocco, the chapter highlights the fact that these intermediaries not only supply and adapt the technology in close association with farmers, but also help manufacturers to provide technology adapted to local demand. Finally, Chapter 18 moves another level away from farmers and deals with the networks and alliances involved in establishing and managing subsidy schemes. It shows how the state of Gujarat (India) set up a business-like semi-autonomous public company to handle subsidy delivery. It highlights that the positive image of drip irrigation in the State is more related to the amount of subsidies distributed and to the surface area equipped with drip irrigation than to the actual use of it and to who benefitted from the subsidy programme.

Conclusion: bringing common sense back about drip irrigation?

To conclude, this book challenges conventional wisdoms about drip irrigation. It is about understanding how drip irrigation has come to be seen as a 'Panda': something that everybody loves and that therefore becomes almost unquestionable. The book does so through an empirically grounded analysis of the multiple realities of drip irrigation-in-use. It first uses case studies and analyses to critically question the likelihood that the introduction of or a conversion to drip irrigation will lead to water savings. Second, it discusses the social and environmental desirability of the types of agricultural modernization that drip irrigation forms part of or helps promote, also exploring alternatives. And third, it revisits the attractiveness of low-cost drip irrigation technologies as a tool for alleviating poverty, also using the findings to raise some larger questions about the practices of international development cooperation. Telling the untold stories about the everyday and often messy realities of drip irrigation is also laying bare some larger challenges of water governance, challenges about how to democratically organize and account for water uses and interventions when these escape public forms of control. It shows how strong political choices favour certain agricultural development models and questions the related increasing influence of global private actors and the impact it has on smallholders.

As a whole, the book underscores that the enthusiasm that surrounds drip irrigation as a technology is primarily based on a theoretical construct, that of its 'potential' and 'promises', rather than on what it achieves and does in farmers' fields or river basins. Yet, many public, civil society and private sector actors strategically make use of stories about this potential to negotiate funds, mobilize political support or justify re-allocations of water. This, together, with the difficulty to voice alternative stories, sometimes makes the Panda look like a Hydra. Those responsible for promoting drip are seldom, if ever, held accountable when the promises of drip irrigation are not met. Instead, the shortcomings of on-going initiatives are used as further justification to increase support for the technology: failures thus become incentives to invest even more in the creation of the necessary conditions needed to make drip irrigation perform well.

Finally, our work suggests that the journey of drip irrigation is not over. The hardware, the discourses and the success stories will continue to travel to other destinations, other users and uses, and other political arenas. We hope that some of the common sense that prevailed when drip irrigation was not yet the global success it is now portrayed to be, can be regained, also to direct much needed attention to the wider processes of environmental and social change that drip forms part of or helps promote.

Notes

1 E.g. https://en.wikipedia.org/wiki/Drip_irrigation, accessed October 19, 2016.
2 Interestingly, there is a direct link between drip irrigation and the WWF panda, as WWF itself also promotes drip irrigation (see http://wwf.panda.org/wwfnews/?159841/Modern-irrigation-techniques-could-save-Turkeys-water, accessed October 31, 2016).
3 According to the International Commission for Irrigation and Drainage, about 11.1 million ha are now under drip irrigation worldwide. This figure is likely to be largely underestimated (see also the FAO Aquastat database, and Chapter 1 of this book). While the area under drip irrigation is considerable, and the associated investments substantial (at about 5,000€ per ha, the aggregate investment cost realized to date would amount to 55.5 billion € – not accounting for the necessary maintenance and replacement of the system), this should be compared to the total surface area under irrigation, which is about 300 million ha (Siebert et al., 2010).
4 Or perhaps 'fluidity', see De Laet and Mol, 2000 or Chapter 2.
5 The hydra is a Greek and Roman multi-headed mythological figure guarding one of the entrances to the underworld. It was invulnerable as soon as it retained a head (and several heads grew back each time a head was cut). It was eventually killed by Heràkles.

References

Benouniche, M., Kuper, M., Hammani, A., and Boesveld, H. (2014). Making the user visible: Analysing irrigation practices and farmers' logic to explain actual drip irrigation performance. *Irrigation Science* 32(6): 405–420.

Burt, C.M., Clemmens, A.J., Strelkoff, T.S., Solomon, K.H., Bliesner, R.D., Hardy, L.A., . . . Eisenhauer, D.E. (1997). Irrigation performance measures: Efficiency and uniformity. *Journal of Irrigation and Drainage Engineering* 123(6): 423–442.

Garb, Y., and Friedlander, L. (2014). From transfer to translation: Using systemic understandings of technology to understand drip irrigation uptake. *Agricultural Systems* 128: 13–24.

Hillel, D. (1988). Adaptation of modern irrigation methods to research priorities of developing countries. In G. Le Moigne, S. Barghouti, and H. Plusquellec (Eds.), *Technological and institutional innovation in irrigation. Proceedings of a workshop held at the World Bank*, April 5–7, 1988. Washington, DC: The World Bank, World Bank, Technical Paper No. 94, pp. 88–93.

Huffaker, R., and Whittlesey, N. (2000). The allocative efficiency and conservation potential of water laws encouraging investments in on-farm irrigation technology. *Agricultural Economics* 24(1): 47–60.

Kijne, J.W., Barker, R., and Molden, D.J. (Eds.). (2003). *Water productivity in agriculture: Limits and opportunities for improvement*. Colombo: IWMI and CABI.

Kulecho, I.K., Weatherhead, E.K. (2005). Reasons for smallholder farmers discontinuing with low-cost micro-irrigation: A case study from Kenya. *Irrigation and Drainage Systems* 19(2):179-188.

Laet, M. de, and Mol, A. (2000). The Zimbabwe Bush Pump: Mechanics of a fluid technology. *Social Studies of Science* 30(2): 225–263.

Lamm, F.R., Ayars, J.E., and Nakayama, F.S. (2006). *Microirrigation for crop production: Design, operation, and management* (Vol. 13). Elsevier: Amsterdam.

Li, T.M. (2007). *The will to improve: Governmentality, development, and the practice of politics*. Durham, NC: Duke University Press.

Mosse, D. (2005). *Cultivating development: An ethnography of aid policy and practice*. London: Pluto Press.

Perry, C.J. (1999). The IWMI water resources paradigm – definitions and implications. *Agricultural Water Management* 40(1): 45–50.

Polak, P. (2008). *Out of poverty: What works when traditional approaches fail*. San Francisco, CA: Berret-Koehler Publishers.

Postel, S., Polak, P., Gonzales, F., and Keller, J. (2001). Drip irrigation for small farmers: A new initiative to alleviate hunger and poverty. *Water International* 26(1): 3–13.

Riedl, M.O., and Young, R.M. (2010). Narrative planning: Balancing plot and character. *Journal of Artificial Intelligence Research* 39(1): 217–268.

Röling, N., and Maarleveld, M. (1999). Facing strategic narratives: In which we argue interactive effectiveness. *Agriculture and Human Values* 16(3): 295–308.

Seckler, D.W. (1996). *The new era of water resources management: From "dry" to "wet" water savings*. Research Report No. 1. Colombo: International Water Management Institute.

Siebert, S., Burke, J., Faures, J.M., Frenken, K., Hoogeveen, J., Döll, P., and Portmann, F.T. (2010). Groundwater use for irrigation – a global inventory. *Hydrology and Earth System Sciences* 14(10): 1863–1880.

Spielman, D.J., Ekboir, J., and Davis, K. (2009). The art and science of innovation systems inquiry: Applications to sub-Saharan African agriculture. *Technology in Society* 31(4): 399–405.

Venot, J.P. (2016). A success of some sort: Social enterprises and drip irrigation in developing countries. *World Development* 79: 69–81.

1 From obscurity to prominence

How drip irrigation conquered the world

Jean-Philippe Venot

Introduction

Over the last 60 years, drip irrigation has evolved from an *experimental technology* tested in a few research stations in a limited number of countries to a truly *global phenomenon*. This chapter recounts the journey of how drip irrigation has become one of the most popular technologies in the field of irrigation and agriculture for professionals and the wider public alike.

This evolution, the chapter will argue, partly hinges on the ability of drip irrigation to lend and reinvent itself in line with continually evolving dominant environment and development discourses. This 'malleability' was not a given and emerged progressively as ever more people and organizations became interested and involved in the use and promotion of drip irrigation. These actors now form a loosely bound but wide-reaching coalition through which drip irrigation artifacts (plastic pipes and emitters) are attributed inherent characteristics such as that of efficiency, productivity and modernity. The malleability of the technology and its systemic nature, in turn, allow multiple actors (including farmers; see Chapters 15 and 16 for instance) to engage with it, adapt it to their context of use and specific needs, thus giving weight to far-reaching claims of environmental preservation, agricultural modernization and poverty alleviation.

This chapter's objective is not to assess whether those claims are 'real' (or not) but to understand how they have come into being; neither do we aim at a detailed ethnography of the way drip irrigation systems have evolved over time. Rather, we recognize their diversity but use the term *drip irrigation* to designate a concept (the frequent application of small quantities of water directly at the root zone of crops) and a suite of objects through which this concept is put into practice (a system of perforated plastic pipes, emitters and ancillary devices).

The chapter adopts a diachronic and geographic perspective and identifies 'significant trends' regarding the way drip irrigation has been, and still is, envisioned. This is done on the basis of an extensive review of documents published since the 1970s (grey literature, journal articles, policy documents), an analysis of wide audience media pieces (news articles, blog posts, public statements) and key informant interviews with irrigation professionals who have been active in the field for several decades. The analysis also benefitted from the multiple interactions the author had with colleagues when preparing this edited volume.

The chapter begins by showing how data on drip irrigation, and the way they are discussed and presented, contributes to establishing drip irrigation expansion as a global imperative, notably pursued through far-reaching public subsidy schemes (section 1). The chapter then turns toward identifying the main discourses that have been associated with drip irrigation since the 1960s, and how these play out today in different regions/countries of the world (see the color section for a spatial illustration). Section 3 highlights that such drip irrigation discourses are not 'de-incarnated' but have been actively shaped by multiple actors whose influence has waxed and waned with time. We notably show the importance of 'mythical figures' to bring an 'expert technology' out in the open, and the fact that agricultural and irrigation engineers working in public organizations have progressively lost ground vis-à-vis private manufacturing companies in shaping the debate around drip irrigation. A short conclusion recaps the findings.

Uncertain data but a general consensus to promote drip irrigation

As pointed out by Venot et al. (2014), a first challenge in attempts to understand the dynamics of drip irrigation relates to the availability and reliability of data. Most current drip irrigation development takes place outside large-scale public irrigation systems (whose extent is relatively well known) and is largely driven by private initiatives, which makes monitoring more difficult for governmental and international organizations.[1]

The first multi-country assessment of the extent of drip irrigation use dates back to 1975 (Shoji, 1977). The International Commission for Irrigation and Drainage (ICID) conducted its first worldwide micro-irrigation survey in 1982 and has regularly updated it since (data is obtained from national irrigation and drainage committees). These surveys, however, remain largely incomplete (only 25 countries were listed in 1982 and 45 countries are listed in the most recent 2012 survey).[2] The surveys as well as the literature highlight that large-scale use of drip irrigation started in the 1970s in countries such as Israel, the United States, and South Africa. Countries in the south of the European Union joined the trend in the 1980s (Bucks, 1995; Reinders, 2006). Since the mid-1990s, most drip irrigation development is occurring in emerging economies; notably on the southern shores of the Mediterranean Sea, Latin America and Mexico, and above all China and India (which together are estimated to account for more than one third of the entire area irrigated with drip worldwide; see Box 1.1 for a brief description of trends in China).

Box 1.1   China: a (not so) hidden giant?

In 2012, ICID evaluated at 1.6 million hectares the drip irrigated area in China, and some manufacturers do not hesitate talking of a growth rate of

20% a year,[3] but it remains very difficult to assess the extent to which drip irrigation is used in China, especially for non-Chinese scholars.

What is sure, however, is that China has emerged over the last few decades as a drip irrigation powerhouse, both in terms of manufacturing capacity (alongside Israel and India) and drip irrigated area (alongside India). This has notably happened because of the strong support of the Chinese government whose priority is to support an agricultural boom in the country but is increasingly concerned by the unsustainable rate of groundwater abstraction in some of the most productive regions of the country (the northwestern plains). The commitment of the Chinese government to promote drip irrigation has to be understood in this context and was clearly illustrated by the No. 1 document of the Chinese government of 2011. This high level policy document that identifies the government priorities focused on irrigation and identified 'water-efficient irrigation technology' as a way to pursue the policy goal of 'water conservancy' (USDA, 2011; Burnham et al., 2015). This priority notably translated in the form of subsidies to farmers for purchasing drip irrigation equipment (of more than 50% of the equipment cost), as well as low interest loans and tax reductions or exemptions for manufacturing companies.

If drip irrigation systems have long been used by smallholders in 'traditional' Chinese greenhouses for the production of vegetables, it is now mainly used for cotton cultivation, most notably in the semi-autonomous arid region of Xinjiang in the extreme west of China, but also for sugarcane and maize production. Apart from the Xinjiang region, the semi-arid provinces of the northwestern plains (such as Gansu, Shaanxi, Ningxia) are a hot-spot of drip irrigation use as well. It is no surprise that the major equipment manufacturers are located there.

Drip irrigation was introduced by governmental research institutes in the 1970s but it is only in the 1990s that the drip equipment manufacturing sector got structured. The three biggest equipment manufacturing companies were set up in the late 1990s (Xinjiang Tianye Water Saving Irrigation and Dayu Conserving Water Group in 1999 and Gansu Yasheng Industrial Group in 1998); since the mid-2000, they are registered on the Chinese stock market. In addition to these giants, there are probably over 500 small- and medium-scale manufacturing companies. Taken together, the manufacturing capacity of Chinese companies would reach about 2 billion meters of drip line a year (that is, if a spacing of one meter is considered, the equivalent of 200,000 hectares/year) (Tianye, on its website, announces that the total cumulative irrigated area that could be equipped with the equipment it manufactured reaches 4 million hectares in China – this includes other equipment than drip irrigation such as sprinklers and pivot www.tianyejieshui.com.cn). Chinese manufacturing companies conduct most of their business in China but some also try to enter the international market – notably through acquiring foreign companies. On the other hand, many large non-Chinese drip irrigation companies have offices and distributors in China. They tend to highlight the quality and 'high-tech' nature of their products as most Chinese companies still manufacture relatively simple (and cheap) products. Entering the Chinese market for international actors does not seem to be an easy task, as can be illustrated by the history of Netafim in the country. The company first entered China in 1994; in 2000s it partnered with a Chinese company and built two plants there. These were shut down in 2008. Only in 2016 did Netafim open a new plant in Ningxia, hoping to benefit from the booming wine industry of the region. The company is also looking at partnering with Chinese fertilizer companies to expand its business, communicating on the potential of drip irrigation in terms of input optimization through fertigation (the use of soluble fertilizers) (www.netafim.com/news-item/231841; www.ft.com/intl/cms/s/2/24e45b40–3b00–11e2-bb32–00144feabdc0. html#axzz303cS9mWm; accessed October 23, 2016)

In 2012, and still according to ICID, India, China, Spain, the USA, Italy, Korea, Brazil, South Africa, Iran and Mexico were, in that order, the 10 countries with the largest area under drip irrigation (ICID, 2012). Middle-eastern countries (United Arab Emirates, Israel and Jordan) had the largest drip irrigated area when expressed as a percentage of their national irrigated area (ICID, 2012) (see Figure 1.1). The statistics as well as scholarly analysis further show that drip irrigation is mostly used for vegetable crops (in open fields or greenhouses), fruit trees (including vines), as well as industrial crops such as cotton and sugarcane (Ayars et al., 2007; Narayanamoorthy, 2004). Indeed, countries in which such crops are common are also the countries in which drip irrigated areas are largest (see the list above).

Although data are few and unreliable, there is a striking commonality in how the data that are available become part of larger narratives. For scholars, governments and the wider public, the tremendous growth observed over the last few decades (although it is not supported by consistent statistics) serves as proof of the multiple benefits that drip irrigation can bring to farmers and the society as a whole. Indeed, why would it spread so quickly if it was not good? Most scholars and policy documents further highlight, however, that drip irrigation **still only** represents a marginal share of all irrigation worldwide (about 5%; ICID, 2012), especially in emerging economies and developing countries (see GoI, 2004; Narayanamoorthy, 2008 on India and Zhu et al., 2013 or Zou et al., 2012 on China). They also stress that drip irrigation is concentrated in a limited

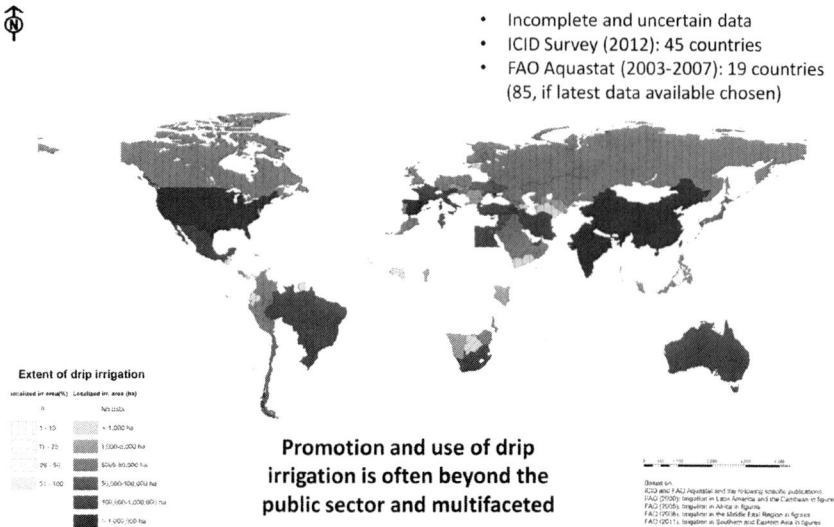

- Incomplete and uncertain data
- ICID Survey (2012): 45 countries
- FAO Aquastat (2003-2007): 19 countries
  (85, if latest data available chosen)

Promotion and use of drip irrigation is often beyond the public sector and multifaceted

Figure 1.1 A worldwide map of the importance of drip irrigation: data shortcomings.

number of countries (when taken together, the United States, Spain, China and India account for two thirds of all drip irrigated areas) and lament the fact that its benefits are still largely eluding smallholder farmers notably because of high investment costs (Hillel, 1988; Postel et al., 2001).

Contrasting the tremendous expansion with the limited actual extent of drip irrigation use is not a neutral way of telling the drip irrigation story. It serves as a justification for massive public investment and lending schemes in support of the promises of drip irrigation for a better future. This happens even if a large share of the world's irrigated food crops – rice, wheat, maize – are little adapted to existing drip irrigation systems. Indeed, many governments have engaged in aggressive policies to increase the share of the irrigated area under drip irrigation.

Public policies often take the form of subsidy programs that are, depending on the country, said to contribute to broader agricultural development and/ or water resources management policies. They are also invariably justified on the ground that high initial investment costs still constitute a major barrier for farmers to adopt drip irrigation (drip kits put aside), especially in emerging economies and developing countries (Postel et al., 2001; Ayars et al., 2007).

The map in the color section spatially represents the main discursive frameworks that underpin public policy and international development programs promoting drip irrigation around the world (the level of analysis is the country; the map is based on a review of available policy and development projects documents). In India, for instance, federal level financing schemes to promote drip irrigation have been launched in the early 1980s and have remained uninterrupted since then, culminating in the National Mission on Micro-Irrigation

(NMMI) in 2010 (Venot et al., 2014; Chapter 18). Drip irrigation was first seen as a way to promote a 'modern' (i.e. productive) agricultural sector; over the last two decades, however, the government primarily justified subsidy schemes on the basis that extending the area under drip irrigation (a technology that allows efficient water use) would alleviate the current pressure over ground-water resources, which are unsustainably tapped in many parts of India. Subsi-dizing drip irrigation is also presented as a way to include smallholders (who have limited financial resources) in the on-going efforts to build a modern and intensive agricultural sector (see Chapter 18 for further discussion on a subsidy scheme adopted in one particular state of India, Gujarat). In the same way, in Morocco, drip irrigation has been at least partly subsidized since 1986 onward (Venot et al., 2014). The recent national plan for water saving in irrigation (PNEEI) as well as the Green Morocco Plan (the national agricultural policy) have led to a massive conversion and extension of areas under drip irriga-tion. In Morocco, subsidies schemes are said to serve the objectives of preserv-ing groundwater resources and modernizing low performing public irrigation schemes (see Chapters 5, 6 and 10, this volume).

Among the 10 countries with the largest drip irrigated areas in the world (see above), Mexico (since 2003; see Chapter 9), Iran (since 1990), and China also subsidize the installation of drip irrigation systems. In Mexico and Iran, subsidizing drip irrigation is part of broader irrigation modernization projects and is also said to contribute to water preservation (for Iran) and agricultural intensification (for Mexico). In China, the primary stated objective to subsidize drip irrigation is 'water conservancy' (see Box 1.1). Other countries such as Chile, Iraq, Oman, Pakistan, Peru (through the irrigation sub-sector project, which started in 1996), the UEA, Syria, Tunisia, and Turkey provide direct sub-sidies to the installation of on-farm drip irrigation systems.[4] Broadly speaking, governments in South America generally justify subsidizing drip irrigation as a way to trigger a shift toward a modern and intensive agricultural sector; in the Mediterranean region and the Middle East, the '(ground)water conservation argument' is more prominent. In both regions, and notably in countries with large canal-based irrigated schemes, subsidizing drip irrigation is seen as a way to modernize low performing irrigation systems, which also receive significant attention from international development organizations (this is also the case in Egypt where drip irrigation is seen as a way to extend irrigation in desert areas while improving the management of irrigation systems in the Nile delta).

Support to drip irrigation can also take the form of tax exemption (e.g. Uzbekistan); low-interest loans on drip irrigation equipment (e.g. Iran, Syria, Brazil, micro-credit in sub-Saharan Africa); and subsidized seedlings for crops seen as particularly adapted to drip irrigation (such as date palm trees in Saudi Arabia, where the government also buys dates at a fixed price).[5] Agricultural policies and water laws may also indirectly favor the use of drip irrigation, as for instance is the case in Peru where getting water rights is tied to using 'effi-cient technologies' (see Chapter 8). Finally, in the least developed countries, support to drip irrigation does not take the form of direct subsidy schemes but of projects funded and implemented by public agencies and the third sector (NGOs, development agencies, private foundations and national government; see Chapters 12, 13, 14).

The multiple and reinforcing framing of drip irrigation

The map in the color section provides a snapshot of current discursive framings of drip irrigation, taking countries as the unit of analysis. However, the way drip irrigation has been seen, perceived and discussed in each of these countries has evolved with time, in relation to global discourses in the inter-related sectors of irrigation, agriculture and water management. The current 'multiplicity of drip' was not a given when the first large-scale experiments took place after the Second World War; it emerged with time and several periods can be identified since the 1960s.

The 1960s and 1970s (when experimentations with drip irrigation were increasingly taking place in the United States, Europe and Israel but also in India) are the decades of the Green Revolution, a term that refers to a tremendous increase in the production and productivity of agricultural crops. At the time, it was the *yield increase* allowed by drip irrigation (thanks to better matching crop water requirements)[6] together with the fact that the technology made it possible to *extend irrigation* to areas that could not be otherwise cultivated as well as to *use low quality water* that attracted the most attention. That drip irrigation allowed extending cultivation in arid regions (with sandy and little fertile soils) led to the fact that the technology became closely associated with the attractive image of a 'blooming desert' (see Plates 9 and 10 in the color plate). This was re-highlighted when Daniel Hillel was awarded the World Food Prize in 2012, and also underpins current drip irrigation dynamics in the coastal area of Latin America (see Chapter 8 and the map in the color plate). At the time, the fact that drip irrigation allowed *optimizing fertilizer application* (through fertigation, which consists of applying soluble fertilizers together with irrigation water) and *saving labour* (in developed economies) was the object of much attention, too. In short, in the 1970s, drip irrigation was mostly discussed in relation to *agricultural intensification*. This is still the case, as highlighted above, in developed economies, as well as in many South American countries as well as China.

In the mid- to late-1980s, things started to change. These were the times when irrigation performance started to be questioned and there were increasing calls for irrigation 'modernization', intended as 'a process of technical and managerial upgrading combined with institutional reforms, with the objective to improve resource utilization (labour, water, environment, economy) and water delivery service to farmers' (Facon and Renault, 1999). In some contexts, the modernization debate led to talk of pressurizing open canal systems or building pressurized systems from scratch (United States, Europe but also the Middle East, most noticeably Israel and Jordan and the United Arab Emirates) to increase irrigation efficiency (i.e. limit water losses through evapotranspiration and open conveyance systems). The consequences of shifting from canal to pressurized drip irrigation have been particularly intense in Spain where the government supported massive irrigation modernization programs (see for instance, López-Gunn et al., 2012). In policy discussion, the *high efficiency* of drip irrigation as a *modern technology* was often contrasted with the wastefulness of traditional irrigation methods, and it was this efficiency that took center stage (see van der Kooij et al., 2013 and Chapter 5 for a critical reflection on the notion of efficiency and the uses made of it). Such framing has now 'lost' in importance at a global level but still plays out in countries with large publicly managed irrigation systems such as Turkey, Egypt, Pakistan, Mexico, Brazil, and countries of Central Asia.

The most significant change may have happened in the early 1990s. It can be traced back to the 1992 Earth Summit held in Rio de Janeiro, which established 'sustainable development' (a term first coined in WCED, 1987) as a global policy goal and dominant development discourse. The fact that the global debate on sustainable development influenced the way drip irrigation was perceived is clearly reflected in the 5th International Micro-Irrigation Congress, the first of the series to have a title that reflects environmental concerns: '*Microirrigation for a changing world: Conserving resources/preserving the environment*' (ICID, 1995). Drip irrigation came to be seen through the lens of the environment, and with it *potential water savings* became its most important feature; irrigation modernization programs started to be assessed and evaluated through this lens, most notably in Spain (see above).

The 1990s are also the decade during which irrigation largely fell out of favor with the international community. Returns to public irrigation investments had been notoriously disappointing, whereas the negative consequences of large infrastructural works (notably dams and canals) started raising questions at international level. In addition, the rapid rise in the popularity of neo-liberal ideologies (after the collapse of the Soviet Union and the fall of the Berlin Wall had reduced the appeal of and belief in socialist forms of government) made it increasingly difficult to defend spending so much public money on the construction, operation and maintenance of public irrigation systems. In Against this backdrop, in the late 1990s and early 2000s, the water issue started to be framed as a tension between the environment and agriculture in a context of growing scarcity and competition over water. Agriculture was pointed out as being the single main water user and irrigation as wasting a significant share of it, as many observers emphasized the low efficiency of most irrigation worldwide (for instance, Postel, 1997; Gleick, 2001). The question became: how to save water in agriculture so that it can be allocated to other uses – including the environment.

Despite a growing debate on the water saving potential of drip irrigation in absolute terms (see for instance, Seckler, 1996; Perry, 2007; Molle et al., 2010 as well as Chapter 5, this volume), the alleged *high efficiency* of drip irrigation continued to be a defining element of the discourse. Rather than being framed from an engineering point of view (as had been the case in the 1980s when modernizing irrigation was high on the agenda), the question of efficiency was primarily discussed in relation to the alleged potential that drip irrigation held in terms of *preserving water resources* from overexploitation. This happened at a time when groundwater use in irrigation skyrocketed worldwide (especially in the developing economies of India, China and the southern shore of the Mediterranean) leading to declining water tables (see Shah et al., 2007 for an analysis of groundwater development trends in the agricultural sector). This focus on the potential that drip irrigation would hold in terms of preserving (ground) water is very much to the fore in North African and Middle Eastern countries as well as China and India (see Chapter 5 for a critique of this argument, on the basis of a case study in Morocco).

The mid-2000s, characterized by a renewed concern for food production and food security (notably heightened by the 2008 food crisis) constitute another turning point. Bringing together two distinct narratives, that of *agricultural intensification* (prominent in the 1970s) and of *water use efficiency* (prominent in the 1980s), drip irrigation became the material embodiment of a broad agricultural development discourse stressing the need for 'more crop per drop', which combines calls for intensifying agriculture with the need to preserve natural resources. It became also framed as a technology for adaptation to climate change (UNFCC, 2006; Clements et al., 2011; ADB, 2014).[7] In recent years, in relation with the emergence of the new water-food-energy nexus vocabulary (see Allouche et al., 2015) drip irrigation is also increasingly framed as *an energy saving* technology, notably in India. The energy-saving argument is closely linked to the water-saving idea that underpins many drip irrigation promotion initiatives. Indeed, by supplying water more efficiently to crops, drip irrigation would allow pumping less water per unit area (see Chapter 5), hence using less energy.[8] This is a contentious issue as other scholars highlight that the need to pressurize water to use drip irrigation is a reason for the booming energy consumption in agriculture (see, for instance, Rodríguez-Díaz et al., 2011 in the case of Spain).

At the same time, a parallel discourse the basis of which had been laid out as early as 1988, when Hillel (1988) highlighted that most drip irrigation development eluded small farmers in the developing world, started to take ground. This discourse centered on the potential of drip irrigation as a *poverty alleviation* tool. This new 'feature' of drip irrigation is related to what Postel et al. (2001) identify as the existence of 'a new spectrum of drip systems [that] can form the backbone of a second green revolution, this one aimed specifically at poor farmers in sub-Saharan Africa, Asia, and Latin America'. The late 1990s and early 2000s have indeed been marked by significant efforts from Non-Governmental Organizations (NGOs) (supported by national and international aid agencies), social enterprises, international agricultural research centers, and private irrigation equipment manufacturers to design and disseminate systems that would soon become known under the generic term of 'drip kit'. Drip kits are widely presented by those promoting them as being affordable, small size, and infinitively expandable. These features make them (seem) particularly suited for meeting the specific needs of smallholders in the developing world. Their small size, low cost and aptitude for irrigating vegetables, in addition, makes it easy to align them with goals of *women's empowerment,* based on the idea that female farmers can make use of the technology in their kitchen gardens (see among others, Polak, 2008; see also Chapters 11, 12, 13 and 14 as well as Plates 12 and 13 in the color plate).[9]

If the main discursive justification of NGOs disseminating these drip kits is to support smallholder farming and contribute to transforming it into an entrepreneurial venture, large-scale equipment manufacturers promote these smallholder drip irrigation systems as part of a broader Corporate Social Responsibility rhetoric[10] but also in an attempt to tap into an important market; the so-called 'Bottom of the Pyramid' (see Venot, 2016 for a critical analysis of these processes). The *poverty alleviation* potential of drip irrigation currently underpins efforts to promote the technology in most of sub-Saharan Africa, some countries of South and Central America (Bolivia, El Salvador, Guatemala, Haiti, Honduras, Nicaragua, etc.) and others in South and Southeast Asia (Bangladesh, Nepal, Myanmar, Cambodia, and Indonesia).

What is striking in this short historical review is that, over time, and at global level, drip irrigation has increasingly come to be discussed in relation to water resources management (and more broadly poverty and the environment) and less and less for its agricultural benefits (higher yields, input optimization, labour savings) that had made its early fame and are still the main reasons why farmers start using it. But what is also central to explain how drip irrigation 'conquered the world' is the fact that one story does not replace the other, rather they 'add-up', reinforce each other, with the emphasis shifting depending on context as shown in Figure 1.1, thus defining a true 'Panda' (see the Introduction of the book). The following section highlights that the discourses and (policy and development) initiatives promoting drip irrigation are not de-incarnated; they have been actively shaped and the trends described also reflect changes in the actors who are most influential in shaping the debate around drip irrigation.

A wide-reaching and multifaceted advocacy coalition

The 'mythical' figures

One of the reasons drip irrigation 'conquered the world' is linked to the fact that the technology found powerful advocates in a number of people from different backgrounds, pursuing different goals, and embedded in different social networks. To use the term of Madeleine Akrich (Akrich et al., 1998a, 1988b), these individuals acted as 'spokespersons' and created 'interessement' in an ever-widening coalition of actors who supported the dissemination of drip irrigation. These individuals came to personify some of the key discursive regimes of drip irrigation; hence lending weight to the latter.

Most authors agree that the concept of drip irrigation (i.e. the slow and frequent application of water to the root zone of crops) is as old as agriculture (e.g. Goyal, 2015), and trace back 'modern day' drip irrigation to experiments with underground clay pipes that took place in the second half of the 19th century

[1860s-1880s] in Germany and the United States (Ayars et al., 2007). The introduction of perforated pipes in the 1920s in Germany constituted a significant breakthrough as did the improvements in plastic technology: during the Second World War, perforated plastic pipes were used for vegetable production in greenhouses in the United Kingdom, France and Germany (Ayars et al., 2007; Goyal, 2015). Experiments in the 1950s followed in the United States and Israel and led to patents on emitters. This is when the personalization of (the history of) drip irrigation really begins.

Simcha Blass, a water engineer also known for having played a key role in the early years of Mekorot (the national water agency of Israel) and the elaboration of a 'fantasy plan' to irrigate the Negev in the late 1930s (a plan that would lay the basis for supplying water to Jewish settlements in the area and also formed the start of the future National Water Carrier),[11] is commonly attributed (together with his son, Yeshayahu) the first patent of a coiled in-line emitter (allowing the slowing of water flow and avoiding clogging) (Shoji, 1977; Ayars et al., 2007). That the role of Simcha Blass in drip irrigation development has attracted significant attention and has found its ways in collective memories can be partly explained by the fact that the contract he signed with the Kibbutz Hatzerim led to the establishment of the Netafim company. This company would later become the largest manufacturer of drip irrigation equipment worldwide (see below) and pursues aggressive communication campaigns. He came to be seen as an example of an Israeli innovator-entrepreneur, who allowed for making the Israeli desert 'bloom'. In the same vein, Daniel Hillel has received recent attention for the role he played in the development of drip irrigation, first in the Negev desert of Israel in the 1960s and in its dissemination in other countries of the world during the following decades. A soil and water scientist who has been affiliated to several Israeli and American universities, he closely worked with international organizations such as the United Nations and the World Bank. He was awarded the 2012 World Food Prize on account of 'his role in conceiving and implementing a radically new mode of bringing water to crops in arid and dry land regions – known as "micro-irrigation".' (www.worldfoodprize.org). Daniel Hillel personally embodies this ambition to make 'the desert bloom' but also an aspiration for making agricultural transformation and modernization benefit smallholders in developing countries – which has been most noticeably associated to other individuals since the 1970s.

Richard D. Chapin, the founder of a small irrigation equipment manufacturing firm in the United States, Chapin Watermatics, is also presented as an individual having been instrumental in developing drip irrigation. He is notably attributed with the development of drip-tape (thin plastic tubing) and early experiments with plastic mulch in the 1950s (Ayars et al., 2007); but also with designing and disseminating the first small-scale drip irrigation systems specifically targeting farmers in developing countries in the 1970s. This was notably done through a dedicated foundation, at the request of the NGO Catholic Relief Services (Keller, 2000; Venot et al., 2014; Chapter 11). In the late 1990s and early 2000s, the poverty alleviation potential of drip irrigation has also been personally embodied by yet another American entrepreneur, Paul Polak, founder of International Development Enterprises (iDE) and by an Israeli researcher, Dov Pasternak. The former attracted attention less for his role in developing drip irrigation systems per se and more for his unapologetic call to adopt a 'business approach to poverty alleviation'. He advocated a system based on the sale of affordable products to farmers. The enterprise would make a small margin on each sale (hence being a profitable venture in itself); that poor farmers bought the product acted as a proof of the added value of the latter and of the fact that farmers derived economic benefits from it. (Polak, 2008; Polak and Warwick, 2014; see Venot, 2016 for a critical analysis). The latter has been pivotal in harnessing support from development aid agencies and private manufacturers of drip irrigation equipment for the dissemination of 'drip kits', notably in sub-Saharan Africa (see Wanvoeke et al., 2015).

Through these individuals and their work, a picture of the different types of actors who have played and still play an active role in shaping the imagery of drip irrigation worldwide emerges, too. The supportive coalition for drip irrigation includes irrigation and agricultural engineers (and their professional organizations); public research centers and universities; private companies; NGOs and social enterprises; and international or national development agencies. When tracing the history of drip irrigation, what becomes clear is that drip irrigation is far from the 'neutral' technology that is often described. On the contrary it has long been, and still is, clearly associated with particular political and ideological movements and agendas (the Kibbutzim in the early years of Israel; the dream to make the desert bloom in the same country but also others such as the United States, Peru and Algeria (see Chapters 8 and 16); the 'will to improve' that drives development agencies; or the search of modernity through technology and entrepreneurship as articulated by social enterprises).

Public irrigation research: highlighting efficiency and productivity

Less publicized but also central in establishing drip irrigation as a technology that would become associated with notions such as efficiency, productivity and input optimization is an 'army' of agricultural and irrigation engineers working in research and development organizations, universities of developed countries, as well as extension services and agricultural ministries. ICID, the main professional body for irrigation practitioners and researchers, constituted the main platform through with international experiences with drip irrigation were shared. Even though most work in the sector still related to canal-based irrigation, an increasing number of actors became involved in drip irrigation research and development during the 1970s to 1980s (in 1985, the contributions of more than a 100 individuals were already acknowledged; Bucks, 1995).

The increasing attention toward drip irrigation among irrigation professionals is illustrated by a series of International Microirrigation Congresses, the first of which was organized in Israel in 1971. From the mid-1970s to the mid-1990s, engineers and researchers from the US Department of Agriculture-Agricultural

Research Services (USDA-ARS) appear to have been particularly active in these events, as shown by the fact that three of the successive four international micro-irrigation congresses were held in the United States in 1974, 1985 and 1995 (in 1988, the congress was held in Australia).[12] Such congresses played a key role in establishing the technical and scientific legitimacy of drip irrigation as a technology allowing to (1) use water (more) efficiently than canal irrigation; (2) save labour in intensive agricultural systems; (3) and improve yields (both in quantity and quality; see, for instance, ICID, 1995). At the time, the debate and discussions revolved on identifying and establishing the conditions for drip irrigation to fulfill a potential that had been demonstrated in experimental plots as well as closely monitored farmers field. It involved significant expert-led refinement of the technological artifacts and discussion around the elaboration of (quality) standards (see Box 1.2) and certification by professional authorized bodies.

The 5th International Micro-Irrigation Congress organized in 1995 marks the height of this era of irrigation engineering research. At the time, research focused on (1) determining water requirement and irrigation scheduling; (2) development in nutrient management, fertigation methods, and water quality issues, (3) system design and uniformity; (4) the development of decision support tools and simulation models (Phene, 1995). From 1995 onward, the influence of ICID as a global knowledge broker on drip irrigation progressively waned. Three subsequent international micro-irrigation congresses were organized in South Africa (2000), Malaysia (2008) and Iran (2011). These attracted less interest and attention than the 1995 congress (a much lower number of papers were presented during these meetings), but provided room for discussing drip irrigation in the context of emerging and developing economies.

That ICID became less pivotal perhaps can be attributed to the fact that drip irrigation became increasingly discussed in relation to water rather than agriculture while its members are first and foremost aging (male) agricultural and irrigation engineers. It is also linked to a widespread feeling among these individuals that the potential of drip irrigation (in terms of efficiency, productivity and scope of use) had been clearly demonstrated and that little interesting work (read: technical oriented research) remained to be done, as well as to dwindling investment (at least in Europe and the United States) in training, research and education in the field of irrigation engineering.

Box 1.2   An elusive search for standards?

At a time where certification schemes (built on a sets of ideas and standards; see box in Chapter 8 for instance) are gaining in importance in the agricultural and water sectors, the drip irrigation sub-sector stands aside and is conspicuous by the near absence of standards.

The issue of elaborating standards for drip irrigation systems has, how-ever, long been debated among irrigation professionals as illustrated by the existence of an ISO subcommittee on 'irrigation and drainage equip-ment and systems' since the late 1970s (Baudequin and Molle, 2003).[13] The 'proponents' of elaborating standards argue that the latter can act as a guarantee of 'quality' (material and installation), which they see as a pre-requisite to fulfill the potential of the technology (be it in terms of water saving or yield improvement). The 'opponents' to elaborating standards, on the other hand, stress that, given the systemic nature of drip irriga-tion systems (they are, by essence, networks), this is an impossible task. Indeed, standards could not possibly account for the diversity of on-farm situations and would act as impediments to innovation by users (see, for instance, Solomon and Dedrick, 1995). The opponents to 'strict standards' see the boom of drip irrigation development as a proof that the latter are not needed. They however also stress that many systems observed function in a sub-optimal way, and attribute this to the fact (1) that the systems are not designed or used according to 'good engineering practice' or (2) that the actual discharge of the drippers differ from the nominal value displayed by the manufacturer.

These discussions over standard development mostly take place among private manufacturers, industry associations whose *raison d'être* is to fur-ther develop and promote irrigation businesses (such as the US-based irrigation association and the EU-based European Irrigation Associa-tion), national and international standard agencies, and public testing laboratories. Farmers (who use drip irrigation systems and are presented by irrigation professionals as being the ultimate beneficiaries of standard development) are rarely contributing to these debates and meant to be represented by research laboratories – though the latter appear to be far from farmers' realities (see box on field-measurement performance level in Chapter 5).

Against this backdrop, and since its creation, the ISO sub-committee has endorsed over 30 international standards. A similar number of stand-ards have also been developed by the American National Standards Insti-tute (ANSI) and the European Committee for Standardization (ECS). Several countries such as Morocco and India also have their own stand-ards, which need to be followed by farmers if these want to benefit from the public subsidies.

Most existing standards at international level are 'design standards' for specific pieces of equipment (notably ancillary devices such as filters or pumps) and 'procedural standards' that set the conditions under which irrigation equipment shall be tested. Most commonly, drip irrigation pipes and emitters are tested for (1) the degree of variation the material has when compared to manufacturing specifications (notably in terms of the relation between water flow and pressure), and (2) the level of uniformity across emitters. How this relates to water use efficiency, the quantity of water used or yield levels is never discussed – these are indeed seen as beyond the realm of the tests and solely attributed to farmers' practices rather than to the equipment.

There is a conspicuous lack of 'performance standards' and, despite the name of the subcommittee, no 'system level' standard according to which systems need to be designed and their performance measured (again, this is 'justified' by the fact these are seen as depending on designers and farmers practices rather than to the equipment).[14] In such conditions, it is difficult to hold private manufacturers accountable to farmers or policy makers when promises of improved efficiency or productivity are not met. They can, as a result, continue to promote their products on the basis of their water savings and efficiency potential irrespective of whether this potential is ever realized. Very much aware of this discrepancy between efficiency claims and the ways the irrigation equipment works in farmers' fields, and in a context in which they are sidelined from the processes of standard development, several testing laboratories (among which the IRSTEA-France, the INCTEI-Brazil, the UCLM-Spain, and CIT-USA; see Plate 4 in the color plate for an illustration of a drip irrigation test bench) have formed the International Network of Irrigation Testing Laboratories (INITL). They aspire to conduct independent cross tests of drip irrigation equipment to form an alternative informal certification mechanism among peers.[15] Yet, the laboratories are limited in terms of the tests they can conduct and are also in a delicate situation as they derive part of their funding from private manufacturers (when the latter need a certification of their material to qualify for public tenders notably), who may (threaten to) sue them if they do not comply with internationally standardized tests.

Jean-Philippe Venot and Bruno Molle

When private companies gain in importance

The lower influence of networks such as ICID and public research organizations happened as the private sector, partly structured through industry associations, played an increasingly pivotal role in developing drip irrigation systems and supporting their dissemination through aggressive marketing policies – including lobbying national governments and organizations such as the EU so that they would consider subsidizing so-called 'water efficient' technologies.

The first drip irrigation manufacturing companies were set up in the 1960s (e.g. Chapin Watermatic in 1960 and Netafim in 1965). In the 1980s and 1990s, several companies that had been active in the field of agriculture and/or irrigation (such as Jain irrigation in India and Toro in the United States)[16] specifically started manufacturing drip irrigation equipment. In countries such as Cyprus, Greece, France, Italy and Spain a multitude of smaller scale manufacturers also emerged.

From the 2000s onward, two different trends can be identified. First, the emergence of a local manufacturing capacity and private companies in countries such as Turkey, Jordan, Morocco and evidently Korea and China, which have seen a tremendous increase in the use of drip irrigation. Second, a tendency toward a consolidation of the drip irrigation equipment market, which goes hand in hand with a financialization of the sector (see Box 1.3). Through a series of acquisition (Chapin Watermatic in 2006 and the Israeli Naandan in two stages, in 2007 and 2012), Jain irrigation for instance emerges as a second powerhouse alongside the Israeli company Netafim (both companies are valued at more than 1 billion US$). Alongside this trend, the profile of the decision makers of these companies has also evolved. While the directors and CEO of the 'early age' drip irrigation equipment companies had often been trained as agriculture and/or irrigation engineers, this is less and less the case today; they rather have MBA training or the like. This has come together with a change in the way drip irrigation is discussed: there is far less attention to the technical aspects of the technology (the technical potential of drip irrigation is seen as a given) and much more emphasis on overarching claims of water preservation and agricultural productivity that are difficult to actually grasp and assess.

Box 1.3   Global finance and drip irrigation

That drip irrigation has entered the world of 'big-business' is evidenced by the publication of reports on market trends and potential, which are sold for hefty sums (for instance Sustainable Asset Management, 2004; Transparency Market Research, 2013).

Maybe more significant is the fact that pension funds such as the London based private equity firm, PERMIRA, acquired a majority share of 61% of Netafim in 2011 for $850 million (Kibbutz Hatzerim and Kibbutz Magal remain minority shareholders with 33% and 6% respectively) and justified this investment on the ground that '*Netafim [was] the undisputed market leader and [. . .] ideally positioned to be the market maker in one of the most attractive segments of the agriculture industry worldwide*' (Permira, 2011a). In February 2013, PERMIRA further reported on its webpage that '*Through the cycle, Netafim has delivered strong double digit revenue growth and it targets a similar growth rate over the coming years. It will continue to pursue a growth strategy built around enhancing its leadership position as a market maker in drip irrigation. Specific growth opportunities include a significant expansion in emerging market businesses, particularly India and Latin America, entrance into the Chinese market and further penetration of large commodity crops*'. For the CEO of Netafim, '*having the backing of an international investor like the Permira funds [was] invaluable*' to accelerate the company's international growth (Permira, 2011b). This is not the only example of the involvement of global finance in the drip irrigation equipment sector. FIMI, one of the largest Israeli private equity funds also acquired John Deere Water in 2014 (which had itself been set up in 2006 and acquired Plastro and T-Systems, two other drip irrigation companies, in 2008).

What is striking in the two cases here is the terminology used. The focus is on economic growth and profit making; the technology – let alone the farmers – are conspicuously absent. There are tentacular implications to such trend. Netafim has 28 subsidiaries and 16 manufacturing plants worldwide, and employs 4,000 workers, some of which commented during interviews that the sales target for regional managers increased after the acquisition of PERMIRA to support the double-digit growth objective of the pension fund, while field workers benefit-sharing schemes were cut. More surprising, maybe, is the fact that Dutch civil servants (including contributors to this book) are part of the story. Indeed, PERMIRA is financed by other funds among which ABP (the Dutch pension fund for civil servants) through Alpinvest.

Jeroen Vos and Jean-Philippe Venot

Conclusion

Over the last 60 years, drip irrigation has been envisioned according to a multiplicity of lenses, which tend to reinforce each other in a never ending process. It was first seen as a way to extend agricultural land and a means to increase agricultural production. It then became associated with the idea of irrigation modernization and improved efficiency and productivity. Agricultural and irrigation engineers working in public and/or international agencies started losing their influence from the mid-1990s onward as drip irrigation travelled from experimental sites to farmers' fields, the political sphere, and global water arenas. In the latter, and in line with the dominant discourse of sustainable development, drip irrigation started to be discussed as a technology to preserve water resources and later as an adaptation strategy to climate change, including by and for the benefits of small farmers in the developing world. As the technology travelled from the realm of self-proclaimed irrigation experts and moved into a broader public sphere but also the world of big businesses and international water arenas, caution regarding the prospects it held largely gave way to definitive claims on the potential of the technology. This is partly because of the ever higher political and economic stakes linked to the promotion of drip irrigation.

What is striking in this 'traveling process' is that new sets of characteristics ascribed to the technology do not replace earlier ones; they rather add up to form a mythical artifact, which we called a Panda or a Hydra in the Introduction.

As such, drip irrigation has the characteristics of a boundary object (Star and Griesemer, 1989), that is, an artifact that is multivalent in character, can be cast in different ways to speak to various social worlds and communities of practices, and embodies multiple discourses in the process acting to mediate between different actors.

A multiplicity of drip irrigation narratives co-exist, which, when taken together, make the technology seem capable of addressing a number of current challenges in order to contribute to the shaping of a better future. Beyond the technical attributes commonly associated to the artifact (efficiency, modernity, productivity), it is this widely shared positive imagery that explains the attraction that drip irrigation exerts among a wide variety of actors.

Acknowledgments

The author would like to thank the four reviewers and his fellow co-editors for their valuable comments on earlier versions of this chapter. I also greatly acknowledge the insights from the many people interviewed in the course of this research project who shed light on the multiple aspects and dynamics of drip irrigation worldwide.

Notes

1 In countries where the installation of drip irrigation is subsidized, consistent monitoring of subsidy schemes at the national level could be a proxy to assess the extent of drip irrigation development. It would, however, overlook the fact that many farmers (often smallholders) purchase drip irrigation equipment on the market, without resorting to any subsidy scheme (see, for instance, Chapters 5, 7 and 16 in this volume). Furthermore, volumes of subsidy or sales do not always reflect actual use by farmers, as demonstrated in Chapter 13.

2 FAO Aquastat gives data on 'localized irrigation' (meant to include drip and sprinkler irrigation) for 85 out of 199 countries, with some records dating back to over two decades (FAO, 2013; Venot et al., 2014).

3 'China buys Israeli smart irrigation tech company for $20 million' (Anav Silverman, 2014); published in 'Breaking Israel News'; www.breakingisraelnews.com/21745/china-buys-israeli-smart-irrigation-tech-company-20-million.

4 In Algeria drip irrigation was heavily subsidized in the early 2000s but subsidies have been progressively cut back since then.

5 From the 1970s onward, Saudi Arabia pursued an elusive objective of food sufficiency and, in order to do so, heavily subsidized its agriculture sector, notably wheat production under pivot irrigation (Elhadj, 2004). Since the early 2000s, wheat subsidies have been phased out in part due to concern over groundwater over abstraction and the government is now prioritizing vegetables under greenhouses, using drip irrigation (FAO/GIEWS, 2015).

6 Many experiments consisted in assessing the yield-response of different crops under drip.

7 IPCC (2014) identifies 'efficient irrigation' as an adaptation strategy, without mentioned drip irrigation explicitly.

8 This concern over the energy use of (drip) irrigation has led some organizations to test associating drip irrigation systems with solar pumping together.

9 Interestingly, small drip irrigation systems such as drip kits are also promoted in the United States, though in specific contexts, that of the reservations of Native Americans. There, drip kits are promoted on the ground they could contribute to better health through improved diets and physical activity (vegetable gardening) in a context of very high incidence of diabetes.

10 Netafim, the largest manufacturer of drip irrigation equipment worldwide, for instance widely advertises that (1) by selling drip irrigation systems, the company contributes to preserving water resources worldwide and that (2) by partnering with NGOs, research centers and international development organizations in projects promoting drip kits, it contributes to poverty alleviation worldwide (see www.netafim.com/corporate--responsibility).

11 See www.mekorot.co.il/ as well as Siegel (2015). These early plans to develop the Negev desert were part of a Zionist strategy that aimed at influencing the drawing of the boundaries of the future State of Israel. The National Water Carrier plays a key role in the Israeli identity as does drip irrigation repeatedly presented as one of the innovation Israel 'brought to the world'.

12 Between 1979 and 1997, ICID had a dedicated working group on 'micro-irrigation', which would then be renamed 'on-farm working group' in 1998 (www.icid.org). Irrigation and agriculture engineers from Israel, South Africa, several European countries as well as India were also particularly active during these two decades.

13 Currently, the ISO/TC23/SC18 counts 11 participating and 18 observing members (www.iso.org) among which the countries where drip irrigated areas are the largest and the irrigation industry well developed (among the 10 countries with the largest drip irrigated areas, only South Africa and Mexico do not contribute). Solomon and Dedrick (1995) highlight that Israel, France, the United States, Canada, Spain and Italy are the most active countries.

14 A noticeable exception of 'performance standard' is a US standard on the performance of drip irrigation tape. Due to the opposition of Israeli manufacturers to the idea of standardizing tape, there is no equivalent ISO standard (US and Israeli manufacturers are competitors on the drip irrigation tape market). The difficulty to develop (material) performance standards at international level can also be illustrated by the 'journey' of a proposition that aimed at developing a standard on drippers' sensitivity to clogging, which, after 10 years, has been registered as a 'technical report' rather than a standard (hence is not bounding). In terms of system standards, the American Society of Agricultural and Biological Engineers (ASABE) has devised an 'Engineering Practice' which specifies practical engineering methods for field evaluation of existing microirrigation systems. India and Morocco have their own (national level) quality and dimensioning standards that farmers need to adhere to if they want to benefit from public subsidies. Whether the material and the system respect the standards is assessed by certified companies and engineers (the equipment of all major drip irrigation manufacturers is eligible to the subsidy).

15 Many of these laboratories are not certified by their national standard organizations, let alone by ISO, due to the costs this would imply for them in regard to a rather limited 'market' for testing drip irrigation equipment.

16 TORO is specialized in turf and landscape management. It entered the irrigation equipment sector in 1962 (turf and landscape) and expanded into drip irrigation for agriculture in the mid-1990s through acquisition. Rainbird, set up in 1933, is another major player in the irrigation equipment sector – specialized on sprinklers.

References

ADB (Asian Development Bank). (2014). *Technologies to support climate change adaptation in developing Asia*. Mandaluyong City, The Philippines: ADB.

Akrich, M., Callon, M., and Latour, B. (1988a). A quoi tient le succès des innovations? 1: L'art de l'intéressement. *Gérer et comprendre, Annales des Mines* 11: 4–17.

Akrich, M., Callon, M., and Latour, B. (1988b). A quoi tient le succès des innovations? 2: le choix des porte-parole. *Gérer et comprendre, Annales des Mines* 12: 14–29.

Allouche, J., Middleton, C., and Gyawali, D. (2015). Technical veil, hidden politics: Interrogating the power linkages behind the nexus. *Water Alternatives* 8(1): 610–626.

Ayars, J.E., Bucks, D.A., Lamm, F.R., and Nakayama, F.S. (2007). Introduction. In F.R. Lamm, J.E. Ayars, and F.S. Nakayama (Eds.), *Microirrigation for crop production*. Vol. 13: Development in agricultural engineering. Elsevier: Amsterdam, pp. 1–26.

Baudequin, D., and Molle, B. (2003). Is standardization a solution to improve the sustainability of irrigated agriculture? Improved irrigation technologies and methods: Research, development and testing. *Proceedings ICID International Workshop*, Montpellier, France, September 14–19, 2003, pp. 1–12.

Bucks, D.A. (1995). Historical developments in microirrigation. In ICID, *Microirrigation for a changing world: Conserving resources/preserving the environment. Proceedings of the fifth international microirrigation congress*, April 2–6, 1995, Orlando, FL, pp 1–5.

Burnham, M., Ma, Z., and Zhu, D. (2015). The human dimensions of water saving irrigation: Lessons learned from Chinese smallholder farmers. *Agricultural and Human Values* 32(2): 347–360.

Clements, R., Haggar, J., Quezada, A., and Torres, J. (2011). *Technologies for climate change adaptation: Agriculture sector*. Roskilde, Denmark: UNEP Risø Centre.

Elhadj, E. (2004). *Camels don't fly, deserts don't bloom: An assessment of Saudi Arabia's experiment in desert agriculture*. SOAS Occasional Paper No 48. London: SOAS, University of London.

Facon, T., and Renault, D. (1999). Modernization of irrigation system operations. *Proceedings of the 5th ITIS Network International Meeting*, Aurangabad, October 28–30, 1998.

FAO/GIEWS. (2015). *Saudi Arabia country brief*. Available at www.fao.org/giews/country-brief/country.jsp?code=SAU (Accessed November 23, 2015)

Food and Agriculture Organization (FAO). (2013). *AQUASTAT database*. (Accessed November 25, 2015)

Gleick, P. (2001). Making every drop count. *Scientific American*, February 2001.

GoI (Government of India). (2004). *National task force on micro irrigation in India*. New Delhi: Ministry of Agriculture.

Goyal, M.R. (Ed.). (2015). *Sustainable micro irrigation: Principles and practices*. Toronto and Waretown, NJ: Apple Academic Press.

Hillel, D. (1988). Adaptation of modern irrigation methods to research priorities of developing countries. In G. Le Moigne, S. Barghouti, and H. Plusquellec (Eds.), *Technological and institutional innovation in irrigation. Proceedings of a workshop held at the World Bank*, April 5–7, 1988. World Bank Technical Paper No. 94. Washington, DC: The World Bank, pp. 88–93.

ICID. (1995). Microirrigation for a changing world: Conserving resources/preserving the environment. *Proceedings of the fifth international microirrigation congress*, April 2–6, 1995, Orlando, FL.

ICID. (International Commission on Irrigation and Drainage). (2012). *Sprinkler and micro irrigated area*. Available at www.icid.org/sprin_micro_11.pdf (Accessed November 25, 2015).

IPCC. (Intergovernmental Panel on Climate Change). (2014). *Climate change 2014: Impacts, adaptation, and vulnerability. Part A: Global and sectoral aspects*. Contribution of Working Group II to the Fifth Assessment Report of the Intergovernmental Panel on Climate Change, Cambridge University Press, Cambridge and New York, 1132 pp.

Keller, J. (2000). *Gardening with low-cost drip irrigation in Kenya: For health and profit*. Technical Report prepared for International Development Enterprises (IDE). Available at siminet.org/fs_start.htm (Accessed November 25, 2015)

Kooij, S. van der, Zwarteveen, M., Kuper, M., and Errah, M. (2013). The efficiency of drip irrigation unpacked. *Agricultural and Water Management* 123: 103–110.

López-Gunn, E., Mayor, B., and Dumont, A. (2012). Implications of the modernization of irrigation systems. In L. De Stefano and M.R. Llamas (Eds.), *Water, agriculture and the environment in Spain: Can we square the circle?* Madrid: Fundación Botín, pp. 241–255.

Molle, F., Wester, P., and Hirsch, H. (2010). River basin closure: Processes, implications and responses. *Agricultural Water Management* 97(4): 569–577.

Narayanamoorthy, A. (2004). Impact assessment of drip irrigation in India: The case of sugarcane. *Development Policy Review* 22(4): 443–462.

Narayanamoorthy, A. (2008). Drip and sprinkler irrigation in India: Benefits, potential and future directions. In U. Amarasinghe, T. Shah, and R.P.S. Malik (Eds.), *India's water future: Scenarios and issues.* Colombo: International Water Management Institute, pp. 253–256.

Permira. (2011a). *Permira funds complete acquisition of Netafim.* Permira Press release 21/12/2011.

Permira. (2011b). *Investing in growth- annual review 2011.* Permira: London.

Phene, J.J. (1995). Research trends in microirrigation. In ICID, *Microirrigation for a changing world: Conserving resources/preserving the environment. Proceedings of the fifth international microirrigation congress,* April 2–6, 1995, Orlando, FL, pp. 6–16.

Polak, P. (2008). *Out of poverty: What works when traditional approaches fail.* San Francisco, CA: Berret-Koehler Publishers.

Polak, P., and Warwick, M. (2014). *The business solution to poverty: Designing products and services for three billions new customers.* San Francisco, CA: Berret-Koehler Publishers.

Postel, S. (1997). *Last oasis: Facing water scarcity.* New York: W.W. Norton & Co.

Postel, S., Polak, P., Gonzales, F., and Keller, J. (2001). Drip irrigation for small farmers: A new initiative to alleviate hunger and poverty. *Water International* 26(1): 3–13.

Reinders, F.B. (2006). *Micro-irrigation: World overview on technology and utilization.* International Commission on Irrigation and Drainage. Keynote address at the opening of the 7th International Micro-Irrigation Congress in Kuala Lumpur, Malaysia, 13-15 September 2006.

Rodríguez-Díaz, J.A., Pérez-Urrestarazu, L., Camacho-Poyato, E., and Montesinos, P. (2011). The paradox of irrigation scheme modernization: More efficient water use linked to higher energy demand. *Spanish Journal of Agricultural Research* 9(4): 1000–1008.

Seckler, D. (1996). *The new era of water resources management: From "Dry" to "Wet" water savings.* International Irrigation Management Research (IIMI) Report No. 1. Colombo: IIMI.

Shah, T., Burke, J., and Villhoth, K. (2007). Groundwater: A global assessment of scale and significance. In D. Molden (Ed.), *Water for food, water for life: A comprehensive assessment of water management in agriculture.* London: Earthscan and Colombo: International Water Management Institute, pp. 395–423.

Shoji, K. (1977). Drip irrigation. *Scientific American* 237(5): 62–68.

Siegel, S.M. (2015). *Let there be water: Israel's solution for a water-starved world.* New York: Thomas Dunne Books and St. Martin's Press.

Solomon, K.H., and Dedrick, A.R. (1995). Standards development for microirrigation. In ICID, *Microirrigation for a changing world: Conserving resources/preserving the environment. Proceedings of the fifth international microirrigation congress,* April 2–6, 1995, Orlando, FL, pp. 1–5.

Star, S.L., and Griesemer, J.R. (1989). Institutional ecology. "Translations" and boundary objects: Amateurs and professionals in Berkeley's museum of vertebrate zoology. *Social Studies of Sciences* 19(3): 387–420.

Sustainable Asset Management. (2004). *Precious blue: Investment opportunities in the water sector.* Zurich, Switzerland: SAM.

Transparency Market Research. (2013). *Micro irrigation systems market -global industry analysis, size, share, trends and forecast, 2012–2018.* Transparency Market Research.

UNFCC (United Nations Framework Convention on Climate Change). (2006). *Technologies for adaptation to climate change*. Bonn, Germany: UNFCC Secretariat.

USDA. (2011). *Peoples republic of China- agricultural policy directive: Number 1 document for 2011*. Global Agricultural Information Network Report No. CH 11024. Washington, DC: USDA.

Venot, J.P. (2016). A success of some sort: Social enterprises and drip irrigation in the developing world. *World Development* 79: 69–81.

Venot, J.P., Zwarteveen, M., Kuper, M., Boesveld, H., Bossenbroek, L., Kooij, S. van der, Wanvoeke, J., Benouniche, M., Errahj, M., Fraiture, C. de, and Verma, S. (2014). Beyond the promises of technology: A review of the discourses and actors who make drip irrigation. *Irrigation and Drainage* 63(2): 186–194.

Wanvoeke, J., Venot, J.P., Zwarteveen, M., and Fraiture, C. De. (2015). Performing the success of an innovation: The case of smallholder drip irrigation in Burkina Faso. *Water International* 40(3): 432–445.

WCED (World Commission on Environment and Development). (1987). *Our common future*. Oxford: Oxford University Press.

Zhu, X., Li, Y., Li, M., Pan, Y., and Shi, P. (2013). Agricultural irrigation in China. *Journal of Soil and Water Conservation* 68(6): 147A–154A.

Zou, X., Li, Y., Gao, Q., and Wan, Y. (2012). How water saving irrigation contributes to climate change resilience – a case study of practices in China. *Mitigation and Adaptation Strategies for Global Change* 17: 111–132.

2 Decentering the technology

Explaining the drip irrigation paradox

Margreet Zwarteveen

When I was a teenager, there was a riddle that my friends and I liked telling each other. It went more or less as follows: what allows you to swim, skate, ride horseback, bike, run and dance? The answer was a particular brand of tampons (I forgot which), and it referred to the television commercial that was used to promote it: it showed a happy, care-free, young and pretty woman who was swimming, skating, riding horseback and so on. While the commercial, of course, aimed to suggest that the tampons would make the menstrual period pass almost unnoticed, its message could also be read as saying that it was the tampon itself that produced the capacity to swim, ride horseback, etc., in the young woman.

The results of our studies of drip irrigation remind me of this riddle. When reading the various articles, books and documents or watching the many short videos about drip irrigation on the Internet, one gets an impression of drip irrigation as an almost magical device, something that will increase water use efficiencies, production and incomes while also saving labour and energy. These claims of its efficiency and productivity increasing potential motivate substantive efforts to promote the adoption of or conversion to drip irrigation. In countries as diverse as Spain, Peru, Morocco, Mexico or India, national governments actively encourage the use of drip irrigation through targeted subsidy programs. International donors and research centres such as the World Bank, the FAO or the Japanese International Cooperation Agency (JICA) are sponsoring and technically supporting the transfer from surface to drip irrigation. The policy story behind this is that irrigation is one of the largest consumers of water, while competition with other uses – and most notably drinking water in cities – is increasing. This is why efforts to generate an intensification of farming (needed to feed an ever expanding population) need to be accompanied with deliberate attempts to use water more efficiently and productively. On paper, a large-scale shift to drip irrigation thus serves the purpose of further increasing the area irrigated (until the 'potential' is reached), while at the same time making significant water savings and reversing the trend of groundwater overexploitation. Drip also figures prominently in attempts to better account for or create awareness about the water used in the production of crops: many of the certificates, trademarks and labels developed to inform customers that what they buy has been produced in water-wise ways use drip irrigation as the only or main indicator for this (see Box 8.1 in Chapter 8; Vos and Boelens, 2014).

In parallel to these efforts to promote drip irrigation as an intrinsic part of larger processes of green modernization, a range of companies and social enterprises have developed and are developing smaller scale low-cost and low-pressure systems for use by poorer farmers (see Chapters 11 to 14 in this volume). These aim making the benefits of drip irrigation also available to farmers who cannot afford conventional drip systems – which tend to be associated with the more well-to-do farmers. The efforts of iDE-International Development Enterprises are perhaps best known in this respect. iDE indeed found influential spokespeople (see Akrich et al., 1998a, 1988b) for spreading their almost evangelical message about the benefits of low-cost drip kits, for instance in Sandra Postel (Postel et al., 2001).

In analogy to the tampon riddle, drip irrigation would thus be the possible answer to the question: what allows conserving aquifers, alleviating poverty, and producing more food with less water? However, while the suggestion that a tampon has the capacity to do many things is hilarious – that was precisely why the riddle was funny – the suggestion that drip irrigation has the ability to do many things does not provoke laughter. On the contrary, it is a serious proposition, one that is oft rehearsed and repeated in professional meetings and texts and one that figures prominently in the policies of many governments. The popularity of drip has a self-reinforcing effect. It has turned the technology into a brand, with its overall positivity rubbing off on anyone who associates with it, almost irrespective of how it is used or how it works. Conspicuously displaying one's affiliations with drip irrigation – through installing a drip system on one's field, or by promoting it in policies of agricultural modernization or water saving – therefore often also serves the symbolic function of showing one's own modernity, technological sophistication or environmental awareness.

The seriousness of the proposition that drip has the capacity to do many good things is not founded on much evidence of the technology actually delivering on its promises. Our own research findings, for instance, show that drip irrigation does not automatically or necessarily bring about water savings in farmers' fields. Maya Benouniche's measurements of what happens when farmers start using drip irrigation systems suggest that many use more water than needed with drip irrigation not because they are not able to operate the technology, but rather because water saving is not a priority for them (Benouniche et al., 2014a). Harm Boesveld's observations (based on his wealth of field evidence in advising farmers on how to use drip, see also Chapter 3), on the other hand, suggest that few farmers master the difficult art of precisely tuning water gifts to crop requirements.

Even though governments actively promote drip irrigation as a means to use water more efficiently, many farmers seem to adopt drip not because of a desire to save water, but because it facilitates the ease of operations; because it allows irrigating on slopes; because it makes fertilizer applications more precise and efficient or again because it allows them to renegotiate their social status or tap into wider support networks. Similarly, the evidence Jonas Wanvoeke collected in Burkina Faso shows that there is a distinct mismatch between the enthusiasm of governments, donor agencies and NGOs about low-cost drip kits on the one hand and that of farmers on the other (Wanvoeke, 2015; see also Chapter 13).

What explains the continued enthusiasm for drip irrigation, in spite of little evidence that it realizes expectations? What, in other words, is the basis for the seriousness – and durability or tenacity – of the proposition that drip irrigation has the capacity to save water, improve productivity or alleviate poverty? It is not as if there is a lack of information or knowledge about the often disappointing results of drip-in-use. The influential and widely available United Nations Water Development Report of 2009, to name just one source of such information, for instance clearly states:

> Localized irrigation (micro-irrigation) for example, has a limited impact on water depleted by evapotranspiration in the field and chiefly reduces return flows. Thus water 'saved' by upstream irrigators can come at the expense of downstream users, allowing upstream users to expand their cultivation. This may be desirable from the perspective of the upstream farmer, but the result is increased water depletion.
>
> (WWA, 2009: 156)

In Morocco, engineers of the public agricultural services will also readily acknowledge that drip irrigation may not result in water savings. Many Indian water scholars and practitioners are likewise well aware that drip irrigation will not necessarily lead to higher water use efficiencies. As one of them, the reputed water scholar Tushaar Shah, wrote in response to our question to him why the government of India was so keen to promote drip irrigation:

> The claim that this will raise water use efficiency to 95% is untenable, even disingenuous. The real losses – that is evaporation and not seepage which is much needed groundwater recharge – are from the open canals which will remain open even after this project. (...) So farmers are unlikely to collaborate in improving water use efficiency – which is a meaningless concept in Indian systems. I doubt if this scheme will deliver efficiency.
>
> (Shah, personal communication, 2013)

Likewise, the lack of enthusiasm of farmers for low-cost drip kits in several African countries as well as in India and Nepal is diligently reported in articles and reports. Yet, rather than dampening the enthusiasm for and belief in drip irrigation, this evidence only works to re-invigorate efforts to better market and disseminate it, train farmers in how to use it, improve support services or create a better enabling environment, or find a different problem to which it can serve as a solution (such as improving yields or alleviate poverty). The technology itself remains beyond doubt and question.

In what follows in this chapter, I propose one possible way of explaining what we – as the project team involved in studying drip irrigation – have come to refer to as the 'drip irrigation paradox': its continuously reified status as a solution to problems of water, food security and poverty even when there is little evidence to support this status and much to challenge it. In the explanation, I draw on some of the Science and Technology Studies (STS) insights that we have used to frame our overall project, making use of the opportunity to also briefly introduce and explain these insights to those who are less familiar with this body of scholarly work.

From DRIP to drips: decentering the technology

An important pillar supporting the proposition that drip irrigation systems can help solve contemporary water and food challenges is science: it is because the design and functioning of drip technologies is based on the universal laws of nature that drip systems are (or appear) as true and are believed to be capable of fulfilling expectations. In the case of drip irrigation, these are first of all the laws of hydraulics. Hydraulics is an applied science and a form of engineering dealing with the mechanical properties of liquids or fluids. It thus explains, predicts and allows manipulating the behaviour of water through canals, pipes and tubes. Next to hydraulics, also meteorology, agronomy and soil science – in particular, how these come together in the calculation of crop water requirements (i.e. the amount of water a plant needs in different stages of maturation for optimal growth and yields; see for instance the different FAO publications on this topic; e.g. Allen et al., 2006, Doorenbos and Kassam, 1979 and Doorenbos and Pruitt, 1977) – provide the scientific foundation for the claim that drip works. Together, insights from these sciences form the basis for the design of drip irrigation systems as tools that allow the precise administration of water to plants, making sure that water gifts accurately match requirements for optimal growth, yields and indeed profits.

The firm scientific anchors of the technology work to place its functionality beyond doubt. Hence, if actual performances in for instance farmers' fields are suboptimal, this raises questions about 'the context of use' – and particularly about the behaviours and practices of those operating the system – rather than about the (basic design of the) technology or the engineers behind it. While suggested solutions to problems of underperformance may include some adaptations to the technology – for instance, to make it fit better to particular soil-weather-crop conditions – most remedial proposals concern changes in the conditions of its use, better adapting those to the optimal functioning of the technology. These adaptations, for instance, include teaching farmers how to (better) operate their systems; devising tools and applications that allow irrigators to more precisely determine when to give how much water (see Chapter 3); or setting up support services.

In this line of thinking, the technology – the drip irrigation system – is at the centre. It is what it is and does what it does because of its intrinsic characteristics – characteristics that have been purposively designed by trained experts based on scientific insights. This line of thinking, also, tends to be technologically optimistic and linear: the drip irrigation system itself is or becomes the emblem of irrigation wisdom; it embodies, helps bring about or even enforces wise (modern, rational) water behaviour.

An important preoccupation of Science and Technology Studies (STS) is and has been to challenge such technocentric lines of thinking by literally 'decentering the object in technoscience' (Law, 2002). STS replaces an understanding of society-technology interactions from the normative and prescriptive perspective of the engineer (or the technology) with one that is more descriptive and agnostic about what technologies are and do, approaching these interactions as happening in complex and dynamically evolving settings that are often full of surprises and contingencies. Attempts to 'decenter the technology' took inspiration from and evolved alongside a large body of critical scholarship on the 'socialness' of technology and science. Important here was and is the argument that the divide between the natural and the social – the divide that provides the ultimate anchor for scientific truth claims in positivist or more populist views of science – is not something that exists prior to the investigation or prior to being named by people. Instead, this separation – and by implication the one between 'science' and 'politics' (the anchor of everyday definitions of objectivity; see Latour, 2004) – is the *effect* of particular ways of naming, defining, categorizing and mapping.

For drip irrigation, it is not difficult to show how its performance as a rational water user is indeed *the effect* of particular ways of aligning it to water sources, soils and plants, rather than the outcome of its intrinsic characteristics. A review of the scientific literature on drip irrigation (van der Kooij et al., 2013) revealed that most studies that test (and most experiments to improve) the water use efficiency of drip irrigation are done in experimental plots. Most research work consists of meticulous attempts to isolate the drip experiment – and thus measurements of the performance of the technology – from any outside 'interferences', creating a laboratory-like setting (that is, at least in theory, replicable or reproducible elsewhere). When reporting the results of these experiments, the constructed and purposively manipulated experimental setting comes to represent 'the natural' (that which can be objectively known and therefore represents a universal truth) and indeed 'the real', thereby serving as the anchor for the claim that drip irrigation works. Unlike suggested, this 'natural' or 'real' is clearly not an 'untouched' reality that existed before the investigation or experiment, but something that is consciously constructed and manipulated by research engineers. Through the reduction of an overwhelmingly complex world to an isolated laboratory-like setting, they create a closed system-model in which a small number of controllable variables determine water flows. Some drip irrigation studies, for instance, conduct their measurements in closed containers to avoid disturbances from the outside world (see Goodwin et al., 2003; see also Plate 4 in the color section), others go through great efforts to avoid what they consider 'interfering' water flows, such as runoff, drainage, capillary rise. Such disturbances and interferences are of course impossible to avoid on farmers' fields. To express the implication of this in STS terms: when drip irrigation 'saves water' in experimental settings, this is not just the result of the technology itself, but the performative effect of a careful process of aligning the technology to water flows, pumps, crops, people and stories (words, meanings, ideas). The water use efficiency of drip irrigation is therefore only real within settings that resemble these experiments: only when it is similarly aligned to words, tools, resources and people can it be expected to perform similarly.

This theoretical move to decenter the technology destabilizes the belief in the intrinsic 'goodness' (modernity, wisdom, rationality) of drip irrigation, thereby also troubling questions of whether it works. The move turns normal concerns about the objectivity of science – centred on the question of how to keep politics out – upside down. The explicit acknowledgment that science (or technology) and society are always deeply entangled is precisely the starting point of STS studies. Technology is, as a consequence, no longer a more or less objective given, something that exists outside and independent of society or history. Rather than emanating from universal 'natural' laws, (the design of) a technology becomes something that happens (or is done) in specific cultural and social settings, by people with specific identities. Their ideas and practices are importantly shaped by traditions, interests, (geo-)politics and institutional path-dependencies.

The popular and dominant engineering definition of drip irrigation sees it as a single and stable object that does what it does because of its intrinsic material characteristics – let's refer to it as DRIP. STS replaces this with a definition of drip irrigation as something that comes into being in and through day-to-day sociomaterial practices, engineering practices among them (cf. Mol, 2002: 6). In this definition, what drip irrigation does and how is not pre-given by its design characteristics, but is a performative effect of aligning the technology to a context in a specific configuration. This does not mean that design characteristics do not matter. They do, because they importantly confine the number of possible configurations of technology with context: what a drip irrigation system can do is clearly co-determined and limited by its specific material properties. One implication of the technology-in-context definition that STS studies propose is that there are many possible drip irrigation systems: the definition replaces DRIP with drips. After all, different contexts each produce a slightly different version of the technology. Hence, the engineering version of drip irrigation – let's call it e-drip (with 'e' referring to engineering and efficiency, and to experiment) – is only one possible version of the technology. Instead of the one and only DRIP, e-drip thus becomes just another drip which is not a priori granted a privileged status or considered normatively superior to other possible drips. Another implication is that the question whether drip works, or whether it performs well, can no longer be answered with a clear-cut 'yes' or 'no': there are many shades and grades of working (de Laet and Mol, 2000; van der Kooij et al., 2015).

In the next sections, I present and discuss some of the drips that we have come across in our research project. To do this, I have made a broad categorisation of two main worlds in which drip is entangled: the world of farming, and the world of policymaking. As I show, in farming, drip is so successful and popular because it indeed allows for many things: there are multiple *drips*. In its entanglement with policymakers, what drip irrigation does instead depends on DRIP – the belief in e-drip as the only right and true version of drip irrigation. I use a discussion of the overlaps and frictions between these two worlds as a device to shed light on the 'drip paradox.'

Drips in farming: fluidity and multiplicity

To most farmers and the engineers directly working with them, the statement that there are many different drips will not be surprising or shocking at all. In Chapter 3 of this book, Harm Boesveld for instance clearly argues from a practical engineering perspective that it is misleading to talk about or present drip irrigation as one standardized technological package. Although any drip irrigation system is built up from more or less similar technical elements – including a pump, a filtration system, (sub) mains and laterals, emitters and valves – there are many different ways to combine these different elements to make a particular drip irrigation system. In addition, there is a wide range of materials available for every individual component of a drip irrigation system, further increasing the number of possible systems. Boesveld gives the example of emitter types: these range 'from highly sophisticated pressure compensated emitters to a simple small spaghetti tube with a knot that is inserted into a pinched hole in the lateral' (Boesveld, Chapter 3 of this volume). In the same way, pumps may range from sophisticated electrical pumps to simple treadle pumps from local materials.

Making drip irrigation systems work (which may mean a range of things, as I discuss below) in a farmer's field is far more complex than simply following standard guidelines. It requires fitting the technology to the specific needs and capabilities of the farmer (in terms of knowledge, the availability of funds and labour), the characteristics of the farming system, water availability, water quality, and field, soil and crop types. As Boesveld's chapter also shows, drip system operators face many challenges when working with the technology. These challenges include emitters being moved by strong winds; rats biting through drip lines to quench their thirst; and drip lines that start to wander because of how they shrink and expand with differences in temperature. To fit the system to one's needs and abilities and to deal with these challenges, drip irrigation operators (often together with the design engineers asked or assigned to help them) have to be inventive and creative: they need to continuously make technological or managerial adaptations to meet their requirements for use and indeed make the drip irrigation system work for them.

Benouniche et al. (2014b) provide some telling illustrations of such technological and managerial adaptations by smallholder farmers – which they themselves refer to as *bricolage* – in the regions of the Saïss in Morocco (see also Chapter 15, this volume). For instance, because it is difficult for many of them to mobilize the necessary capital to purchase a full-fledged drip irrigation system with all its components, many start their engagement with drip irrigation by buying secondhand material; drip lines and emitters that were discarded by larger agro-export farms. Another way to save on initial investment costs is the use of a *roubiniyatte,* or valve system. This relatively cheap and simple system is composed of only the basic components of a drip system: engine, pump, tubing and valves. In general, there is no filter system or a basic fertigation unit. The system is particularly suitable for plots where the ground is not even, as the different drip lines can be opened and closed according to the pressure available. Farmers who have problems of water shortage also use the system, as it allows them to irrigate the different parts of a plot consecutively. However, the system has a number of limitations (clogging of emitters, poor fertigation, labour requirements), which is why farmers consider it a transitory system: they hope that it will allow them to make enough profits to buy a more sophisticated system in the future. Interestingly, some farmers in their efforts to improve their drip irrigation systems to local conditions and requirements collaborate with local craftsmen and welders, who as a result also become actively enrolled in the larger project of 'making drip work' for this specific region and this specific group of farmers. A particularly noteworthy achievement here is their fabrication of a new and sturdier version of the hydrocyclone filter (Benouniche et al., 2014a).[1]

As part of the same study, but published in a different article, Maya Benouniche et al. (2014b) further elaborate how different farmers use drip for different purposes. This is a telling illustration that a drip irrigation system can perform differently depending on specific farm contexts. What is perhaps most striking in this analysis, in addition to showing that few farmers use drip for water-saving purposes, is that their reasons to adopt drip irrigation not only relate to how it allows changing irrigation practices but also include poetic, cultural or symbolic motivations. Many smallholders for instance adopt drip irrigation to increase or improve their social status. Taher is one of them. He got to know drip irrigation when working as a labourer on a large farm. An important reason for him to install drip irrigation on his farm was to improve his standing within family and community networks. His uncle, with whom he had a long-standing conflict over a land inheritance question, had been the first to install drip irrigation in the community – earning him the status of a modern and progressive farmer. Taher recounts: '[T]he day I installed drip irrigation, I was proud of myself. I finally had drip irrigation, there was no longer any difference between me and my uncle' (Taher, 31 years old, 2011). Taher's mother even compared the installation of the drip irrigation system on their field to 'a wedding party. I invited my neighbours, and I organized a small party. Oh yes, on that day I was happy, it felt right, now we had drip irrigation like the large-scale farmers. I showed the drip irrigation system to my neighbours. Afterwards, everybody wanted to install drip irrigation. Our neighbour Mounir, for example, came over next day and asked Taher to install drip irrigation on his farm' (mother of Taher, 62 years old, 2011). (Benouniche et al., 2014a: 215).

Likewise, based on the study that Saskia van der Kooij conducted in a farmer-managed irrigation system in the same region of Morocco, we identified four possible performances – or enactments – of drip irrigation, four versions of drip that co-existed (and partly overlapped) in this specific context: (1) drip as a modernizer of farming; (2) drip as a device to make farming cleaner; (3) drip as an emancipator of sharecroppers; and (4) drip as an alliance builder with the State (van der Kooij, 2016). Here, we just briefly discuss the first two, using these to further illustrate that there are many possible versions of drip irrigation which all do something else, or indeed work differently. The first version of drip irrigation in the farmer-managed irrigation system Khrichfa directly stems from its overall image as a modern technology, modern both in the sense of being more advanced and sophisticated than surface irrigation systems (i.e. according to engineers) and in the sense of being associated with more entrepreneurial and intensive forms of farming (the first drip irrigation systems in Morocco were used by large capital-intensive farms, often on the coast). The association of drip irrigation with modernity is actively reproduced by the Moroccan Government in its Green Morocco Plan (see Chapter 10). For many, especially young male farmers, drip irrigation has therefore become an important symbolic marker of more modern ways of farming and rural life. The image of modernity is not just symbolic, however, but also stems from how drip irrigation allows making the activities of farming and irrigation cleaner and easier (the second performance). With drip, irrigation only requires someone to open the valve, walk around the plot to check whether all emitters work properly, and then close the valve again. In addition, drip irrigation keeps the irrigator and his clothes literally cleaner than surface irrigation – which involves going into the muddy fields to open and close field channels with shovels, thereby unavoidably becoming soiled with muddy water. Here, the way in which drip allows changing irrigation practices – making them easier and cleaner – thus overlaps with and reinforces the more symbolic association of drip irrigation with progressiveness and modernity.

A second way in which drip irrigation makes farming into a cleaner profession is that, in Khrichfa, drip marks a change from tobacco to onion cultivation. Tobacco is considered dirty in a number of ways. First, tobacco is *haram* (religiously forbidden). In addition, and as government campaigns have made clear, the fact that the smoking of tobacco is unhealthy make its production into something to be almost ashamed of. Also, the drying of tobacco leaves is a laborious and not much-liked activity, often done by children. While lacing the tobacco leaves, the sticky juice that oozes from the leaves marks one's hands with brown stains that are difficult to remove. Onions are a cleaner and more positive crop. Growing them is a source of pride, as onions are a prominent part of the national dish *tajine*. Yet, to be able to grow onions – which need to be irrigated once every two days during peak demand to get an optimal yield – with surface water (most onion cultivators in the region use groundwater), the Khrichfa water users had to adapt their rotational water distribution – which provided water every five to seven days. They did this by smartly increasing the irrigation frequency and by installing small reservoirs (van der Kooij et al., 2015) that made their access to water less dependent on what happened in the rest of the irrigation system. The introduction of drip irrigation was a further innovation that contributed to making the growing of onions possible. When taken together, the changes associated with or facilitated by drip irrigation – in crops, technologies, cultivation methods – allowed young farmers to position themselves as cleaner, more modern and entrepreneurial farmers than their fathers.

The evidence of low-cost drip irrigation (see Wanvoeke, 2016; Chapter 13) likewise illustrates that the popularity of drip irrigation among farmers not always stems from how it changes irrigation or farming practices. In Burkina Faso, most smallholder farmers agree to adopt drip irrigation systems because it usefully connects them to development cooperation networks. Hence, in the arena of the development project and for farmers, micro-drip kits are defined by the side benefits that accompany their introduction, such as motorized pumps, free inputs, the promise of credit, or the prospect of acquiring social prestige and forging new alliances.

All this shows that the generic label DRIP is a broad term to refer to a wide range of systems, practices and performances that are only loosely associated with particular ways to guide water from a source to the root zone of plants, or indeed with e-drip. There are at least two 'ranges' of drip irrigation systems, two ways in which drip systems pluralize. The first range has to do with what Marianne de Laet and Annemarie Mol, when discussing the Zimbabwe bush pump, refer to as *fluidity*. Talking about how the pump changes form and function depending on where and by whom it is used, they observe: 'the boundaries are vague and moving, rather than being clear or fixed' (de Laet and Mol, 2000: 225). Even more than the Zimbabwe bush pump, drip irrigation systems come in many different shapes and sizes. Because they are composed of different components all of which are available in a range of different dimensions, qualities and materials, the number of material forms that a drip irrigation system can take is very large. And similar to the Zimbabwe Bush Pump, drip systems lend themselves to tinkering by farmers and engineers: when installing or using drip systems, farmers or operators often make smaller or larger adaptations to the system to better fit it to their particular preferences or conditions. Indeed, and as also Andy Keller suggests in his foreword to this book when praising drip irrigation's 'flexibility to fit nearly any shape field; its scalability; its adaptability to nearly any topography, soil and crop', it may well be that it is partly this *fluidity* of drip irrigation systems that makes it into such an attractive and popular technology.

A second range, or way in which drip irrigation systems pluralize, is that they are used in a variety of contexts, and for a variety of purposes. Some farmers install it to use water more efficiently, as when drip irrigation allows irrigating a larger area with the same amount of water. Others choose drip because it facilitates irrigation operations, making irrigation into a cleaner job, or because drip helps them to improve their yields. In Burkina Faso, farmers agree to adopt the technology not because of any intention to actually use it but because it usefully connects them to wider flows of development cooperation support and benefits. Drip can also perform in more poetic or symbolic ways: in the Saïss in Morocco, having a drip system for instance confers modernity and progressiveness onto the owner – a capacity that directly stems from the successful way in which it is promoted and positively branded by an alliance of drip equipment manufacturers, public or private design engineers (who are often taught that DRIP is a more advanced form of irrigation) and government agencies. This *multiplicity* of drip, its capacity to do or perform many different positive things may be another reason to explain its popularity among farmers.

An important implication of the fluidity and multiplicity of drip irrigation is that the relation between e-drip – the scientific and engineering version of drip – and drips-in-use cannot be assumed: the mere presence of the technology says little to nothing about how farmers use water or farm.

DRIP in policymaking: plasticity and black-boxing

By invoking the terms Panda and Hydra, the Introduction of the book signalled what it referred to as the *plasticity* of drip: 'it can be mobilized as a "solution" to a range of ever-changing and conjectural international policy issues, including environmental conservation, food security, gender equity, poverty alleviation, agricultural modernization'. A possible way to think about this is to conceive drip irrigation as a boundary object (see also Chapter 1). Boundary objects are objects that are both plastic enough to adapt to the local needs and constraints of the several parties employing them, yet robust enough to maintain a common identity across sites. They have different meanings in different social worlds but their structure is common enough to more than one world to make them recognizable, a means of translation (see Star and Griesemer, 1989; Star, 2010).

This boundary object capacity of DRIP – the version that postulates e-drip as the only true and therefore normatively superior version of drip irrigation, one that will perform similarly anywhere provided conditions and contexts are (made) suitable – is one reason why DRIP is a useful political device: it can fuse diverse political projects, social visions, ecological concerns, cultural imaginaries, discursive formations, institutional practices and economic strategies of global competitiveness together (Swyngedouw, 2013). In Morocco, to once again refer to the country about which we have the most information, the solution of drip irrigation thus usefully allows the Ministry of Energy, Mines and Water (in charge of the health and longer term conservation and sustainability of water resources) and the Ministry of Agriculture to find each other: DRIP promises to simultaneously help modernize agriculture and to save water. Similarly, and most strikingly in sub-Saharan Africa, in its reliance on manufacturing and design companies, low-cost drip irrigation fits the contemporary policy rhetoric of combining aid with trade. It allows private companies to flourish, and helps a range of development actors (NGOs, funders) to remain in business. In how it promotes the intensification of farming, low-cost drip irrigation neatly fits currently popular poverty alleviation strategies in agriculture that are predicated on turning smallholder subsistence farmers into entrepreneurs.

Drip irrigation – that is the DRIP version of it – may serve as a life buoy for public irrigation engineering, which has been and is under severe pressure because of the worldwide decline in public support for irrigation that characterized the last decades. The explicit mention of DRIP on political agendas provides public irrigation engineers in different countries with new projects and with a new sense of mission and urgency, while providing those in charge of water conservation with a credible storyline. Simultaneously, DRIP creates new employment opportunities for young irrigation engineers in the private sector, as government support for DRIP often entails the active welcoming and involvement of private irrigation companies, both to provide the equipment as to help design and install systems on farmer's fields.

A second but related way in which DRIP does important political work is in how it allows black-boxing or circumventing contentious questions about water (re-)allocations. Its promise of water savings makes it seem possible to continue or even increase water diversions or extractions, as it creates the suggestion that there will be more of it to share among diverse competitors (see also Chapter 4). As the chapters about Morocco, Peru, Mexico, Chile and India in this volume suggest, DRIP nourishes the belief in the possibility to combine economic growth – increases in agricultural productivity or modernization – with environmental conservation, expecting (or calling on) the private sector and (quasi-)market mechanisms to make this happen. It is also in this capacity that DRIP figures prominently in water sustainability certification schemes (Vos and Boelens, 2014; also see Box 8.1 in Chapter 8). Its promise of wise water use allows farmers (big and small, although it is mostly the larger agro-export companies that will go through the effort of certification) to continue abstracting water from aquifers for the cultivation of water-intensive export crops.

Another way of saying this is that in usefully allowing to combine three important features of contemporary development strategies – economic growth, environmental conservation and active private sector participation – DRIP gives business as usual (intensive farming) either an attractive green gloss, or – by linking it to poverty alleviation – provides it with an attractive 'social responsibility' image. The argument can also be reversed: the combination of neo-liberalism with environmentalism that characterize today's favoured solutions for tackling economic as well as environmental problems provides a perfect policy context for DRIP to flourish. In this sense, the spread and success of drip irrigation marks the transition from a hydro-structural to a decentralized, but still definitively state-led, market environmentalist framework (see also Introduction and Chapter 9; Swyngedouw, 2013). Indeed DRIP makes it seem as if there will be no losers in the modernization project: no one has to limit agricultural production and neither will incomes of farmers decrease (cf. Allan, 1999).

Both the plasticity of drip irrigation – its ability to function as a boundary object – as its capacity to black-box contentious allocation questions crucially hinge on its DRIP version: the version that postulates e-drip as the only true and therefore normatively superior version of drip irrigation.

Conclusion: the success of drip irrigation as a network-effect

Dividing the presentation and discussion of many drips in two 'worlds', the world of farming and the world of policymaking, provides a good starting point for solving the drip paradox. Drip in the world of farming is about *drips*: it is about a plurality of drip systems that perform in many different ways. Indeed, its fluidity and multiplicity make drip irrigation into a beautiful and popular device among farmers. What it does, and how, can however not be assumed; its performance is the contingent and sometimes unexpected outcome of its interactions with irrigators, soils, water sources and flows, funding streams, markets, etc. The technology co-determines water-society relations, but does this in ways that are difficult to predict or plan. Rather than neatly following the designs and intentions of its inventors (engineers) or of policymakers, drip irrigation shifts shape and identity depending on where and by whom it is invoked or used. The relation between e-drip and drips, therefore, is never a given.

Drip in the world of policymaking is about DRIP – the e-version of drip irrigation elevated to universal status through the use of specific notions of science and technology as existing outside of history and society. DRIP functions as a useful device to bring together different interests and political projects, while also allowing to depoliticize contentious questions of water re-allocation. The widespread policy popularity of drip irrigation is only loosely related to its popularity among farmers and has little to nothing to do with how it changes farming and irrigation practices. Instead, the policy popularity of drip irrigation has to do with how it *promises* to change these practices. This promise, as I have shown, appears scientifically founded and true as it is based on e-drip – the one enacted by research engineers on experimental plots in controlled conditions. Yet, through DRIP – the specific definition of technology that ascribes its performance to its intrinsic characteristics – drips become an as-yet not realized but eventually realizable form of DRIP, an immature version of it. In this way, farm realities do not challenge DRIP. This allows the very presence of a drip irrigation system to become the indicator of wise water use, improved productivity or poverty alleviation. The effective performance of drip irrigation as a boundary concept or depoliticizer crucially hinges on this belief in e-drip as the only true drip (DRIP).

Bruno Latour famously proposed that realities (and knowledges of realities) 'depend on practices that include or relate to a hinterland of other relevant practices' (Law, 2009: 241). Sustainable knowledge rests in, and reproduces, more or less stable networks of relevant instruments, representations and the realities that these describe. This is what makes realities – together with the techniques and representations that enact them – seem stable, durable and reliable (Law, 2009: 241–2). The implication is that realities are only real within particular networks (of people, devices, funds) or systems of circulation. Truths – in our case DRIP – therefore, are not universal (Law and Mol, 2001); they are only 'realized' in definite form within the networks of practices that perform them. The question of how this happens then becomes important: this is a question of power, interests, traditions and culture.

Following this line of argumentation allows to show that the wide circulation and 'success' of drip irrigation is the effect of powerful networks and alliances which have made the technology central to contemporary policy proposals for solving problems of environmental sustainability. The DRIP alliance not just consists of governments, agro-export companies seeking cheap access to water, and drip equipment manufacturing companies looking for new markets for their products but also counts on the active support of international environmental organisations (the World Wildlife Foundation for instance); international water policy networks (such as the Global Water Partnership or the Stockholm International Water Institute that awarded the Stockholm Water Prize to NETAFIM, a major manufacturer of drip equipment, in 2014) and diverse policy-science initiatives that have taken up water as their cause (such as the Water Footprint Network – see http://waterfootprint.org/en/). The alliance notably includes researchers, whose funds may depend on promises of the technology's success. All these actors have strongly invested in DRIP, and have a stake in perpetuating the belief that the technology performs. Their power and funds also attract and enroll many others, even those who do not believe in DRIP. Many irrigation engineers and researchers who know very well that DRIP will never fulfil its promises have become cautious and reluctant to openly admit this, as they themselves have come to depend on DRIP – for their jobs, careers, status. In its popularity, DRIP has come to confer an overall aura of positivity: progress, greenness, advancement, entrepreneurialism (and what not) on those who associate with it, including smallholder farmers.

Few of those who have invested in DRIP are interested in finding out about drips. This is both dangerous and ethically wrong, as it means that huge amounts of (public and private) money are spent without water being saved or poverty being alleviated. Part of the funds that are currently spent on promoting the adoption or conversion to drip irrigation would be better spent on in-depth investigations of how different farmers actually use water to irrigate their crops, either with or without drip irrigation, and with what environmental and social effects. Rather than from a normative and managerial identification with engineers that pre-define wise water use and wise water users, such investigations should attempt to make sense of water and farming practices from the perspective of those who engage in them. Such a serious engagement with the messy and dynamic practices of farmers forms the only viable starting point for identifying strategies for changing their irrigation practices and improving the sustainability and equity of water governance.

Acknowledgments

I thank all members and contributors to the Drip Irrigation in Perspective team. This chapter reflects the many discussions and interactions we have had over the years. Special thanks to Jean-Philippe Venot and Marcel Kuper for their useful comments and suggestions.

Note

1 In this book, the chapter of Mohamed Naouri et al. (Chapter 17) provides yet another account of the innovativeness of farmers in appropriating and adapting drip systems to their specific needs and demands, while Chapter 15 provides a more general account and discussion of *bricolage*.

References

Akrich, M., Callon, M., and Latour, B. (1988a). A quoi tient le succès des innovations? 1: L'art de l'intéressement. *Gérer et comprendre, Annales des Mines* 11: 4–17.

Akrich, M., Callon, M., and Latour, B. (1988b). A quoi tient le succès des innovations? 2: le choix des porte-parole. *Gérer et comprendre, Annales des Mines* 12: 14–29.

Allan, T. (1999). Productive efficiency and allocative efficiency: Why better management may not solve the problem. *Agricultural Water Management* 40: 71–75.

Allen, R.G., Pereira, L.S., Raes, D., and Smith, M. (2006). *Crop evapotranspiration (guidelines for computing crop water requirements)*. FAO Irrigation and Drainage Paper No. 56. Rome: FAO.

Benouniche, M., Kuper, M., Hammani, A., and Boesveld, H. (2014a). Making the user visible: Analysing irrigation practices and farmers' logic to explain actual drip irrigation performance. *Irrigation Science* 32(6): 405–420.

Benouniche, M., Zwarteveen, M., and Kuper, M. (2014b). Bricolage as innovation: Opening the black box of drip irrigation systems. *Irrigation and Drainage* 63(5): 651–658.

Doorenbos, J., and Kassam, A.H. (1979). *Yield response to water*. FAO Irrigation and Drainage Paper No. 33. Rome: FAO.

Doorenbos, J., and Pruitt, W.O. (1977). *Crop water requirements*. FAO Irrigation and Drainage Paper No. 24. Rome: FAO.

Goodwin, P.B., Murphy, M., Melville, P., and Yiasoumi, W. (2003). Efficiency of water and nutrient use in containerised plants irrigated by overhead, drip or capillary irrigation. *Australian Journal of Experimental Agriculture* 43(2): 189–194.

Kooij, S. van der. (2016). *Performing drip irrigation in the farmer-managed Seguia Khrichfa irrigation system, Morocco*. PhD thesis, Wageningen University, Wageningen.

Kooij, S. van der, Zwarteveen, M., Boesveld, H., and Kuper, M. (2013). The efficiency of drip irrigation unpacked. *Agricultural Water Management* 123: 103–110.

Kooij, S. van der, Zwarteveen, M., and Kuper, M. (2015). The material of the social: The mutual shaping of institutions by irrigation technology and society in Seguia Khrichfa, Morocco. *International Journal of the Commons* 9(1): 129–150.

Laet, M. de, and Mol, A. (2000). The Zimbabwe Bush Pump: Mechanics of a fluid technology. *Social Studies of Science* 30(2): 225–263.

Latour, B. (2004). *Politics of nature: How to bring the sciences into democracy*. Cambridge, MA and London: Harvard University Press.

Law, J. (2002). Aircraft stories: decentering the object in technoscience. Durham, NC: Duke University Press.

Law, J. (2009). Seeing like a survey. *Cultural Sociology* 3(2): 239–256.

Law, J., and Mol, A. (2001). Situating technoscience: An inquiry into spatialities. *Society and Space* 19(5): 609–621.

Mol, A. (2002). The body multiple: Ontology in medical *practice*. Durham, NC: Duke University Press.

Postel, S., Polak, P., Gonzales, F., and Keller, J. (2001). Drip irrigation for small farmers: A new initiative to alleviate hunger and poverty. *Water International* 26(1): 3–13.

Star, S. L. (2010). This is not a boundary object: Reflections on the origin of an object. *Science, Technology, & Human Values* 35(5): 601–617.

Star, S. L., and Griesemer, J. R. (1989). Institutional ecology. "Translations" and boundary objects: Amateurs and professionals in Berkeley's museum of vertebrate zoology. *Social Studies of Sciences* 19(3): 387–420.

Swyngedouw, E. (2013). Into the sea: Desalinisation as hydro-social fix in Spain. *Annals of the Association of American Geographers* 103(2): 261–270.

Vos, J., and Boelens, R. (2014). Sustainability standards and the water question. *Development and Change* 45(2): 205–230.

Wanvoeke, J. (2015). *Low cost drip irrigation in Burkina Faso: Unravelling actors, networks and practices*. Thesis submitted for degree of Doctor at Wageningen University, Wageningen, The Netherlands.

Wanvoeke, J., Venot, J.-P., Zwarteveen, M., and de Fraiture, C. (2016). Farmers' logics in engaging with projects promoting drip irrigation kits in Burkina Faso. *Society and Natural Resources* 29(9): 1095–1109.

World Water Assessment Programme. (2009). *The United Nations world water development report 3: Water in a changing world*. Paris: UNESCO and London: Earthscan.

3 The practice of designing and adapting drip irrigation systems

Harm Boesveld

Introduction

Drip irrigation is a method of frequent, slow application of water to the crop through physical devices, called emitters, at selected points along water delivery lines. Drip irrigation is based on the concept of irrigating only the root zone of the crop and maintaining the water content of the root zone at near optimum levels (Keller and Bliesner, 1990; James, 1988; Jensen, 1983). Over the past decades, irrigation engineers have developed various drip irrigation systems ranging from high tech, sophisticated technological systems to relatively simple, low-cost systems. Any drip irrigation system is built up from more or less standard technical elements. These building blocks include, among others, a pump, a filtration system, (sub) mains and laterals, emitters and valves. However, standard drip irrigation systems do not exist. The challenge for the engineer and the farmer is to balance farmer's needs, farming system characteristics, water availability, water quality, field and crop types, and farmers' capabilities with the technological options available. Even if this is done with great care, many irrigation practitioners experience problems in operating drip irrigation systems. Farmers and extension workers continuously encounter specific problems and are inventive in making adaptations to solve these. This chapter describes the basic layout and design aspects of a drip irrigation system and describes the most commonly encountered problems at farm level and the adaptations made. These adaptations can be a response to flaws in initial design, ill-adapted choice of components, improper installation or end-users skill set.

Drip irrigation systems discharge water close to plants or plant rows in small amounts. Terms like drip irrigation, trickle irrigation, localized irrigation and micro irrigation are also often used to refer to the same method of irrigation, which can cause some level of confusion. They all refer to the slow application of water in the form of drops close to each plant. Water can be applied at a single point on the land surface through devices called emitters. This creates a wetted area around each plant. It can also be applied through a line source from either closely spaced emitters or tubes with continuous or equally spaced openings that discharge water a drop at a time. Discharge rates for emitters are generally between 1 and 12 Liters per hour (L/h) for online emitters and 1 and

6 L/h for inline emitters. In this chapter the term 'drip irrigation' is used to refer to the aforementioned methods.

Mostly, emitters are placed on the ground, but they can also be buried in subsurface irrigation systems. Subsurface irrigation involves the use of emitters to apply water below the soil surface, directly into the root zone. Discharge rates for subsurface irrigation are generally in the same range as classic drip irrigation rates. Since the laterals are buried, little visual control is possible on the functioning of the emitters.

Basic composition of a drip irrigation system

Water flows through valves, filters, mainline, sub mains and laterals before it is discharged into the field through emitters. There are many handbooks and publications that elaborate on the components and design aspects of drip irrigation systems (Burt and Styles, 2007; Reinders et al., 2012; Keller and Bliesner, 1990). The typical components of drip irrigation systems are shown in the Figure 3.1 and briefly introduced below.

• *A pump unit or pressure supply.* All drip irrigation systems require energy to move water through the pipe distribution network and discharge it through the emitters. In some instances, this energy is provided by gravity as water flows downhill through the delivery system. However, in most irrigation systems, pumps are used to pressurize irrigation water so as to impart a head (flow and pressure) to the water for transport and uniform distribution over the fields to the crops. In low-cost drip irrigation systems

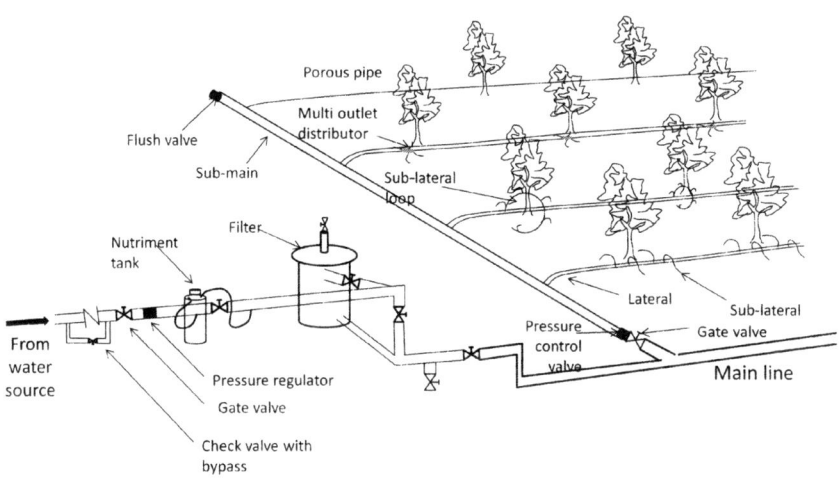

Figure 3.1 Typical layout and components of a drip irrigation system.

Source: Drawing by Jean-Philippe Venot, adapted from Brouwer et al., (1990).

the pressure is often provided by a reservoir at an elevation of a few meters high. The reservoir can be filled by hand with buckets, a simple (treadle) pump or even a motor pump (Chapter 11).

- *A check valve.* A check valve is a type of valve that allows fluids to flow in one direction and closes automatically to prevent flow in the opposite direction and drainage of the system when the pump is shutdown. In some countries, a separate backflow protection device is required by law when water is obtained from a domestic water supply. This prevents water containing substances, like fertilizer or pesticides, from flowing back and contaminates the water source. A check valve is usually not a legally approved method of backflow prevention.

- *A nutrient tank (fertigation).* Water soluble fertilizers can be applied through the drip irrigation system; this is a system of fertilization called 'fertigation', which allows the application of soluble plant nutrients directly to the root system of the crop. Drip irrigation systems may include injection equipment from a nutrient tank. Injection is located after the pump since fertilizers and chemicals may be corrosive to the pump.

- *Filters and pressure gauges.* Filters have the function to remove suspended solids from the water. Irrigation water is rarely without suspended materials and removal of these particles is usually required for enhancing drip irrigation performance and avoiding emitters clogging (this is often one of the main problems encountered with low-cost drip systems; see Chapter 11). Filter systems are selected depending on the type of materials in the water. Settling basins, sand filters, screen filters and centrifugal separators are the main devices to remove suspended material. Pressure gauges before and after the filter are used to determine the timing of cleaning the filters. When the pressure difference between the gauge before and after the filter is increasing, it is usually time to clean the filters, as this is a sign of solids accumulating in the filter. Cleaning can be done manually or be fully automated.

- *Water meter.* The water meter registers the amount of water that is flowing into the system. It is not an essential part of a drip irrigation system but it is highly recommended because it can provide valuable information on the operation and functioning of the system. It is a useful instrument to monitor the amount of water applied and can provide early warning of clogging or leaks.

- *Main line and submains.* The main and submains are the pipelines that convey and distribute the irrigation water to the plots. The main is the largest diameter pipeline of the network capable of conveying the full flow of the system. The submains are smaller diameter tubes which extend from the mainline to divert the flow to the various plots. Often the main and the submains are permanent pipelines that are buried into the soil. In most cases (sub)mains are buried and the cheaper polyvinyl-chloride (PVC) is used. In case the (sub)mains are above ground the materials is black high density polyethylene (HDPE) as PVC will become brittle when exposed to sunlight.

- *Laterals or drip lines.* These are the smallest diameter pipelines with the emitters that are placed along the plant rows to release the water to the root zone of the crop. Black polyethylene (PE) is generally used as the laterals are usually above ground. The emitters can be inline or online. Inline emitters are inserted in prefabricated drip lines with emitters incorporated in the PE drip line at regular intervals. Online emitters are mounted on the PE lateral through punched holes in the lateral and placing the emitters at the desired position. The lateral's downstream end is plugged with a piped fitting or simply crimped over. These should be opened periodically for flushing the lateral lines; flushing is a critical management action to maintain a drip system. Dripline with automatic flush ends are available on the market as well, these open automatically when the pressure drops under a certain threshold allowing water and dirt to drain from the lateral. Equipment such as laterals and emission devices typically remain in one place during the growing season. Systems are usually permanently installed for trees and vines and row crops. For other field crops they may be portable and moved to a different field after the irrigation season is complete.
- *Emitters or drippers.* The emitters are the devices that release water from the lateral into the environment in a controlled way. There are many types and designs of point-source emitters available commercially. The main characteristic is a low discharge often in combination with a long or extended irrigation time. Emitters come in different forms varying from very simple devices to highly sophisticated ones.
- *Pressure regulators* reduce the water pressure to the required operation pressure. Most drip irrigation systems operate at a relative low pressure and the pressure regulator provides a relatively constant water pressure at the inlet to the irrigation system even if the incoming pressure may vary. A pressure regulator can only reduce the water pressure and not increase it.
- *Valves* turn on and off the water to drip lines or sections of the irrigation system so different sections can be irrigated in sequence. The control valves may be manually operated or fully remotely operated. Manual control is simple, the valve has a handle to open or close the section. The remotely operated valves are mostly electrical valves that are switched on and off by a timer.

Designing a drip irrigation system

There are two primary types of designers, those designing the system components and those preparing the hydraulic design of the system.

The component designers design the individual irrigation materials like emitters, tubes, pumps, etc. This type of design mostly takes place in the private sector of equipment manufacturers. Every separate piece of material has been designed and is the result of the vision and assumptions of the designer on how the piece of equipment will be used. Therefore, a wide range of materials is available for every individual component of a drip system. An example are the wide range of emitters type from highly sophisticated pressure compensated emitters to a simple small spaghetti tube with a knot that is inserted into a pinched hole in the lateral. The expensive former one comes with complex designed labyrinth to regulate discharge. Within a range of pressure variation the emitters provide an equal discharge. The designer focused mainly on diminishing the variability of discharge (in relation to pressure fluctuations) at the emitter level, so as to allow for high discharge uniformity and was less concerned with the cost of the device. The latter, a simple spaghetti tube with a knot, is the extreme on the other side of the spectrum. The designer made every effort to keep the price low and accepted lower discharge uniformity. The discharge of this device is very much dependent on the operating pressure. These tube-emitters are often chosen if the investment cost of a system is a major issue and are often observed in many low-cost drip irrigation systems disseminated in the developing world (see Chapter 11). In between these extremes there is a wide variety of emitters to choose from. In the same way, pumps range from sophisticated electrical pumps to simple treadle pumps from local materials.

The system designers are those who plan and assemble, from separate irrigation material on the market, a complete operational irrigation system at field level that meets the requirements of use of the farmer. The system designers are often private consultants, public extension agents or employees of equipment manufacturers. Usually, the design procedure starts with a consultation with the farmer to determine the requirements for use of the drip system and an inventory of available data. Availability of water sources, soil data, crop data, peak water requirement of the crops cultivated by the farmer and socio-economic data is the basis for the system design choices. This will result in a preliminary layout of the system and choices with regard to the different irrigation materials but the actual designing and dimensioning of a drip irrigation system for a given plot is an iterative process.

There is a vast amount of literature and handbooks dealing with the design aspects of drip irrigation systems at field level (Burt and Styles, 2007; Lamm et al., 2006; Hoffman et al., 2007; Keller and Bliesner, 1990). We will touch only on the key elements that influence the use and performance at farmers' fields (for details, one can refer to these publications: distribution uniformity, peak crop water requirement, water availability, cropping pattern, soil types and water quality are important criteria in the design of a drip irrigation system.

Distribution uniformity

For system designers, distribution uniformity (DU) is a key factor in designing a drip irrigation system. DU is a term that relates to the evenness of water application throughout the field (Burt at al., 1997). A low DU is a sign of under irrigation and/or over irrigation at different places in the field. This will lead to suboptimal crop growth in parts of the field. DU should not be confused with irrigation efficiency (IE), seen here as the ratio of the volume of irrigation water beneficially used and the volume of irrigation water applied (Burt at al., 1997; see van der Kooij et al., 2013 for a discussion on the multiple ways the term efficiency is used). Although IE relates to irrigation practices and DU to hardware status, the two parameters are often interacting. A farmer with a system having distribution problems (and thus a low DU) may be attempted to compensate by providing more water to the entire area, resulting in over irrigation on a part of the plot, and thus a reduction in the irrigation efficiency. On the other hand, irrigation may be uniform but if the irrigation timing and duration are not appropriate it may cause run-off and deep percolation losses resulting in low irrigation efficiency. However, in general a high IE can only be achieved if the DU is high.

The selection of drip irrigation materials and the configuration of a drip system is often a trade-off between DU and economics. Drip irrigation systems have been improved over the last few decades with a lot of effort from Western irrigation equipment suppliers who aimed at meeting the demands of commercial agricultural enterprises for higher irrigation efficiency and DU. Expensive and sophisticated drip equipment were designed to reach these goals but did not meet the widespread need for inexpensive, divisible irrigation systems for poor farmers cultivating small plots. For these, affordability and rapid payback are necessary and this means cheaper and simpler drip equipment and accepting lower DU performance.

There are a number of factors and components that influence the overall system DU and are important considerations in the design of a system for a given situation. Pressure differences between emitters will cause flow rate differences (especially for non-pressure-compensated emitters). The pressure difference is influenced by the length and dimension of the tube system transporting the water to the emitters. Varying topographical conditions also have an effect also on pressure differences within the system. Other factors like (partial) clogging, wear and manufacturing variation can also play a considerable role. Burt and Styles (2007) showed that in California about half of the non-uniformity is caused by pressure differences and around half by a combination of clogging and wear and to a minor extend to manufacturing variation.

Peak water requirements

Crop evapotranspiration (ETc) is the combined process of evaporation from soil and plant surfaces and transpiration from the plants. An irrigation system is often designed on the premise that it should be able to supply enough water to meet the peak water requirement of crops (this is generally when the crop is growing vigorously and the weather is the hottest and driest). This means that the system is oversized for most of the year. To limit oversize, most systems are designed with several blocks that can be irrigated on a rotation basis. This results in lower pump capacity, smaller dimensioning of the tubing system, and therefore in lower investment and operation costs. Some drip systems are designed to irrigate the whole field at once. This requires a large pump capacity and bigger tubing.

Maximum ETc will be the basis for determining the required net application depth per plant, tree or irrigated area taking into consideration the partial wetting pattern. After this, the dimensioning (pump flow rate, length and diameter of pipes) of the system can be done so that the system can provide the minimum pressure requirement of the selected emitter located the farthest away from the water source (hence, the required depth of water at this point). In the rest of the system, the pressure variation should stay within the required operating pressure of the emitters. The pressure loss in the system is determined mainly by three factors: a) the amount of the water that needs to flow through the system (Q), b) the length of the tubing system (L) and c) the diameter of the tubing system (D). A desired plot layout will determine the discharge Q and the length L of the system. Through iteration the optimal diameter D can be calculated. Prefabricated drip lines with inline emitters often come with a table with the diameters of the drip tubing and the flow and head loss per unit of length. These tables can be used with the field size for selection of the combination of length and tubing size. This design process needs also to take into account changes in elevation.

Besides this straightforward procedure of system design, Benouniche et al. (2014) give examples of 'bricolage' in which designers, farmers and local craftsmen adapt and modify standard design. Through bricolage, local actors effectively share responsibilities of the design process with engineers adapting it to their own specific needs (see Chapter 15).

Cropping patterns and soil types

Variation in crop types and different plant spacing may result in different numbers of emitters per unit area in the field. However, most drip lines are designed with specific emitter spacing and these are generally acceptable for a range of crops. Crop rotation over various cropping seasons might lead to suboptimal emitter spacing. Designing a system with a fixed and equal number of emitters per unit area will equally lead to suboptimal placement of the emitters to the plants. This might be partly compensated for by applying water in different planting zones for different durations.

Soil types influence the infiltration and wetting pattern. Sandy soils generally have generally a higher infiltration rate and will result in a fast vertical infiltration resulting in a narrow drip cone. Clay soils have more lateral movement and will result in a larger wetted area. Soil characteristics will therefore influence the number and position of the emitters in a field. A designer will work on a targeted percentage of wetted area to effectively wet a sufficient portion of the root zone of the crop. The crop characteristics of root growth and plant spacing determine this targeted percentage of wetted area. Since the area wetted by each emitter is a function of the soil hydraulic properties and discharge of the emitter, one or more emission points per plant may be necessary for widely spaced crops. A low discharge will result in a narrow drip cone with high vertical infiltration while a high discharge will result in more lateral spreading of the water and a larger wetted area.

Water quality

Information on water quality is important in the design process. Drip irrigation systems require clean water to avoid clogging of the emitters and a system
design will need provisions to avoid clogging. Drip irrigation systems have
low flow rates and extremely small passages for water. These passages are easily clogged by mineral particles (sand, silt or clay), organic debris carried in
the irrigation water, and by chemical precipitates and biological growth that
develop in the system. Carbonates may accumulate within the drip system and
cause clogging when high levels of calcium, magnesium and bicarbonates are
present in the irrigation water. Clogging can also occur when dissolved iron and
manganese in well water precipitates when exposed to the atmosphere. Warm
and high nutrient conditions within a drip system can cause rapid growth of
several species of algae and bacteria. These microorganisms often become large
enough to cause complete clogging. The corrective treatment for controlling
clogging depends on the type of clogging agent. Many physical agents can be
removed using settling ponds, water filtration and periodic flushing. Injections
of acids, oxidants, algaecides and bactericides are treatments used to control
chemically and biologically caused clogging. This will need the inclusion of
treatment facilities in the design.

Encountered problems and irrigation practitioners responses at field and farm levels

*Adaptation of the fully developed root system of perennial crops
to limited soil wetting*

Generally, if drip irrigation is installed at the moment of planting, tree seedlings and the root system develop in relation to the wetting pattern of the drip
system with intensive root development in the drip cone. However, in many
parts of the world, orchards have long been irrigated through surface irrigation
and the shift to drip irrigation takes place as the trees are already old and fully
developed. Many farmers complain and express their worry that only a limited
part of the widely developed root zone of older trees is wetted and that trees
will suffer from water stress. In the village of Senyera in Spain, the community
switched to drip irrigation in 2005 in existing citrus plantations but maintained
the surface irrigation system for several years to apply supplemental surface
irrigation one or two times per season to their older citrus trees. Gradually, the
older trees adapted with a more intensive and active root system in the limited
wetted area and a full transition to drip irrigation was done. Other farmers
increased the number of emitters around the trees by placing a sub-lateral in a
loop around the trees to secure sufficient coverage of the root zone. It is also
common to observe old citrus trees receiving water from several lines of laterals. Alternatively, the old trees can be heavily pruned to reduce their water

requirement to facilitate their adaptation to the new drip irrigation system. In any case, most trees will need multiple emitters; with new orchards these emitters can be added as the trees develop.

Clogging of emitters

One of the main concerns in designing a drip irrigation system is to avoid clogging of emitters. Great care to water quality and the filtration system is key to avoid clogging. Still, clogging is a common phenomenon and one of the main reasons for uneven application of water to the crop and even for abandoning a drip irrigation system. Frequent maintenance and rinsing of the irrigation system and backflushing of the filter system is crucial. It is however observed that manual backflushing is often little done and only when pollution is severe and clogging problems already observed. Ideally, a filtration system should have an automatic backflushing system that automatically rinses the system when needed. Such a system has a pressure gauge before and after the filter and if the pressure difference exceeds a certain level (because of pollution) it will backflush. Despite this, clogging often occurs. Tapping the emitters with a piece of wood to loosen the coagulated minerals or deposited solids in the emitter followed by flushing the irrigation system sometimes helps to unclog the emitters. Often, this unclogging is just temporal and the frequency to tap and rinse the system increases over time. When individual emitters are clogged some farmers cut out the emitters and reconnect the drip line (see Plate 2 in the color plate).

Another way of dealing with clogging is to adapt the filtering system. Benouniche et al. (2014) describe problems with filtration systems after the introduction of drip irrigation in Morocco. Farmers were not satisfied with the standard filtration systems; they experienced problems with leakages in the tight welding of the hydro-cyclone filtration system that were manufactured. This was due to fluctuating pressure as the pumps were directly connected to the drip irrigation system without pressure regulator. Local welders redesigned the hydro-cyclone filtration to meet the requirements for use and robustness required by the farmers. This led to a blossoming of a small local industry manufacturing filtration systems for the local market.

Dealing with iron and manganese in groundwater

Water quality and water treatment are closely related to the problems of emitters clogging. Water analysis is often done in laboratories to check the quality and suitability of the water source for drip irrigation. Guidelines in handbooks on threshold and methods of treatment are abundant. Many problems related to clogging of emitters are caused by high levels of iron and/or manganese. Iron and manganese levels over 1.5 mg/L are a serious threat to the drip system. In groundwater, these levels are in a complete soluble form and the water looks perfectly clean but as soon as it comes into contact with oxygen it will oxidize and form a solid deposit. This typically happens at the outlet of the emitters immediately blocking it. Farmers in Myanmar and Thailand have a quick method to analyze the water for these substances. They add this water to green tea and it will turn black instantly if the water contains high levels of iron and/or manganese. Villagers cleaned the water by letting it pass through a container with burned rice husks. High levels of iron in the water were removed instantaneously as water passed through the container. It is believed that the high alkaline nature of the rice husk in combination with the large surface area of the rice husk are the main mechanisms for the quick oxidation of the soluble iron in the water (Rowell and Soe, 2015). Elderly farmers in the Netherland relate similar experiences with adding water to common tea and it would turn black and form a film on the tea surface. Passing water through a reservoir with wheat straw cleaned the water to acceptable levels.

Dealing with high salinity levels in irrigation water

The most active root system is in the wetted soil volume with a higher concentration of active roots in aerated upper soil layer. These active roots can be hampered by the accumulation of salt in this zone. Dissolved salts accumulate at the perimeter of the wetted zone and particularly at the soil surface where the water content of the soil is generally lower. Farmers sometimes build ridges around or along their crops to lead the main salt concentration to these ridges and away from the active root zone. The ridges need to be leached outside the main cropping season in a period of abundant natural rainfall or leaching irrigation. Farmers continue irrigation during this leaching through rain to prevent the salts from leaching into the active root zone. In parts of the world farmers changed from salt sensitive crops to less sensitive crops. For example, in Greece, farmers near the coast and with problems of salt intrusion into the root zone changed from the salt sensitive culture of cucumber to a more salt tolerant tomato production.

Resolving the problem of 'wandering' drip lines

Due to changes in temperature between day and night, the length of the drip line enlarges and shrinks. The longer the drip line, the larger the expansion and shrinking during an irrigation event. This can result in emitters moving away from their initial position, close to the plant. To deal with this issue, farmers place pegs into the ground at the end of the drip line and tie the drip lines firmly to the pegs during an irrigation turn. This does not prevent drip lines from swaying when the temperature rises. However, as soon as the irrigation water flows and cools down the drip line, the latter will shrink again and put itself in the original position again. In some cases farmers stick the drip line down to the soil with U shaped wire at regular intervals. Low-cost drip irrigation systems, such as those disseminated in Zambia are often made of lightweight tape as laterals because it brings the cost down. However, lightweight drip lines often get blown away from their correct position due to strong winds.

Some farmers placed wood pegs at the end of the laterals. Others replaced the flat lightweight tape by heavier and more expensive commercially available drip lines (Tuabu, 2012). Hybrid systems, made of low-cost drip material and commercial available drip irrigation equipment, were assembled to keep the system running as farmers desired. To connect these different types of material, farmers used rubber bands as the available connecters could not do the work. Also shoveling some dirt over the lines at regular interval is practiced to keep the drip lines in position.

The seasonal removal of drip lines for the harvest and soil preparation of seasonal crops

Temporarily removing and storing drip lines to allow land preparation or harvesting is laborious and can easily cause damages to the drip lines and the emitters. Drip irrigation is suitable for perennial crops when drip lines can stay in position during a number of cropping seasons. In annual crops like vegetables, there is generally a need for land preparation, planting, sowing, etc. An intensive network of drip lines hampers these activities and moving them too often may lead to significant damages, hence low performance. Farmers have developed innovative simple devices to facilitate the frequent installation and removal of drip lines and limit damages to the latter (see Plate 3 in the color plate).

Mechanical weeding

An issue that farmers face with drip irrigation is the fact that drip lines hamper mechanical weeding. In drip irrigation systems, weed growth is generally most vigorous near the drip lines, which makes mechanical weeding a challenge because it can damage the drip lines. To solve this problem, it is common that farmers with orchards hang the drip lines in the trees, way above the ground. However, if the drip line hangs with loops there is a danger that water from several emitters flow along the drip line to the lowest place of the loop and water of several emitters is concentrated on a single spot creating uneven distribution uniformity. This can be avoided by a tight wiring and clipping of the drip line to the wire on short interval. In vineyards the trellis system is use to hang the wire and the drip tube.

Dealing with crop water requirements: irrigation scheduling for drip

Achieving high application efficiencies with drip irrigation systems requires fine monitoring of crop water requirements and, as these vary over a growing season, adjusting water application and irrigation scheduling. Irrigation scheduling has been an important topic in agricultural research for decades and several irrigation scheduling computer models have been developed; many of these being based on water budgeting. However, few studies have linked these irrigation scheduling tools with actual on-farm irrigation practices. Both large commercial farms and smallholder farmers make little use of scheduling tools for various reasons: they find them overwhelming and lack the skills or funds necessary to install, operate and troubleshoot them. Smallholders in developing countries (but also in developed economies) need simpler, cheaper and more comprehensive support tools to achieve improved irrigation management at the farm level. One example of such a tool is the Drip Planner Chart (DPC) developed by Wageningen University and Research Centre (WUR). The DPC is a manual disk calculator to calculate daily irrigation requirements based on the FAO irrigation requirement calculations (Allen et al., 1998) and has been introduced in various countries like Zambia, Nepal, Tunisia, Myanmar, and the Netherlands (Boesveld et al., 2012; Rowel and Lar Soe, 2015).

Dealing with fertilisation for organic crop production

Organic certification schemes value drip irrigation for its assumed water-saving capabilities; on the other hand, drip irrigation systems are sometimes designed with fertigation units in such a way that artificial soluble fertilizers can be mixed into the irrigation water. This raises a problem as the latter are not acceptable under organic certification schemes. To benefit from an organic certification, farmers need to use organic fertilisers that are not completely soluble and lead to emitters clogging. In several countries (Morocco and Zambia), farmers have been experimenting with a 'teabag method' whereby organic fertilisers are drenched in water for several days in jute bags. The extract is then enriched with nutrients and can be injected into the drip irrigation system provided it is well filtered. Alternatively, farmers can apply organic fertiliser/manure to the crop and soil directly and use the drip system to apply water only.

The Acequia Real de Jucar is a surface irrigation system in Spain of more than 20,000 ha that is in the process of shifting to drip irrigation. The shift occurs progressively with sectors of 400–600 ha being gradually changed to drip irrigation and working as separate irrigation units. Each sector has its own infrastructure for pumping and applying fertilisers to the water. Water from the system is always enriched with artificial fertilisers and therefore unacceptable under any organic certification. Farmers who want to produce in an organic manner are left with no other option but to dig their own well and operate an individual system within a large communal system. This situation is also a challenge and hurdle for 'traditional' farmers who might want to change to organic farming at some stage. In Biskra, Algeria (Chapter 16), local manufacturers and farmers designed simple fertigation units at the level of the greenhouse, thus leaving the decision to apply nutrients or solely water to the farmers hence creating much more liberty to operate the system according to the farm's needs. Some farmers in Morocco also adjusted their fertigation unit by replacing the large nutrient tanks by small nutrient tanks to make it easier to calculate the right dose of fertilisers.

Conclusion

Many drip irrigation systems do not function according to the wishes of farmers, despite the care and effort designers have put into their design, leading the former to display much innovativeness in using the systems. Any irrigation system is assembled from separate pieces of equipment that are used as building blocks. For every building block, there is a wide variety of components available to choose from but, in all of them, the designer included his/her ideas of how the material would be used as a component of a full drip irrigation system. In the same way, the system designer, who plans and assembles the individual pieces of equipment into a full operational system, has his/her own ideas of how the system should function and how it will be used. This does not always fully reflect the farmer's realities or needs. Farmers and extension workers further adapt and tinker with the irrigation systems in the field to operate it under their own specific farming conditions. Modifications may relate to shortcomings in the initial design, poor installation of the irrigation system, choice of irrigation components or the skill set and capabilities of the farmers who operate the system.

Acknowledgments

The author would like to thank two anonymous reviewers and the editors for their valuable comments on earlier versions of this chapter.

References

Allen, R.G., Pereira, L.S., Raes, D., and Smith, M. (1998). *Crop evapotranspiration: Guidelines for computing crop water requirements*. FAO Irrigation and Drainage Paper 56. FAO: Rome.

Benouniche, M., Zwarteveen, M., and Kuper, M. (2014). Bricolage as innovation: Opening the black box of drip irrigation systems. *Irrigation and Drainage* 63: 651–658.

Boesveld, H., Zisengwe, L.S., and Yakami, S. (2012). Drip planner chart: A simple irrigation scheduling tool for smallholder drip farmers. *Irrigation and Drainage Systems* 25(4): 323–333.

Brouwer, C., Prins, K., Kay, M., and Heibloem, M. (1990). *Irrigation water management: Irrigation methods*. FAO Training Manual No. 5. FAO: Rome.

Burt, C.M., Clemmens, A.J., Strelkoff, T.S., Solomon, K.H., Bliesner, R.D., Hardy, L.A., Howel, T.A., and Eisenhauer, D.E. (1997). Irrigation performance measures: Efficiency and uniformity. *Journal of Irrigation and Drainage Engineering* 123(6): 423–442, November.

Burt, C.M., and Styles, S.W. (2007). *Drip and micro irrigation design and management for trees, vines and field crops,* ITRC: San Luis Obispo, CA.

Hoffman, G.J., Evans, R.E., Jensen, M.E., Martin, D.L., and Elliott, R.L. (2007). *Design and operation of farm irrigation system*. ASAE Monograph. ASAE: St. Joseph, MI.

James, L.G. (1988). *Principles of farm irrigation systems design*. New York: Wiley & Sons.

Jensen, M.E. (1983). *Design and operation of farm irrigation systems*. American Society of Agricultural Engineers. Monograph Nr 3. ASAE: St. Joseph, MI.

Keller, J., and Bliesner, R.D. (1990). *Sprinkler and trickle irrigation*. New York: Chapman and Hall.

Kooij, S. van der, Zwarteveen, M.Z., Boesveld, H., and Kuper, M. (2013). The efficiency of drip irrigation unpacked. *Agricultural Water Management* 123: 103–110.

Lamm, F.R., Ayers, J.E., and Nakayama, F.S. (2006). *Micro irrigation for crop production*. Developments in Agricultural Engineering 13. UK: Elsevier Science Technology.

Reinders, F.B., Grové, B., Benadé, N., van der Stoep, I., and van Niekerk, A.S. (2012). *Technical aspects and cost estimating procedures of surface and subsurface drip irrigation systems* (Vols. 1, 2, 3). ARC-Institute for Agricultural Engineering. Water Research Commission Report No. TT 524/12.

Rowel, B., and Lar Soe, M. (2015). Design, introduction, and extension of low-pressure drip irrigation in Myanmar. *HortTechnology* 25(4): 422–436.

Tuabu, O.K. (2012). *Configuration of KB drip irrigation systems by different socio-economic, biophysical, technical and organizational/institutional conditions in Zambia*. Minor MSc thesis International Land and Water Management, Wageningen University.

4 Re-allocating yet-to-be-saved water in irrigation modernization projects

The case of the Bittit irrigation system, Morocco

Saskia van der Kooij, Marcel Kuper, Charlotte de Fraiture, Bruce Lankford and Margreet Zwarteveen

Introduction

While the world faces a growing demand for food, water availability is limited. One much proposed and advocated (among others by international donors) solution for producing more food with less available water is the so-called modernization of irrigation systems (Playán and Mateos, 2006; World Bank, 2006; EEA, 2009; OECD, 2010). The term modernization refers to the 'upgrading' (Burt, 1999: 15) of irrigation systems through the introduction of new management arrangements and technologies that stimulate efficient water use (van Halsema, 2002). Increases in the efficiency of irrigation systems are expected to result in increases in the productivity of the irrigated sector, as for instance expressed in terms of more production per hectare or more production per cubic meter of available water. Modernization of irrigation is also often associated with a larger agricultural intensification discourse, with farmers becoming more competitive, a liberalization of markets and a reduction in subsidies (Lecina et al., 2010).

Although many would agree (see for instance Burt and Styles, 1998) that modernization cannot be achieved through improvements in the hardware only, planned efforts to increase the efficiency of irrigation systems often centre on the introduction of more efficient irrigation technologies such as drip irrigation or the lining of earthen canals. Several reasons explain why decision makers prefer a technological solution to dealing with problems of scarcity or distributional dilemmas (see Playán and Mateos, 2006). In this article we zoom in on how technological solutions are attractive because they allow avoiding making hard choices about decreasing the water demand of (some) users and changing water allocation patterns. As also Allan (1999) suggested, with technological solutions there appear to be no losers: no one has to limit agricultural production and neither will incomes of farmers decrease. In comparison, soft solutions that aim at reducing water use through a change in management – for example through water pricing – are much more directly and openly political and are therefore more likely to be contested.

Yet, although seemingly neutral and presented as a win-win situation, the introduction of efficient technologies clearly also involves political choices about the allocation of water (Lankford, 2012a). These choices, however, are much less open and transparent and therefore less likely to be questioned or contested. The suggestion that water will be 'saved'[1] by new technologies provides the opportunity to solve distributional dilemmas in an elegant manner: it creates the suggestion that more water will be created, satisfying increased demand or demands of new users without having to take water away from others. This part of the re-allocation process is deliberate and open; on the face of it, it simply consists of allocating the 'gained' or previously 'wasted' water to old or new claimants. Yet, these allocations of yet-to-be-saved water in actual practice often hide re-allocations of water. First, this occurs because it is not certain that the anticipated 'savings' will indeed be attained; it is not sure whether the new technology will indeed be as efficient when used as expected. Although there is remarkably little evidence of the actual water gains obtained by introducing drip irrigation, there is an interesting study by Benouniche et al. (2014) which shows how the actual irrigation efficiency and uniformity of drip irrigation depends on farming practices. Where some farmers achieved high efficiencies, on the fields of others drip irrigation was even less efficient than surface irrigation. The study shows that farmers can indeed attain a 90% efficiency with drip irrigation, but not all of them have the ambition to do so (Benouniche et al., 2014; see Box 4.1). Also Wolf et al. (1995) mention disappointing efficiencies with drip irrigation. When anticipated efficiencies are not reached, this also means that the water 'gains' that were already allocated will not become available. The promised extra water thus has to come from other destinations.

Box 4.1. Water saving is only one of the motivations of farmers using drip irrigation

In the literature, the irrigation efficiency of drip irrigation is typically announced to be around 90–95% (van der Kooij et al., 2013). Such values are obtained in experimental stations or in laboratories. However, there is not much information available about actual irrigation efficiencies obtained by farmers. Uneven topography, clogging of drip lines, imperfections of the equipment or in the design of the system, lack of time and irregular water supply are some of the constraints farmers handle on a daily basis. Irrigation performance also depends on farmers' practices and skills, and on farmers' strategies. Actual irrigation efficiencies show, therefore, more contrasted figures than those presented in the literature.

We conducted field measurements on 22 farms in the Saïss plain in Morocco (see Benouniche et al., 2014). The results show that five farmers obtained efficiencies close to those obtained in experimental stations, i.e. from 90 to 107%. These are generally small-scale horticultural farmers aiming for 'good' irrigation as they have limited water supplies. Some even under-irrigate, as compared to irrigation requirements. On the other extreme, eight farmers obtained efficiencies below 60%, which means that these systems are as efficient (or even less efficient) as gravity irrigation, leading to over-irrigation. Some farmers even apply 2–4 times the required volume. These are often large-scale farmers with subsidized equipment, although such farmers are supposed to be among the most high tech and most efficient in their irrigation practices. In between these extremes, nine farmers obtained efficiencies in the range of 60 to 90%.

So what explains these contrasted efficiency values? In some cases, we observed low values of distribution uniformity (some parts of the field get more water than others), which farmers compensate by over-irrigating to ensure that even the least irrigated part of the field receives enough water. The problem of uniformity is mostly due to design problems (for instance on uneven topography) or to erratic operation and maintenance practices (cleaning of filters, clogging of drippers), in particular in the large-scale farms where the management is delegated to technicians and labourers. In farms where farmers conduct irrigation themselves, the irrigation efficiency is always higher. Some farmers were on a learning curve. They had just started using drip irrigation and encountered difficulties in its use. However, these farmers made quick progress and obtained higher efficiencies within 1–2 years' time.

Finally, and perhaps more importantly, farmers' overall strategies explain most of the contrasted results obtained. Farmers adopted drip irrigation for a variety of reasons. Those confronted to water scarcity, for example a low-yielding well, aimed for very precise irrigation to make the most of a difficult situation. Others installed drip irrigation for reasons of practicality and ease of use, for instance the part-time farmers. For such farmers, the water quantities used (and the associated cost) was much less important than the time gained. A great number of farmers were mainly interested in obtaining high yields. Such farmers systematically over-irrigated to ensure that yields would not be affected by water stress. And some farmers installed drip irrigation for the status it gave to them by becoming 'modern' farmers or by the fact that drip irrigation made farming cleaner, as gravity irrigation is associated with mud, boots and toil.

In all cases, there is scope for higher irrigation efficiencies. This pleads for putting the user central in the debates on water saving. Or in other words, making the user, in all its diversity, visible, as he/she is the one that will make water saving happen or not.

Maya Benouniche, Marcel Kuper, Ali Hammani

Second, it is impossible to just create 'more' water (Perry (2011) refers to the law of conservation of mass). 'Losses', seen from the perspective of the single crop at the plot level, may not be 'losses' when considering the basin scale (Seckler, 1996; Molden, 1997; Molle et al., 2004; Jensen, 2007; Perry, 2011).

Unless these 'losses' were inaccessible or of too low a quality to be re-used – for example when they recharge a polluted aquifer or flow into the sea – it is quite possible and indeed likely that the previous 'losses' were used downstream. The water users dependent on these 'losses' thus lose their access to water when irrigation practices upstream become more efficient through modern irrigation technologies, increasing new water consumption. The introduction of technologies to 'save water' upstream may thus entail a re-allocation of water from downstream users to upstream users. This re-allocation is less open and visible, and may only become apparent after the efficiency intervention is implemented. This re-allocation can be called 'unintended' reallocation (Hooper and Lankford, forthcoming), but as we will show, it is also possible that hiding the re-allocation effects of the introduction of a new technology is a more or less conscious strategy to avoid open conflicts and struggles about distributional priorities.

A well-known example of the re-allocation of water via efficiency interventions is the All American Canal. The All American Canal diverts water from the Colorado River to farmers in the Imperial Irrigation District, which was blamed for its disproportional water use – possibly 436,700,000 m3, thus almost half a cubic kilometre per year could be saved (CGER, 1992). The idea that these 'losses' could be saved, and thus allocated to a new user, resulted in a water-deal: the All American Canal would get lined, financed by the Metropolitan Water District, and the 'saved' water would be used for drinking water. However, in this water deal it did not transpire that the previously 'lost' water used to be re-captured by farmers downstream in Mexico. After lining the canal Mexican farmers dependent on the 'losses' of the All American Canal lost the water that previously percolated through the earthen All American Canal (Jenkins, 2007).

Here we focus on the first part in the re-allocation process: the deliberate allocation of the yet-to-be-saved water. We build further on the suggestion of Allan (1999) that efficient technologies are such attractive tools for modernizing irrigated agriculture, because they apparently do not create explicit losers. Based on empirical results from a case study of the farmer-managed Ain Bittit irrigation system in Northern Morocco, we trace the re-allocations that accompany modernization projects: drip irrigation and the lining of irrigation canals. The recent introduction of drip irrigation in Morocco and more specifically in Ain Bittit illustrates the paradoxical situation that can emerge from modernization interventions: as different parties claim the yet-to-be saved water, it seems that more rather than less water is used for irrigation.[2] The analysis of the past lining projects of Ain Bittit sheds further light on how efficiency projects effectively allow to depoliticize contentious questions of water allocation. By making it seem as if only the 'saved water' is allocated to specific users, it appears that other users will not notice or be affected by changes in discharge. Yet, in Ain Bittit, new technologies created increases in water abstractions, further reducing the availability of water for the command area: only one fourth of the command area can be irrigated.

Methods

We base our study on fieldwork carried out by the first author, who followed the irrigation practices in the secondary canal Khrichfa of the Bittit irrigation system during two irrigation seasons (2012 and 2013). To explore the reconfigurations of water distribution that accompanies the drip irrigation projects, 15 drip irrigation users of the Khrichfa canal or near the Khrichfa canal were interviewed. For tracing how efficiencies shaped the history of the Ain Bittit irrigation system, interviews were carried out with water users (ca. 40); (ex-)officials of government institutes (10) and various actors that have been involved in the modernization projects in Bittit. In addition, use was made of documents obtained from government institutes and water users.

The Ain Bittit irrigation system

Ain Bittit is a small-scale, farmer-managed irrigation system in Northern Morocco covering approximately 2000 hectares cultivated with cereals, forage crops, tobacco, horticulture (onions and potatoes) and fruit trees. Its irregular layout and scattered plots mark its community-built origin, while frequent state-interventions have influenced the lined irrigation infrastructure and the management of the system. The Ain Bittit water (*Ain* meaning 'spring' in Arab) is shared between the state, which obtained since 1929 the right to 60% of the springs' discharge – used by the drinking water company of Meknes – and two rural communities, Ait Ouallal and Ait Ayach, which are each allocated 20% of the springs' discharge. Here, we focus on the secondary canal Khrichfa within the Ait Ouallal community (see Figure 4.1). Khrichfa's share of the springs' water is ca. 100 l/s and as an additional water source, the Khrichfa water users make use of a collective tube-well that releases ca. 40 l/s to the Khrichfa canal during the peak demand of crops. In total, the water availability in Khrichfa is only sufficient to irrigate one fourth of its total command area.

Over the past two decades, several farmers in the Khrichfa region introduced drip irrigation. The drip projects are based on surface water from the Khrichfa canal, sometimes in combination with groundwater pumped by tube-wells. Drip irrigation can irrigate the steep and undulating land above the Khrichfa command area, which increases the value of these previously rain-fed lands. Investors from the nearby cities bought plots in the rain-fed zone, where they drill tube-wells and install drip irrigation, thus increasing the pressure on groundwater (although the Bittit springs' discharge is not influenced by the tube-wells). Khrichfa water users, having both land in the command area and rain-fed land, also convert the rain-fed land into irrigated lands with drip irrigation. They do so by using their water rights from the Khrichfa canal: they store the surface water in a storage reservoir, from which they pump the water into their drip irrigation system. Some farmers strategically combine surface water and groundwater in their drip irrigation projects.

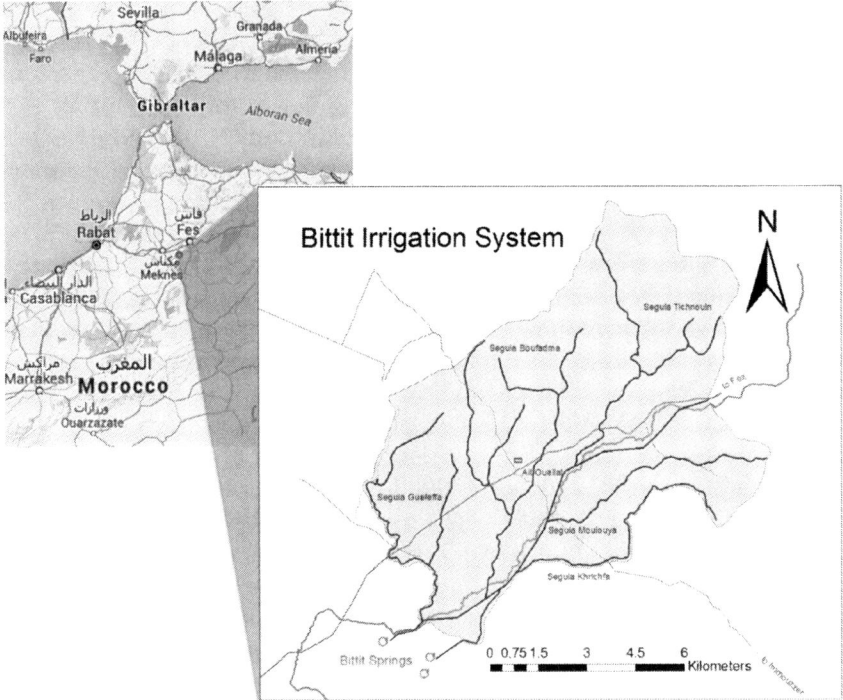

Figure 4.1 The Bittit irrigation system.

Source: the authors.

Results

Modernization 1: drip irrigation in Morocco for sustainable extension of agriculture

Morocco faces declining groundwater levels, negative water balances and half-full dams which figure prominently in the discourses of the Ministry of Agriculture and the water basin authorities to call for modernization of the irrigated sector (El Meknassi, 2009; El Haouat, 2012). In reaction to the alarming figures, the Moroccan government subsidizes the conversion from surface to drip irrigation with subsidies that cover 80–100 % of the investment costs. Indeed, the drip irrigation subsidies are a prominent part of the National Irrigation Water Saving Plan of 2007 (Plan National d'Economie d'Eau en Irrigation, PNEEI) to counterbalance the deficits in the water sector on the basis of theoretical efficiencies of 90% of drip irrigation. More specifically, the PNEEI aims to 'save' 826 Mm^3 per year, of which 514 Mm^3 will be 'saved' in the large-scale irrigation systems. The water 'saved' in the large-scale irrigation projects will be used to compensate structural deficits within the irrigation systems, thus increasing agricultural production while diverting the same amount of water from the dams to the large-scale irrigation systems. The PNEEI expects to 'save' 312 Mm^3 per year within the private irrigation sector, which mainly uses groundwater. As this water will not be pumped by farmers, the PNEEI argues, the 312 Mm^3 is expected to remain in the aquifer thus containing the over-exploitation of groundwater (Ministère de l'Agriculture du Développement Rural et des Pêches Maritimes, 2007). Later interpretations of the PNEEI are even more optimistic about the possible water savings. Referring to the private irrigation sector, Arrifi, 2009 and Belghiti, 2009 (both associated with the Moroccan Agricultural Ministry) expect that 740 Mm^3/year (Belghiti, 2009) or 700 Mm^3/year (Arrifi, 2009) will be saved and will thus remain in the aquifer. For the large-scale irrigation systems, they think 700 Mm^3/year (Belghiti, 2009) or more than 750 Mm^3/year (Arrifi, 2009) can be recovered to fulfil the irrigation requirements that currently cannot be satisfied. The different estimates of the water savings from different authors indicate that although the precise amount of water to be 'saved' with drip irrigation is open for interpretation, the 'saved' water is already allocated to two (competing) destinations: increased agricultural production and protection of the aquifer.

The PNEEI of 2007 is encapsulated in the Green Morocco Plan, which has guided Moroccan agricultural development since 2008. The Green Morocco Plan has a strong focus on increasing agricultural development, both in economic terms and in terms of social welfare. With this strong focus on increasing agricultural production, the goal of the PNEEI seems to also be shifting toward using the 'saved' water for increasing production (as was already indicated for the large-scale irrigation systems in the PNEEI), rather than for also safeguarding the aquifer: 'The National Water Program . . . aims at filling the water gap which is considered the principal limiting factor in improving the agricultural productivity' (www.agriculture.gov.ma/pages/economie-de-leau, accessed on 29–05–2015). In the efforts to make Moroccan agriculture competitive, the Green Morocco Plan stimulates farmers not only to shift to drip irrigation but also to produce high value crops, which often demand more water, and to extend the irrigated area.

While the Green Morocco Plan thus interprets efficiency as more production per cubic metre of water, the Basin Agencies, responsible for safeguarding Morocco's water resources, are concerned about the negative water balances and interpret efficiency as less water diverted to agriculture and thus remaining in the aquifer. In their efforts to safeguard the aquifer system under the Saïss plateau (where Ain Bittit is located), the Sebou River Basin Agency aligns with the National Water Plan and the efforts of the agricultural ministry to stimulate the shift from surface irrigation to drip irrigation, based on its efficiencies which would lead to less abstraction of groundwater. They do so by adding training, subsidies and pilot technologies to the efforts of the agricultural ministry in reconverting Morocco's irrigation areas to drip (El Haouat, 2012).

'Efficient' drip irrigation makes it possible for the basin agencies and the agricultural ministry to work together toward an apparently shared future: one in which water is used 'efficiently', even while their goals compete. In the field, it becomes clear that the water to-be-saved is claimed by different parties who each have a different destination for the to-be saved water. The ministry of agriculture intends to use the water saved for increasing agricultural productivities; the farmers hope to use the water on newly developed land; and the Basin Agency aims to keep the saved water in the aquifer. Paradoxically, the introduction of drip irrigation in Bittit leads to an increased abstraction of water, as also observed in other irrigated areas where drip irrigation leads to an extension of the irrigated area (López-Gunn et al., 2012; Berbel et al., 2013; Benouniche et al., 2014).

Drip irrigation in the Khrichfa area in Bittit: increased
groundwater use with tube-wells

The massive introduction of drip irrigation in the Saïss area started outside of the surface irrigation systems, with individual groundwater users who wanted to economize on pumping cost, intensify agricultural production or irrigate more surface area with the available water (Ameur et al., 2013). In the area near the secondary canal Khrichfa in the Ain Bittit irrigation system, most of these individual drip irrigation projects are located on previously rain-fed lands (Table 4.1). Two main conclusions can be drawn from Table 4.1: 1) most drip irrigation projects rely on groundwater, abstracted via tube-wells, and 2) most drip irrigation projects extend the irrigated area and do not relate to a conversion of existing irrigated land, as the official policy proclaims, but can rather be characterized as extensions of the irrigated area.

New settlers in the Bittit area, attracted by the availability of rain-fed land above the irrigation system with a high potential for finding groundwater, consider tube-wells as part of the drip-package, the set of technologies that accompany drip. They rely on groundwater as the main source for their drip irrigation

Table 4.1 Location and water source of the drip irrigation projects in the Khrichfa area

Water source \ Location	Within command area	Agricultural Extensions	Overlapping command area and agricultural extensions	Total
Surface water (via storage reservoir)		2		2
Groundwater (via tube-wells)	2	6	1	9
Groundwater + surface water		2	2	4
Total	2	10	3	15

Source: the authors.

installation. A tube-well secures access to water, making the investment in drip irrigation less risky than with a springs' variable discharge, and is readily available at hand enabling the farmer to irrigate whenever required.

Right holders to Khrichfa's water who want to use surface water as the main source for their drip irrigation project secure the specific access to canal water that a drip irrigation system requires by ensuring good relations with the WUA and by hiring water. By integrating the use of drip irrigation in a surface canal irrigation system, they challenge the conviction of many engineers that a surface water source like the Bittit springs cannot be combined with drip irrigation. The reasoning of engineers is that drip irrigation will not work in a small-scale, spring-based irrigation system because of the unreliable discharge, the rotational water distribution and a lack of incentives to save water as surface water is free of costs (Ministère de l'Agriculture du Développement Rural et des Pêches Maritimes, 2007; pers. comm. with the engineer in charge of irrigation at the Regional Agricultural Department of Meknes, 28–10–2013). This is also why the Basin Agency, paradoxically, supports the drilling of tube-wells for those farmers with water rights in a surface irrigation system who wish to convert to drip irrigation. One of the new drip irrigation projects in the Bittit irrigation system, a family cooperative of 25 members with plenty of surface water rights, thus installed a tube-well subsidized by the Basin Agency on the rain-fed land uphill from the Khrichfa canal. The family cooperative aims to increase the productivity of their rain-fed land, previously used to cultivate cereals and for pastoral activities. Via the Provincial Department of Agriculture, the cooperative arranged a subsidy of 100% on the drip irrigation installation. The initiators of the cooperative met the Basin Agency at a workshop on drip irrigation organized by an IDRC project from the Al Akhawayn University, which aims to: 'determine whether demand management of water in agriculture can save the future of the Saïss basin' (ACCA, 2010: 1). The Basin Agency told the family cooperative that they stimulated drip irrigation within surface irrigated areas, so they offered to finance the construction of a storage reservoir to stock their surface water from the Bittit springs. Because using drip irrigation in a surface irrigation system is difficult – the Basin Agency argued – in terms of water turn and in terms of reliability of the source, they also subsidized the installation of a tube-well as supplement to the construction of the storage reservoir. The goal of the reservoir was to always ensure access to water. This drip project is now put forward by the Al Akhawayn University and the Basin Agency as a pilot project and an example to follow. Also other drip irrigation users of Khrichfa talk about drilling a tube-well, or have already installed a tube-well on the highest part of their rain-fed parcels. The increased use of tube-wells for drip irrigation thus puts additional pressure on the groundwater resources to support the introduction of so-called efficient irrigation technologies.

The drip irrigation projects in the Khrichfa area show that the Moroccan drip irrigation policies create a paradoxical situation. Drip irrigation's promise of efficiency gave a *carte blanche* to the entrepreneurial farmers with drip, who use drip to extend and intensify agricultural activities, which increases their total water abstraction from the Khrichfa canal and from the tube-wells. While the Basin Agency supports a shift to drip irrigation to safeguard the aquifer, it thus also stimulates an increased groundwater abstraction for agricultural extensions. One could explain this situation as a lack of control and legislation and argue that extensions outside the command area should be prohibited. However, prohibiting extensions or measuring groundwater abstraction with meters on tube-wells would make clear that the Basin Agency and the agricultural ministry have conflicting objectives. A water meter on a tube-well would make the water savings tangible, leading to discussions about how the savings should be divided, which is precisely the discussion that is now avoided by all parties claiming (and if possible using) the expected water savings. The main indicator used by the agricultural ministry to predict and register water savings is the surface area equipped with drip irrigation. To further understand the attractiveness of the promise of water savings from water efficient technologies, we trace the process of re-allocating water that accompanied past modernization projects.

Modernization 2: lining projects in Bittit, (re-)allocating the Bittit sources based on efficiency estimations

The city of Meknes was growing fast in the first half of the 20th century and the city's drinking water source – the river Boufekrane – was not sufficient and of bad quality (Direction des Travaux Publics, 1954). In the search for drinking water for Meknes, the administration during the French Protectorate became interested in the sources southeast of Meknes, amongst which the Bittit sources which were used by the Ait Ouallal and the Ait Ayach tribe. The administrators started to measure the sources' discharge from 1934 onward.

The governors in Morocco could not easily get control over the Ait Ouallal inhabitants. The Ait Ouallals belong to the rebellious Beni M'tir tribe and several Ait Ouallal landowners were part of the political opposition. To settle the Ait Ouallal tribe, the administration during the French Protectorate arranged the Ait Ouallal land rights in 1929. The Ait Ouallal farmers divided the water rights amongst each other, but these were not yet registered officially in the first half of the 20th century. The downstream community Ait Ayach had good relations with the governors (Bazzi, 1987). One of the interests of the Ait Ayach farmers was to ensure a stable water supply for irrigation. They used to get the remaining water of Ain Bittit, which means that the discharge was limited in summer, and (too) high during the rest of the year. To support the Ait Ayach farmers and to gain access to the Bittit sources for drinking water, the administrators at that time decided to register the rights on the Bittit sources. They made an inventory of the water use in 1949 based on the local water rights practiced in Ait Ouallal and Ait Ayach. The report of this inventory (Direction des Travaux Publics, 1949) stated that the farmers in Bittit and Ait Ayach only used 40% of the water from the springs, the rest were 'losses' that were not available for use, unless the infrastructure would be improved:

> [T]he water rights have been calculated according to the discharge that is used in reality ... which is no more than 4/10th of the total discharge of Aioun Sidi Tahar, Sidi El Mir and Ain Sebaa (the three sources that together form the Ain Bittit springs), because of the losses that occur in the existing infrastructure. The committee clearly points out that the fractions of the discharge that represent the losses will not be available ... until the infrastructure is lined and the marshes of the Bittit river bed are drained.
>
> (Direction des Travaux Publics, 1949: 2)

The report proposed that the two communities would be allocated the water rights according to the water they 'really used', while allocating the water 'losses' to the public domain. After describing the shares of the different tribes, the report states:

> [T]he proposals of the inquiry commission are based on the customary water allocation, which is not applied on the total discharge of the springs but on the discharges that are really used. It (the committee) thus registers the losses of the current infrastructure as part of the public domain, as these (losses) were never part of the acquired water rights. But it stipulates that the fractions of the discharges that represent the losses are not available at the springs until the irrigation infrastructure is made water tight and the marshes of the Bittit riverbed are drained. This viewpoint, which is also the viewpoint of the Public Works Department, has been accepted by the users.
>
> (Direction des Traveaux Publics, 1949: 3)

Thus, 60% of the discharge could be assigned to the public domain: 'losses to be recovered by the Public Domain after lining works: 60 %, which is 6/10th of the discharge of the sources' (Direction des Traveaux Publics, 1949: 4).

The state registered 60% of the discharge under the public domain and assigned Ait Ayach and Ait Ouallal both an equal share of the remaining 40%. By referring to the 'losses' as the share of the state, the re-allocation of the Ain Bittit sources was de-politicised. Apparently, no one used the 'losses' and it seemed as if the farming communities would be able to use the same volumes of water as before. To get access to the 60% 'lost' water, the state had to line the Bittit infrastructure. The public works department attracted attention to the water wastage of farmers by mentioning the water accumulation in the tail-end of Ait Ayach, and the problems of water-borne diseases that it caused. This was meant to stress that this water was not used by others, meaning that the state's share amounted to the creation of new water rather than a re-allocation that concerned Ait Ouallal and Ait Ayach.

The administration started the construction work in 1952. At the location of the sources, they constructed a division structure to divide the total flow in 60% for drinking water and 40% for agriculture. Based on the rule of thumb of 200 litres of drinking water per day and an estimation of the population growth, the administration constructed a pipeline with a capacity to divert 400 l/s to Meknes (Direction des Travaux Publics, 1954). Even if the springs' discharge would be lower in summer, 400 l/s is still less than the state's water right of 60%. Meanwhile, the main canal of Ain Bittit got lined (see Figure 4.2 for an overview of the infrastructure). However, the administration did not include the earthen secondary canals in the lining program. They designed the main canal in such a way that the large and political active landowners and some colonial farmers in Bittit got their offtakes directly on the main canal. This satisfied the most influential actors in Bittit both because the lining reached up to their offtakes (thus making optimal use of the lining project), and because having private offtakes meant there was less need to cooperate with other farmers to get access to water. Those Bittit farmers in less favourable positions expected that the state would also line the secondary canals. According to them, these canals were part of the irrigation infrastructure and had to be included to attain the anticipated water savings of 60%.

The construction works had attracted labourers from different regions, and when the infrastructure was ready, the labourers made sharecropping agreements

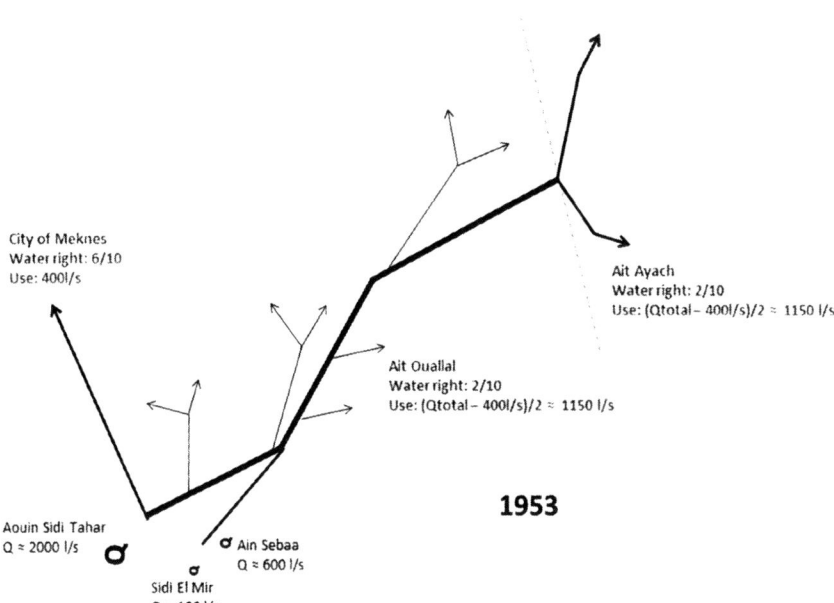

Figure 4.2 Sketch of the sources and irrigation infrastructure in 1953.

Source: the authors.

with the Ait Ouallal landowners to stay in the area. Many newcomers followed. This resulted in the clearing of new land and an increased demand for water within the irrigation system.

A new process of water re-allocations legitimized through lining that happened later, in the 1980s, when both the water demand of Meknes and the water demand of the farmers had grown, resulting in the period between two irrigation turns becoming very long. A rehabilitation project financed by the World Bank created the context for re-negotiating water allocations. Lining the secondary infrastructure again created a promise of 'saved' water, which the drinking water provider of Meknes planned to use. The plan of the drinking water provider of Meknes to construct another pipeline, thus doubling their offtake to 800 l/s, in first instance resulted in protests by the Bittit farmers. However, the Bittit farmers were interested in lining the secondary canals, because it was promised to them in the negotiations over the Bittit springs and because it would speed up the rotation schedule, thus resulting in a shorter period between two irrigation turns (Lankford, 2012b).

The engineers of the rehabilitation project calculated that with the lining of the secondary infrastructure, and minus the additional water use of Meknes of 400 l/s, the Ain Bittit farmers would still gain more water through the lining project than they had before, which could thus be used to irrigate new lands. The calculation was as follows: in total 914 hectare was irrigated with 1282 l/s, thus with an available discharge of 1.41 l/s per hectare, which the engineer considered as too much. As a rule of thumb, he argued, each hectare would need 0.7 l/s, assuming that tobacco would be the main cultivation, as was the case in Bittit at that time. The engineer thus concluded that the system worked at an efficiency of 50% before lining. He estimated that the efficiency of the irrigation system – taking into account the 'losses' in the quaternary canals and at the farmers' fields – could be improved to 65% through lining (Bazzi, 1987). The net efficiency gain from lining the secondary infrastructure would be more than the additional 400 l/s that would be allocated to Meknes (in total resulting in 800 l/s for Meknes). The remaining water could thus be used for new land within the irrigation system, which is how de-stoning land to extend the command area became an integral part of the project (Bazzi, 1987). All users of the Bittit springs seemed to gain a part of the water-to-be-saved with lining. However, after lining the secondary infrastructure and de-stoning of new land, it became clear that the rule of thumb water duty of 0.7 l/s per hectare was not sufficient. Water percolated through the lined infrastructure which was not maintained well enough and the water distribution with orifices in the lined canal went less smoothly than in the earthen canals. As a result, because not all plots could be irrigated, farmers chose which plots they irrigated and which plots they reserved for rain-fed crops.

The examples show how the use of efficiency terminology creates the suggestion that more water is created. The state could allocate this extra water to the drinking water company of Meknes without having to fear the objection of the Ait Ouallal farmers. If we retell the history of allocating the Bittit springs without putting 'efficiency' central, it would become a different story, one about the politics of gaining control over resources. The discharge of the Bittit springs does not change, so it is still the same amount of water that is shared, and over which competition arises. The reference to 'losses to be recovered' gives the impression that this water would otherwise not be used. However, theoretical efficiencies are unlikely to be attained in the field and the 'losses' may well have been used elsewhere when they flow through the Oued Bittit or recharge the groundwater. The history of allocating the Bittit springs is thus also a story of a state taking control over a communities' water source, which they re-allocate to the parties that matter for them: the city, the politically influential landowners, the downstream community.

Conclusion

Modernization projects such as canal lining and drip irrigation in Ain Bittit were initiated because of, or resulted in, a re-allocation of water. In addition, the modernization projects also resulted in increased water consumption. We have shown how policies that promote a shift to technologies with high theoretical efficiencies entail a promise of water 'to be gained' for the ones that make the effort of introducing efficient irrigation technology. Different actors claim and actually use this water for their purposes, even before the water savings are actually attained.

Improving efficiency suggests that water is 'gained' as previous 'losses' are recaptured. The recaptured 'losses' are treated as if these were new, untapped resources waiting to be used (for the ones that make the effort of avoiding these 'losses'). However, local 'losses' were reused elsewhere in the basin, for instance, through the extensive groundwater use in the area and through downstream springs. The improvements in efficiency thus result in limiting the access to water of downstream users. In this line of thought, the implementation of drip irrigation or lining is a re-allocation of water from one use(r) to the other.

The use of efficiency terminology and the discourse about re-capturing 'losses' gives the impression that the 'saved' water can be separated from the actual water use. However, the saved water mixes with the rest of the canal and groundwater, making it difficult to trace how much water is actually 'saved', and by whom it is used. In the case of drip irrigation, the water savings are not just claimed by one user but by several users who were involved in implementing drip irrigation. Each of them contributed to the drip irrigation projects: by providing land and labour, by giving subsidies or training. Different parties assume that the saved water is theirs, resulting in multiple claims on – and actual use of – the savings. This, in turn, results in more, rather than less water abstraction. In this way, modernization projects can effectively provoke an increase in water abstraction at the cost of the aquifer from which more water is pumped.

How efficiently drip irrigation is actually used, or how much water is actually saved at the system or basin level through the efficiency projects is rarely asked: counting on the promises of high efficiencies and thus large water savings is more politically expedient as it allows for (re-)allocating more water. In the history of the Ain Bittit irrigation system, the introduction of technologies to improve efficiencies is rarely accompanied with measures about its impacts in the field (neither before nor after the implementation). The promises of high efficiencies of drip irrigation are only based on experiments in laboratory-like settings, where all variables are controlled (van der Kooij et al., 2013). We suggest that it may not be accidental that efficiencies of drip irrigation in practice are rarely measured. Besides the difficulties that measuring entails (Lankford, 2013), not to measure whether and how much water is 'saved' also conveniently allows avoiding or concealing contentious discussions about how and to whom to allocate the available water. The high promise of drip allows for an allocation of 'yet-to-be-saved water', while other users of the same source seem to maintain access to water just as before. The yet-to-be-saved water can be allocated without difficult questions about whether people lose access to water – which would be the case when water would be re-allocated without a technical efficiency intervention. In the case study of Ain Bittit, no one claims to have lost water because of modernization projects upstream. Aside from the difficulty of measuring, and the strategic aspect of not knowing, questioning modernity and progress is hard. For water users to point at the negative impacts of modernization projects easily results in being blamed for being backward and wasteful, as opposed to modern and efficient.

Acknowledgments

We would like to thank the water users of Ain Bittit for sharing their histories with us. Furthermore, we thank the staff at the Provincial Agricultural Department of El Hajeb and the Regional Agricultural Department of Meknes for sharing their experience and documentation with us. We are pleased with the constructive reviews of Gerardo van Halsema and Chris Perry, which advanced this chapter.

Notes

1   What is perceived as 'losses' or 'savings' depends on the proprietor's perspective (van Halsema and Vincent, 2012) hence such perspective-dependent terms are placed in parenthesis.
2   We focus in this chapter on the notion of efficiency as 'saving water', although 'efficiency' is also tightly linked with increasing productivities, or with issues of equity and reliability, which we do not discuss here.

References

Adaptation aux Changements Climatiques en Afrique (ACCA). (2010). *Des échos du terrain. Réponse équilibrée aux besoins en eaux dans le bassin du Saïss au Maroc.* Rapport annuel 2009–2010. Available at www.idrc.ca/FR/Documents/reponse-equilibree-aux-besoins-en-eau-dans-le-bassin-du-saiss-au-maroc.pdf.

Allan, T. (1999). Productive efficiency and allocative efficiency: Why better management may not solve the problem. *Agricultural Water Management* 40: 71–75.

Ameur, F., Hamamouche, F., Kuper, M., and Benouniche, M. (2013). La domestication d'une innovation technique: la diffusion de l'irrigation goutte-à-goutte dans deux douars au Maroc. *Cahiers Agriculture* 22: 311–318.

Arrifi, E. (2009). L'économie et la valorisation de l'eau en irrigation au Maroc: un défi pour la durabilité de l'agriculture irriguée. *Proceedings of the International Symposium "Agriculture durable en région Méditerranéenne (AGDUMED)"*, Rabat, Morocco, 14–16 May 2009.

Bazzi, A. (1987). *La Petite et Moyenne Hydraulique au Maroc etude de cas: perimetre de Bittit (Meknes), Genie Rural*. Rabat: Institut Agronomique et Veterinaire Hassan II.

Belghiti, M. (2009) Le Plan National D'économie d'eau en irrigation (PNEEI) : une réponse au défi de la raréfaction des ressources en eau. *Hommes Terre Eaux* 143/144, pp.34–36.

Benouniche, M., Kuper, M., Hammani, A., and Boesveld, H. (2014). Making the user visible: Analysing irrigation practices and farmers' logic to explain actual drip irrigation performance. *Irrigation Science* 32(6): 405–420.

Berbel, J., Pedraza, V., and Giannoccaro, G. (2013). The trajectory towards basin closure of a European river: Guadalquivir. *International Journal of River Basin Management* 11: 111–119.

Burt, C. (1999). Current canal modernization from an international perspective. In *Proceedings 1999 USCID Workshop; Modernization of Irrigation Water Delivery Systems*. Denver: US Committee on Irrigation and Drainage, pp. 15–37.

Burt, C., and Styles, S. (1998). *Modern water control and management practices in irrigation: impact on performance*. ITRC Report 98–001. Irrigation Training and Research Centre.

Commission on Geosciences, Environment and Resources (CGER). (1992). California's imperial valley: A "win-win" transfer ? In *Water transfers in the west: Efficiency, equity, and the environment*. Washington, DC: The National Academy of Sciences, pp. 234–248.

Direction des Travaux Publics. (1949). *Reconnaissances des droits d'eau sur les sources de l'oued Bittit (Aioun Sidi Tahar, Sidi El Mir et Ben Sebaa)*. Direction des Travaux Publics. No 2231/46.

Direction des Travaux Publics. (1954). *L'Equipement Hydraulique du Maroc*. Maroc: Publication de la Société d'Etude Economique Sociale et Statistique du Maroc.

El Haouat, S. (2012). *Nappe Fès/Meknès gestion et perspectives*. Powerpoint slides.

El Meknassi, Y. (2009). *Economie et valorisation de l'eau en irrigation au Maroc. Un défi pour la durabilité de l'agriculture irrigué*. PDF document.

European Environment Agency (EEA). (2009). *Water resources across Europe – confronting water scarcity and drought*. EEA Report No 2/2009.

Halsema, G. van. (2002). *Trial and re-trial: The evolution of irrigation modernisation in NWFP, Pakistan*. PhD dissertation Wageningen University, Department of Water Resources, The Netherlands.

Halsema, G. van, and Vincent, L. (2012). Efficiency and productivity terms for water management: A matter of contextual relativism versus general absolutism. *Agricultural Water Management* 108: 9–15.

Hooper, V., and Lankford, B. (2016). Unintended water allocation; gaining share from the ungoverned spaces of land and water transformations. In K. Conca and E. Weinthal (Eds.), *Oxford handbook of water politics and policy*. New York: Oxford University Press.

Jenkins, M. (2007). The efficiency paradox. *High Country News*, February 5, 2007.

Jensen, M. (2007). Beyond irrigation efficiency. *Irrigation Science* 25: 233–245.

Lankford, B. (2012a). Towards a political ecology of irrigation efficiency and productivity. *Agricultural Water Management* 108: 1–2.

Lankford, B. (2012b). Fictions, fractions, factorials and fractures; on the framing of irrigation efficiency. *Agricultural Water Management* 108: 27–38.

Lankford, B. (2013). *Resource efficiency complexity and the commons: The paracommons and paradoxes of natural resource losses, wastes and wastages.* Abingdon: Earthscan Publications.

Lecina, S., Isidoro, D., Playán, E., and Aragüés, R. (2010). Irrigation modernization and water conservation in Spain: The case of Riegos del Alto Aragón. *Agricultural Water Management* 97(10): 1663–1675.

López-Gunn, E., Mayor, B., and Dumont, A. (2012). Implications of the modernization of irrigation systems. In L. De Stefano and M.R. Llamas (Eds.), *Water, agriculture and the environment in Spain: Can we square the circle?* Leiden, The Netherlands: CRC Press/Balkema, pp. 241-255.

Ministère de l'Agriculture du Développement Rural et des Pêches Maritimes. (2007). *Programme National d' Economie d'Eau en Irrigation (PNEEI).* Rabat: Ministère de l'Agriculture et de la Peche Maritime.

Molden, D. (1997). *Accounting for water use and productivity.* IWMI/SWIM Paper No. 1, Colombo: International Water Management Institute.

Molle, F., Mamanpoush, A., and Miranzadeh, M. (2004). *Robbing Yadullah's water to irrigate Saeed's garden, hydrology and water rights in a village of Central Iran.* IWMI Research Report No. 80. Colombo: International Water Management Institute.

Organisation for Economic Cooperation and Development (OECD). (2010). *Sustainable management of water resources in agriculture.* London: OECD Publishing and IWA Publishing.

Perry, C. (2011). Accounting for water use: Terminology and implications for saving water and increasing production. *Agricultural Water Management* 98(12): 1840–1846.

Playán, E., and Mateos, L. (2006). Modernization and optimization of irrigation systems to increase water productivity. *Agricultural Water Management* 80(1): 100–116.

Seckler, D. (1996). *The new era of water resources management: From 'dry' to 'wet' water savings.* Colombo: International Irrigation Management Institute.

Wolf, G., Gleason, J., and Hagan, R. (1995). Conversion to drip irrigation: Water savings, facts or fallacy, lessons from the Jordan Valley. In *Proceedings of the 1995 Water Management Seminar, Irrigation Conservation Opportunities and Limitations*, US Committee on Irrigation and Drainage, Sacramento, CA, pp. 5–7.

World Bank. (2006). *Reengaging in agricultural water management: Challenges and options.* Washington, DC: World Bank.

5 Unraveling the enduring paradox of increased pressure on groundwater through efficient drip irrigation

Marcel Kuper, Fatah Ameur, and Ali Hammani

Introduction: the double groundwater paradox

This chapter investigates the continued promotion of drip irrigation through state subsidies and donor-funded projects, on the basis of claims of water savings that are rarely backed up with evidence, and despite that it enables the intensification and extension of irrigated agriculture and *in fine* actually leads to increased pressure on water resources.

In many cases, groundwater is associated with the rapid extension of drip irrigation and thus bears the brunt of increased water use. This close connection between drip irrigation and groundwater use explains the focus of this chapter.

Groundwater is often portrayed as an overexploited resource, especially through pump irrigation in semi-arid regions (Shah, 2009). Farmers find groundwater attractive as it is 'at hand' through private tube-wells, as opposed to the limited water resources in collective irrigation systems. Obtaining access to groundwater enables farmers to escape a world of water scarcity and water sharing, for a world of plenty with a perceived abundance of water, at least in the short run. Shielding farmers from water scarcity through the access to seemingly unlimited groundwater explains in part why water consumption increases. Flexible use of groundwater allows intensification and diversification of cropping systems, and strengthens farmers' economic conditions. At the same time, it also causes groundwater depletion in many regions around the world and related groundwater quality degradation, land subsidence and ecological impacts on aquatic ecosystems as well as increasing cost in sustaining the use (Llamas and Martínez-Santos, 2005).

This is when drip irrigation comes in the picture. It is often presented as a 'saviour' technology, as it supposedly – through a theoretical efficiency of 90–95% – enables water saving (Gleick, 2003). Increasing the (fraction of) irrigated area under micro-irrigation, in this view, automatically leads to water saving. Gleick (2003) therefore laments: 'As of 2000, however, the area under micro-irrigation worldwide was less than 3 million hectares, only about 1% of all irrigated land'. The way forward in this view is to further increase water use efficiency and shifting the economy away from 'water-intensive production' (Gleick, 2003). However, there has been a strand of critical literature showing that so-called water-saving technology, such as drip irrigation, does not necessarily save water. In fact, these authors state that it is precisely because of the increase in water productivity that farmers are encouraged to intensify agricultural production and thus increase overall water use (Seckler, 1996; Huffaker and Whittlesey, 2000). The debate on this paradox – why is there increased pressure on groundwater despite the use of (purportedly) efficient drip irrigation? – started in the United States in the 1990s and was subsequently extended to different regions of the world, such as South Asia or the Mediterranean region (e.g. Berbel et al., 2013; Batchelor et al., 2014).

What is disturbing in the debate on the supposed positive impacts of drip irrigation on water saving is that the promised water gains are hardly ever quantified, but that drip irrigation support programs continue to roll: 'although water conservation intentions carry considerable political weight, there is all too often little serious evidence on conservation outcomes that would be produced by water conservation programs in policy debates, funding opportunities, and the popular press' (Ward and Pulido-Velazquez, 2008). This reveals a second paradox, which is closely linked to the first one: why do actors continue to use the water saving argument for drip irrigation without ever measuring the claimed water conservation outcomes? Interestingly, a debate building on this paradox is emerging in various fora. The rapporteur of a session on the 'Adoption and Impact of Improved Irrigation Technologies' of the 11th Annual International Water Resource Economics Consortium Meeting, organized at the World Bank in September 2014, thus concludes: 'empirical evidences regarding the effect of these technologies on consumptive water use are not unanimous. There is a need for further analysis of the issue focusing on dispelling the physical, technical, institutional, and policy contexts under which the adoption of new irrigation technologies lead to real water savings'.

This chapter will further explore what explains the continued use of the water-saving argument and how did drip irrigation become a solution of interest to a wide variety of actors. Our argument is not to say that water saving is not possible, but to show how, in many cases, drip irrigation increased water use. Why are none of these actors interested in measuring the outcomes of the expensive investments made? Why is there no water accountability or feedback to the states and donors? In other words, what are the political and technical reasons for not backing up water saving claims with measurements?

Different fallacies in the claims of drip irrigation as a saviour technology

The different critical studies claiming that the introduction of drip irrigation may bring about increased pressure on water resources rather than the proclaimed water savings – the first paradox – advance three main arguments. The first argument refers to the increased water productivity at the farm and the plot levels (more crop per drop; Molden et al., 2003), leading to increased crop production and revenues, thereby encouraging further intensification of agricultural production and thus increased water use (Peterson and Ding, 2005). At the *farm* level, the farmer may increase the fraction of irrigated area (more irrigated plots), increase cropping intensities (more crops per year on a single plot) and introduce high-value crops, which may be more water-demanding. At the *plot* level, the introduction of drip irrigation along with the agricultural intensification decided by the farmer then lead to higher irrigation volumes. In addition, in the Mediterranean region, increased water productivity is often associated with a more inelastic irrigation demand, as agricultural policies advocate permanent crops such as citrus, apples and olives (Berbel et al., 2013). In the recent literature, this first argument is often analysed by referring to the Jevons' paradox, which relates to 'efficiency improvement in the technical process of the use of a resource that ultimately defeats the original purpose through a higher overall use by society' (Dumont et al., 2013). Such unexpected 'rebound effects' may counteract the originally intended water saving: 'In contrast to widely-held beliefs, our results show that water conservation subsidies are unlikely to reduce water use. . . . In fact, water depletion, yields, and acreage are all predicted to increase if total water use is not constrained . . . by the various water authorities' (Ward and Pulido-Velazquez, 2008).

The second argument, which has been much less discussed in the literature, relates to the drip irrigation efficiencies obtained by farmers, as opposed to the theoretical efficiency obtained in laboratory settings (van der Kooij et al., 2013). A recent study in Morocco showed that farmers were able to achieve high levels of efficiency equivalent to the theoretical efficiency of 90%. However, most farmers were simply not interested in water saving and pursued other objectives, such as increased yields or labour saving (Benouniche et al., 2014). Drip irrigation is often associated with high value crops, so farmers do not want to irrigate sparingly to avoid negative effects on crop yields and quality (Benouniche et al., 2014). Most farmers in this study over-irrigated their plots and about one third of the drip irrigation systems observed (8/22) even showed irrigation efficiencies equal or inferior to the presumed efficiency of the much-reviled gravity irrigation system – i.e. around 50% (Benouniche et al., 2014). The results of this study concur with several other field measurements carried out in the Mediterranean (Benouniche et al., 2014).

The third argument, which is probably the most debated fallacy, relates to the distinction between 'wet' and 'dry' water savings by considering the drainage flows at the basin level (Seckler, 1996). According to this author, the scope for improving water use efficiency is low in river basins that are 'closed', i.e. basins where all the 'dry-season flow of usable water is captured and distributed', which is increasingly the case in most semi-arid regions. 'By definition, all of the usable drainage water in closed water basins is already being beneficially used'. A water-saving technique that reduces the amount of drainage water that was used 'beneficially' downstream leads to 'dry' water saving. Only when the drainage water is lost, for example when it goes into a salt sink, water saving will amount to 'wet' savings. By changing from gravity irrigation to drip irrigation, for example, water may be 'saved' at the farm level, as more water is beneficially used inside the farm (typically by increasing the evapotranspiration) while reducing the drainage flow. But if this water was beneficially used downstream, the introduction of water saving technology and expansion of irrigated areas with water-saving technology within the farm will amount to a re-allocation of water from one group of farmers, frequently those situated downstream in the basin, to the group of farmers having installed water-saving technology. This ultimately amounts to 'robbing Peter to pay Paul' (Barker et al., 2000).

As regards the second paradox, three arguments can be advanced to explain why the theoretical irrigation efficiency at plot level continues to be used in justifying drip irrigation projects rather than measured values at the plot, farm and regional levels. First, measuring efficiencies in real-life settings is tedious and requires stringent research protocols to determine these efficiencies and to understand the gap with theoretical efficiencies. The second argument relates to the multi-scalar and thus complex nature of the water-saving debate, as illustrated in the arguments outlined for the first paradox. Although someone like Seckler (1996) showed clear conceptual thinking when distinguishing between 'wet' and 'dry' water savings, the fact that water savings at the plot and farm levels may lead to increased water use at higher levels remains counterintuitive. In such a debate, the use of specific metrics, such as the theoretical efficiency of 90% of drip irrigation at the plot level, leads to what Elliott (2013) called selective ignorance or, in our case, chasing the red herring of dry savings. Third, even when measurements are available they are generally not used to evaluate the impacts of the implementation of drip irrigation. The measured decline in water tables in groundwater-based irrigated areas is strangely enough hardly ever associated with the increased fraction of irrigated area equipped with drip irrigation. In an ironic twist, the argument is generally even inversed: because water tables are declining, more efforts should be made to promote drip irrigation. This raises of course questions so as to why the actual water savings of drip irrigation continue to be ignored. Ignorance has become a well-researched topic 'to explore the social life and political issues involved in the distribution and strategic uses of not knowing' (Gross and McGoey, 2015). This was illustrated for agricultural research by Elliott (2013), who showed how by focusing research on 'maximizing production (often with heavy irrigation, fertilizer, and pesticide use), it can perpetuate a state of relative ignorance about the multiplicity of negative side-effects that accrue on other natural resources'. The way out, suggests Elliot, to challenge 'narrow disciplinary perspectives' is by engaging in 'the thoughtful monitoring of "real-world experiments",' which in our case relates to the observation and analysis of the implementation of drip irrigation in a given case study.

Illustrating the groundwater paradox through a case study

Case study and methodology

To provide an evidence-based illustration of our contention that drip irrigation may lead to increased pressure on groundwater resources, a study area was selected in the rich 220,000 ha agricultural Saïss plain, located close to the large cities of Fes and Meknes (Morocco) (see Figure 5.1). In the past, the region hosted mainly rain-fed agriculture with limited irrigated agriculture around some 100 karstic sources. Since the 1980s a transition to irrigated agriculture took place, taking advantage of the rich aquifer system, consisting of a phreatic aquifer and the Lias confined aquifer. Supported by ambitious agricultural policies, the Saïss is now characterized by a process of groundwater-based agricultural intensification, associated with the rapid extension of drip irrigation. Groundwater use in the Saïss plain is practised through two distinct devices. Shallow wells (up to 40–60 m) rely on the phreatic aquifer and can generally operate only a few hours a day as the recharge rate of the wells is lower than the pump rate. Farmers typically irrigate 1–3 ha with a shallow well. Deep tube-wells (110–160 m) rely on the confined aquifer and can be operated the whole day to irrigate more than 6 ha (Plate 5 in the color plate shows a 'typical' on-farm groundwater abstraction system where a deep tube-well has been installed by deepening the former shallow well).

The study focused on a 4,218 ha area close to the city of El Hajeb, which represents the diversity and dynamics of farming systems in the Saïss. The study area has gone through a series of profound agrarian changes. During the French Protectorate (1912–1956), French settlers occupied the land to grow mainly rain-fed cereals and vineyards and some limited horticulture. The land was progressively nationalized during a 17-year period after Independence. The colonial farms were converted to large-scale state farms that were later on, in the early 1990s, either leased to national agribusiness companies (1,375 ha) or redistributed to smallholders (2,460 ha). The latter were often labourers of

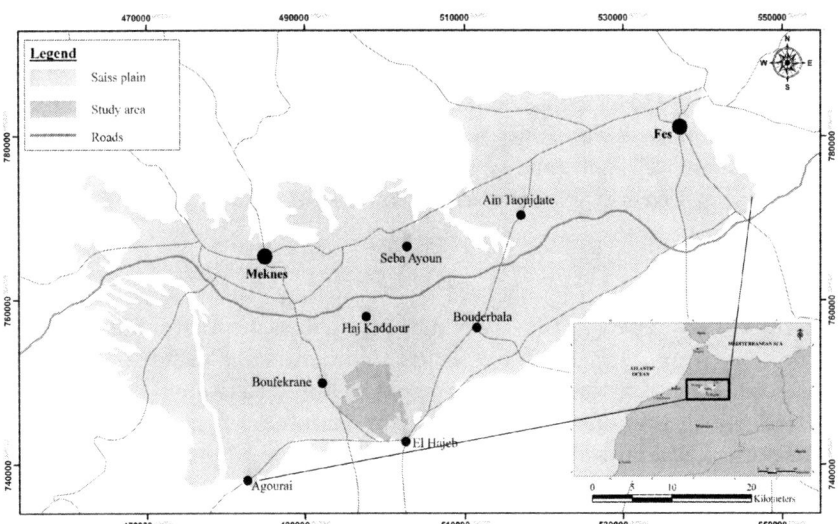

Figure 5.1 Location of the study area within the Saïss Plain.

Source: the authors.

the state farms and were granted access to land (9–13 ha for each beneficiary) through agrarian reform cooperatives. In 2005, the state decided to further liberalize the land of these cooperatives, as these were considered nonproductive. The members of the cooperatives could buy the land they had use-right on until then, but – more importantly – could also sell and rent it out. As many smallholders were indebted, and as the price of land went up to incredible heights (up to 50,000 € per ha), a large number of the members of the cooperatives sold their land or rented it out. This attracted many investors to the area, who – encouraged by generous and ambitious agricultural policies – installed fruit trees and drip irrigation on medium-sized plots (5–10 ha). In 2014, different social categories of farmers co-existed in the study area: 1) two companies conducted arboriculture (citrus, vineyards, mainly) on 1,375 ha; 2) the former members of the agrarian reform cooperatives cultivated 1,396 ha (out of an initial 2,460 ha), mainly with rain-fed crops (cereals, legumes) and some limited horticulture; 3) medium-sized investors now own 981 ha and mainly invested in fruit trees (apples, plums) on farms of 5–10 ha; and 4) lessees rented in 466 ha for highly intensive horticulture (mainly onions, potatoes); their plots in the study area are about 5 ha, but some of them cultivated more than 100 ha in the Saïss.

To assess the agrarian change in the study area, we undertook 30 historical interviews with key witnesses (farmers, local staff from agricultural services). We then interviewed all farmers of the 377 farms in the study area to identify farm profiles according to land ownership, land use and irrigation method for 2013–2014 and to obtain data on how these characteristics had evolved over the period 2005–2014. We also carried out an exhaustive survey of all shallow wells and tube-wells in the study area to determine different attributes (hydraulic characteristics, ownership, access modes).

To determine total groundwater use in the study area, we monitored irrigation applications on 20 selected farms in 2013–2014, and we extrapolated the groundwater use to all identified farms. In order to verify this extrapolation, we compared four methods of extrapolation based on the hydraulic characteristics of the shallow wells and tube-wells, on farming systems (governing irrigation efficiency) and farm type and on actual land use (Ameur et al., 2016).

Groundwater use

The first shallow wells in the study area go back to the 1940s. Most colonial farms (11 out of 13 in 1940) had access to groundwater mainly for drinking water, but some conducted horticulture or fruit tree cultivation on small plots of land. Much later, in the early 1990s, the members of the agrarian reform cooperatives also installed shallow wells to irrigate small plots of horticulture. From then on, the number of wells slowly increased until 2005, when the number of operational wells started to decline as 1) a considerable number of wells ran dry due to the decrease in groundwater levels, and 2) some farmers could no longer afford to continue growing horticultural crops due to volatile market prices. All in all, 96 out of 193 shallow wells were no longer functional in 2014. Some of the former members of cooperatives then invested in tube-wells or in deepening their wells after selling part of the land. However, most of them did not have the means to install a tube-well and conduct the intensive agriculture that could pay back the investment costs. The groundwater abstraction of these farmers thus diminished after 2005 (see Figure 5.2).

The use of tube-wells in the study area goes back to the early 1990s in the large farms of the agribusiness companies. The companies took over existing orchards of the state farms and planted more fruit trees (citrus, mainly) and vineyards. They installed tube-wells and rapidly increased their groundwater use (Figure 5.2). The biggest increase in groundwater use occurred after 2005 with the arrival of investors and lessees. Medium-scale investors invested in fruit trees and installed tube-wells, supported by state subsidies. The lessees conducted intensive horticulture and also installed tube-wells. Both categories are now responsible for 57% of the groundwater use, in addition to the 27% of groundwater use of the two companies, while the former cooperative members only use 16% of the total groundwater volume pumped in 2014, mostly through shallow wells.

Overall, the groundwater use in the study area went up from less than 1 million m^3 in 1990 to 4.8 million m^3 in 2005, and to almost 10 million m^3 in 2014. Most of this use comes from tube-wells (87% of the groundwater use in 2014) that can be operated all day. Meanwhile, the irrigated area went up from only about 400 ha in 1990 (10% of the total area) to 1,580 ha in 2005 (37%) and 2,420 ha in 2014 (57%).

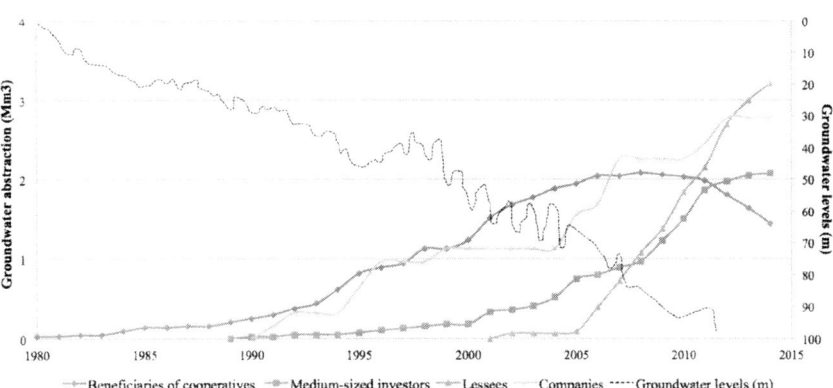

Figure 5.2 Groundwater abstraction by different categories of farmers and decline in water tables in the study area (the authors).

Source: Groundwater abstraction measurements are based on data from Ameur et al. (2016). Groundwater levels are based on reading of the piezometer 290/22 of Haj Kaddour, representing water levels of the Lias confined aquifer (data provided by the Sebou River Basin Agency).

The extension of drip irrigation

Drip irrigation was not common in 2005. It was mostly installed by the two companies in the ex-state farms (600 ha out of 1,375 ha), and – to a lesser extent – by some medium-sized investors (6/25) who equipped 59 ha out of 365 ha. In contrast, only 22 members of cooperatives (out of 261) installed drip irrigation on 38 ha out of 2,460 ha, mainly to irrigate onion nurseries (Table 5.1). By 2005, about 17% of the study area was equipped with drip irrigation (see Figure 5.3).

Drip irrigation accompanied the extension of the irrigated area. It increased rapidly after 2005 with the installation of deep tube-wells, which provided a higher and more reliable water source. In 2014, the situation had dramatically changed: about 48% of the area was now equipped with drip irrigation. The two companies more than doubled the area equipped with drip irrigation and the medium-sized investors multiplied theirs by six (Table 5.1). Both categories benefitted from the public subsidies awarded (80–100% of the investment cost). In all cases, the installation of drip irrigation was accompanied by the installation of tube-wells. Lessees also invested in drip irrigation systems, although they did not receive any subsidies, because they had no formal land titles. Drip irrigation often enabled lessees to double the yields with up to 100 T/ha for onions and 80 T/ha for potatoes. This led to overproduction for horticulture and thus lower prices. The former members of cooperatives, in fact mostly their sons, also installed drip irrigation, but they continued to depend on shallow wells. They installed drip irrigation, without any subsidies, on about 12% of the land. This does not seem much, but as these smallholders only irrigate about 23% of their surface area, drip irrigation covered more than half of their irrigated land.

Table 5.1 Surface area equipped with drip irrigation in the study area for different types of farmers

Actors	Situation 2005			Situation 2014		
	Total area (ha)	*Irrigated area (ha)*	*Area equipped with drip irrigation (ha)*	*Total area (ha)*	*Irrigated area (ha)*	*Area equipped with drip irrigation (ha)*
Former members of cooperatives	2460	378	38	1396	299	167
Agribusiness companies	1375	1092	600	1375	1340	1212
Medium-sized investors	364	100	59	981	404	354
Lessees	19	11	8	466	378	283
Total	**4218**	**1581**	**705**	**4218**	**2421**	**2015**

Source: the authors.

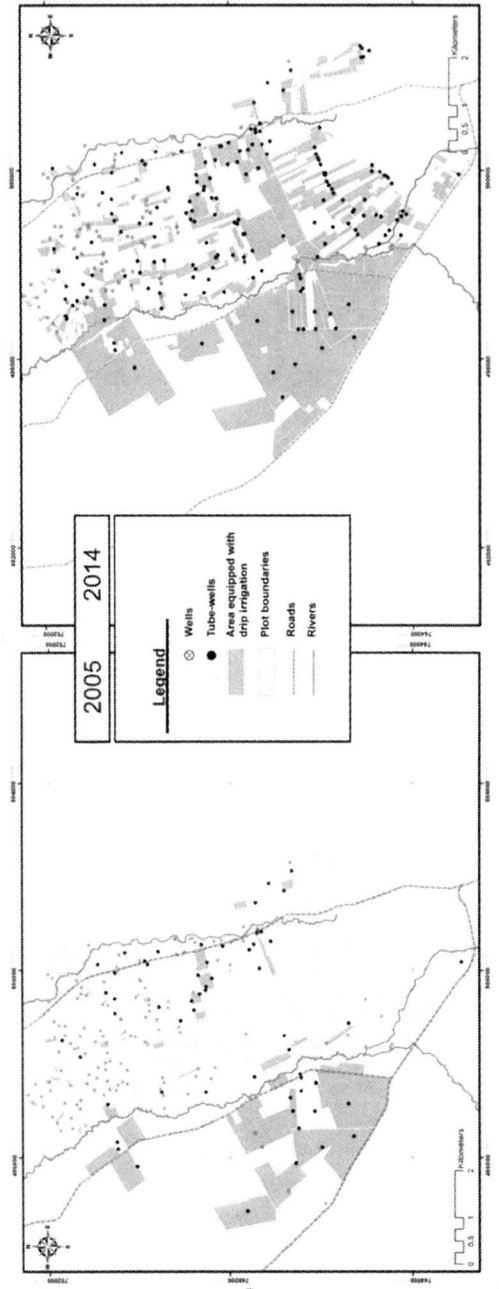

Figure 5.3 The extension of the drip irrigation in the study area.

Source: the authors.

So what about water saving?

We will now verify the three hypotheses contesting that drip irrigation saves water. Drip irrigation may lead to or be accompanied by: 1) increases in the irrigated area, increased cropping intensities and changes to higher water-demanding crops at the farm and plot levels, 2) no water gains at the plot level due to disappointing irrigation efficiencies, and 3) higher consumptive use at the regional scale.

As regards the first hypothesis, Table 5.1 shows that the introduction of drip irrigation was accompanied by an increase in the irrigated area from 2005 to 2014. Farmers converted the existing irrigated area from gravity to drip irrigation: the fraction of irrigated area served by drip irrigation went up from 45% to 83%. However, farmers also extended the irrigated area from 37% to 57% of the cropped area. This is due to the arrival of more entrepreneurial farmers. In 2014, the former members of cooperatives with access to groundwater irrigated only 40% of their farmland, while the companies and investors typically irrigated almost 90% of their farmland (grapes, apple, plums and apricots). The increase in the fraction of irrigated area was accompanied by increased cropping intensities. The lessees, for instance, showed cropping intensities higher than 100% as they grew more than one crop per year. The irrigation volumes doubled in most plots of the lessees, as they shifted from one to two crops per year. In the case of fruit orchards, the irrigation season is extended, as the trees are irrigated for almost the entire year. The introduction of drip irrigation was accompanied by a rapid increase of fruit orchards (1,747 ha in 2013/2014), almost exclusively by the companies and medium-sized investors, and by an increase of horticulture (585 ha in 2013/2014), which are water-demanding crops. The lessees, hardly present in 2005, produced half of the area under horticulture (289 ha). The first hypothesis is, therefore, confirmed. Interestingly, only the former members of cooperatives did not increase their irrigated area during the period 2005–2014, due to problems of water access through shallow wells and of market saturation. In fact, they even decreased the irrigated area, from 378 to 299 ha (table 5.1). However, as the number of tube-wells increases with the arrival of new investors and lessees, the area under drip irrigation will undoubtedly continue to expand, and the risk of continuous decline of groundwater levels is paradoxically linked to the dynamics of drip irrigation extension.

As regards the second hypothesis on field efficiencies, the lessees and the investors, who engage in intensive agriculture, do not want their crops to suffer from water stress. Typically, they over-irrigate by 60% when compared to crop water demands. For instance, an irrigation volume in the range of 7,500–14,500 m^3/ha was determined for onion crops under drip irrigation, whereas the irrigation water requirements are around 4,500 m^3/ha in an average year (Ameur et al., 2016). This leads to efficiencies in the range of 0.3–0.6. These values basically represent the same or even inferior efficiencies as compared to the theoretical value for gravity irrigation (around 0.5). The second hypothesis is therefore confirmed. We also measured the irrigation efficiency for a plot irrigated by gravity irrigation, which showed an irrigation volume of 6,500 m³/ha for onions (irrigation efficiency=0.69). Gravity irrigation is exclusively practiced by former members of cooperatives. They do not have the financial means to cover the higher production costs for more intensive practices and often have shallow wells with low yields. This explains that this field was under-irrigated – i.e. the measured efficiency (0.69) was higher than the theoretical value of 0.5. The financial capital along with the strategic objectives of a farmer thus define the degree of crop intensification, and thus the applied volume at field level and the degree of extension of drip irrigation at farm level. However, this study also shows that 'dry' savings both for drip and for gravity irrigation are possible through careful irrigation practices. Improved irrigation practices enabled, for instance, to reduce irrigation volumes on some fruit orchards to around 6,000 m³/ha, meaning that increased cropping intensities did not always lead to increased water use. In this case, the increase in groundwater use at the farm and regional levels only related to the extension of the irrigated area (first argument).

And finally as regards the third hypothesis, the previous two arguments explain why at the scale of the study area, groundwater consumption doubled between 2005 and 2014: from 4.8 to almost 10 million m³. The increased agricultural production, which was made possible by doubling the irrigation volumes of pumped groundwater, will undoubtedly lead to increased consumptive use and higher depletion levels at the regional scale. This confirms the third hypothesis that there are no 'wet savings'. In addition, the study shows how certain categories of farmers in the study area are excluded from access to groundwater.

Why measure water savings? Historical perspectives and policies of drip irrigation in Morocco

This chapter looks more closely at Morocco, as it is a major protagonist of state-subsidized drip irrigation projects with substantial loans from international donors such as FAO, the African Development Bank and the World Bank. The interest of the state for drip irrigation emerged progressively, and the state is now a major actor in the rapid expansion of drip irrigation. The history of the introduction and the diffusion of drip irrigation in Morocco spans a period of more than 40 years, and can be divided, for analytical purposes, into three main periods. During each of these periods, water saving was one of the stated objectives of at least one of the actors involved in the promotion of drip irrigation (the water-saving argument also justified public support and development of drip irrigation in Spain, see Box 5.1). However, water saving somehow never took any prominence in the implementation and the evaluation of drip irrigation programmes.

Period 1: limited state intervention (1974–2002)

During the first period – from 1974 to 2002 – drip irrigation was mainly a private affair and remained confined to large-scale commercial farmers, growing citrus trees or producing horticulture, including tomatoes, strawberries and bananas (Ababou, 1979; Popp, 1986). These farmers had different reasons for installing drip irrigation: water and labour saving and increased agricultural production, and more generally drip irrigation provided 'ways of intensifying, without much risk, speculative export-oriented crops: citrus fruits, horticulture' (Ababou, 1979). There was also, for some time, the curiosity of a modern technology. Installing it provided a status, that of 'advanced' farmers (Benouniche et al., 2014).

With limited government support, these large growers had installed an estimated 87,100 ha of drip irrigation in Morocco by 2002 (El Gueddari, 2004). During this period, the engineering profession scrutinized this 'unknown' technology and analysed its potential advantages in Morocco (Ababou, 1979). The author indicated somewhat prophetically that drip irrigation was particularly interesting in water-scarce areas where groundwater was used, as drip irrigation could lead to water savings of about 20–30%. This was confirmed by 2002, as it had indeed developed mostly in water-scarce areas relying on groundwater. For instance, in the Souss valley, in the arid south of the country, drip irrigation covered 26,800 ha in 2002 (Baroud, 2002), which constituted 25% of the region's irrigated area and more than 30% of the total area equipped with drip irrigation in Morocco. However, this did not lead to lower pressure on groundwater resources. In the Souss valley, the use of groundwater went up from 205 million m^3 in 1969 to about 615 million m^3 in 2002 (Baroud, 2002).

Box 5.1   Drip irrigation in Eastern Spain. Diverging goals in a converging process

In the last 25 years, drip irrigation has been installed in 1,756,138 hectares in Spain, approximately 53% of the total irrigated area. This was actively promoted through public subsidies in a combined top-down bottom-up process. The administration carried out some large infrastructure projects but most of the investing effort was directed through the provision of generous public grants for drip irrigation implementation. Water Users' Associations (WUAs), which have a prominent role in irrigation management in the country, formally applied for these funds to finance their transformation projects. We analyzed this process in the Valencia Region citrus-based areas, where drip irrigation has acted as catalyst that made the different interests of farmers and planners compatible.

In the Mediterranean Spain, the public administration opted for drip irrigation as a pivotal strategy to reduce agricultural water use, in a context of structural scarcity. Pressure on water resources had markedly increased at the end of the 20th century, and many districts faced important water deficits and coped with recurrent severe droughts. Both the national and regional administrations thus provided legal support and financial resources to boost drip irrigation expansion. Accompanying the legal and economic measures, the State articulated a powerful discourse around drip irrigation, endorsed by international reports, on water savings estimated in scientific works, on the seductive notion of modernity, and also on the necessity to clean up the image of the traditional irrigation sector, frequently stigmatized as water squanderer.

However, water saving or water efficiency were not commonly mentioned when farmers were asked for their reasons to adopt drip irrigation. Factors such as productivity increase, improved fruit quality, labor savings or convenience ranked much higher for them in a context of small landowners, many of them ageing, and the absence of a generational renewal. Actually, most farmers stated that they adopted drip irrigation because they were obliged by the WUAs. They did not directly participate in this decision. Projects were approved by the governing boards and when this issue was voted in the general assembly of the WUAs, the attendance and participation of farmers was low. This is something quite usual in the WUAs in the region, due to the importance of part-time farming, ageing processes and the high level of confidence of farmers in the WUAs' governance. In fact, some governing boards had to organize extraordinary informative meetings to motivate farmers and to seek their support.

The governing boards and managers of WUAs were thus the nexus between different interests that led to widespread adoption of drip irrigation. This was possible because they are a key link in the chain of water management in the region, between water authorities and farmers. On the one hand, they made the decision of introducing drip irrigation, as they were aware of the water deficit problems at basin scale. They understood the strategic importance of drip irrigation as a tool for reducing their own water demand, something that was too far from the individual strategies at farm scale. On the other hand, they were conscious of farmers' necessities, and they believed drip irrigation was a more practical technique for the farmer.

Drip irrigation promotion policies materialized during the period 1996–2008, partially financed by the Valencia Government through 865 public works in the region. In this period, according to data obtained from WUAs, water use decreased approximately by 45% to 55% in citrus based modernized areas, whereas it decreased by 25% to 30% in areas that did not adopt drip irrigation. Impacts on groundwater conservation or adjacent ecosystems are difficult to quantify or to attribute to drip irrigation implementation, due to the parallel trend of land abandonment observed during the same period. However, no signs of rebound effects were observed. As a result of this technological change, during the 2010 decade the administration started a revision of the (generous) water rights in the Jucar basin, in order to adjust the water rights to the real water use.

Thus, there has been a significant decrease in water use, even though the water saving objectives of the administration have become blurred along this top-down chain of development. In fact, farmers and local managers are currently adapting irrigation and fertigation technology to their needs. Some new operating procedures have been implemented to meet goals such as increasing income, reducing labour or 'traditions' that rule practices in the field, regardless of whether these aims conflict with reducing water use or not. For instance, some farmers add fertigation at plot level or complement drip irrigation with gravity irrigation where it is required and still possible. As a result of this flexibility and other advantages of drip irrigation, the satisfaction of most farmers with this technology is generally high in the region. Drip irrigation goes from mouth to mouth, stimulating future conversions.

Mar Ortega-Reig, Carles Sanchis-Ibo, Marta García-Mollá

Period 2: water saving as a new source of irrigation (2002–2007)

During the second period – from 2002 to 2007 – the state progressively became an important player in the promotion of drip irrigation through a series of subsidy programmes. In 2002, the state initiated a comprehensive subsidy scheme, referred to as the National Water Saving Plan for Irrigation (PNEEI). The programme covered 30–40% of the investment cost of farmers and aimed to convert 110,000 ha to micro-irrigation, in particular drip irrigation. According to the ministry, the rate of adoption so far, without state involvement, was too slow: 'this method of irrigation is not widespread enough in view of its numerous advantages' (MADRPM, 1999). The subsidy rate was subsequently increased to 60% in 2006 and even to 80–100% in 2010 in the wake of the second Water Saving Programme of 2007.

To understand why and how the state got involved in drip irrigation development, it is interesting to take a look at the water debates that dominated the late 1990s. First, there was a large international debate on the unfolding global water crisis and the necessity to implement 'water demand management' to "moderate the increase in water demand and even to decrease it by favouring water saving by increasing the water use efficiency" (Margat and Vallée, 1999). Water losses were considered a "new" source of water, 130 km^3/year at the global level (Margat and Vallée, 1999). An understanding emerged in Morocco that, after decades of irrigation development based on 'water supply management' and in the wake of the closing of basins, the major source of extra water could be provided by saving water at the plot level: 'the big source of water saving has been identified within the farm' (El Gueddari, 2004).

Second, a slightly different debate concerned the Integrated Water Resources Management (IWRM) paradigm, which gained much importance in Morocco with the 10–1995 Water Law and the subsequent creation of River Basin Agencies in 2002. Agriculture came under the spotlight as the main user of water resources, for example, during the national water debate of 2006/2007 organized by the ministry in charge of water. Agricultural water use would have to be curbed to cater for uses in other sectors and agricultural users would have to pay for their consumption ('user-pays' principle). Facing increasing scrutiny, promoting the use of water-saving irrigation technology became a perfect means for the ministry in charge of agriculture to regain the upper hand in the national water debate. This led to the formulation of the National Programme on Water Saving in Irrigation (PNEEI) in 2007.

The consultants in charge of preparing this new and ambitious water-saving programme, and more largely the agricultural ministry, were well aware that the change to drip irrigation could possibly mean water reallocation to other sectors. In the programme document, it was thus repeatedly said that the agricultural sector was already facing a water deficit, especially for surface water resources with an insufficient water allocation to the agricultural sector: 'In this context, no release of water resources for other uses, or even to extend the irrigated area, can be envisaged' (PNEEI, 2007). Only in the case of groundwater, it was acknowledged that the agricultural sector was expected to decrease the pressure on water resources by 312 million m³/year. Drip irrigation thus became an interesting tool to further a compromise between a ministry keen to develop irrigated agriculture and the ministry in charge of preserving the water resources. This was testified by the joint presentation of the two ministries on May 15, 2007, in Marrakech during the National Water Debate, which was intended as a contribution to the issue of 'integrated water resources management and the drought'. The presentation pleaded for structural action for water savings, in particular the replacement of gravity irrigation by water-saving irrigation techniques.

Interestingly, the formulation of the National Programme on Water Saving in Irrigation (PNEEI) in 2007 also coincided with critical debates internationally on the use of purportedly efficient drip irrigation: 'Our findings suggest reexamining the belief widely held by donors that increased irrigation efficiency will relieve the world's water crisis (Ward and Pulido-Velazquez, 2008)'. The consultants in charge of preparing the PNEEI – on behalf of the Moroccan government and FAO – were well aware of this international debate: the introduction of water-saving technology may lead to '[a]ccelerating the exploitation of aquifers' (PNEEI, 2007: 52). This is why the consultants proposed to accompany the implementation of the PNEEI by specific measures: 1) promote 'rational' irrigation practices of farmers, 2) constitute a database of all wells and tube-wells, and 3) promote aquifer contracts and install water meters on all tube-wells under the responsibility of the river basin agencies: 'This is a very delicate operation to implement, but it is absolutely essential, because without water metering, reduction of groundwater use is likely to remain wishful thinking' (PNEEI, 2007).

Period 3: water savings became secondary to water productivity

At the beginning of the third period – 2008 to present – the National Water Saving Programme for Irrigation (maintaining the same PNEEI acronym) was absorbed in the larger and very ambitious Green Morocco Plan, which attracted considerable investments to promote intensive irrigated agriculture (Akesbi, 2014). Water scarcity was seen as a potential impediment to agricultural development and drip irrigation became rapidly 'the saviour technology' that could provide a source of saved water to an expanding economic sector. Water saving thus became a secondary objective to increased (water) productivity. This was also evidenced by the fact that even on subsidized drip irrigation systems, no water meters were installed. Even when they were installed, the extracted volumes were not monitored.

The PNEEI also aimed to provide a modern image of agriculture using modern techniques enabling the sector to use water rationally – i.e. increase water productivity, switch to high value-added crops, increase revenues of farmers and promote employment on modern farms. One can add that in financing modern irrigation technology, the ministry of agriculture also modernized its image. The plan was thus also about building the image of the ministry as a modern powerhouse, able to distribute land, water and subsidies.

During this period, the implementation of drip irrigation rapidly increased as the subsidy programme converged with the interest of farmers who had already experimented with the technology in previous periods. The total area equipped with drip irrigation was estimated to be 359,847 ha in 2013 according to the official data of the ministry. It is estimated that the state invested more than 3 billion dirhams (300 million US$) in drip irrigation subsidy programmes between 2010 and 2012 (Boularbah, 2014). In 2012, the subsidies for drip irrigation (which included some subsidies for stone removal) constituted 52% of the amount of the total subsidies for the agricultural sector (Boularbah, 2014).

Discussion and conclusion

Inversing the cause-effect relationship between drip irrigation development and groundwater decline

We have shown how drip irrigation systems contributed to, and were part of, a process of rapid agricultural intensification due to the excellent and reinforcing fit between groundwater and drip irrigation. Declining groundwater levels contradict the frequently posited unmeasured claims that are made about the water-saving potential of new irrigation technologies, such as drip irrigation. The paradox is that declining groundwater levels are often advanced as arguments to subsidize drip irrigation, even though it is increasingly clear that it is precisely the increased water productivity (and associated increases in farmer incomes) allowed by drip irrigation that is responsible for the drawdown of the aquifers.

Different arguments claiming that drip irrigation may increase the pressure on water resources were investigated and quantified for a case study. We showed how drip irrigation contributed to an extension of the irrigated area, to increased cropping intensities, to changes to high water-consumptive crops with relatively low irrigation efficiencies as most farmers were over-irrigating. The announced and non-substantiated water-saving claims were, therefore, not achieved. Following Seckler (1996), we can confirm that the introduction of drip irrigation is not likely to produce any 'wet' water savings and thus results in the reallocation of water to those that have access to groundwater and installed drip irrigation to the detriment of those who do not. In our case, we showed that even 'dry' water savings were often not achieved due to low irrigation efficiencies. This does not mean that drip irrigation cannot achieve 'dry' or 'wet' savings, but both type of savings are unlikely if the total water use is not constrained (Ward and Pulido-Velazquez, 2008) and if there is no change in mind set of the different protagonists, meaning that water saving becomes part of the objectives of actors (Mitchell et al., 2012). This was shown in the case of Valencia (see Box 5.1) where 'wet' water savings were obtained in a context of a declining agricultural sector and ageing farmers.

At first sight, drip subsidy programmes seem to be part of a 'win-win' scenario of increased agricultural productivity and water saving, but a more careful analysis shows who is losing out. The first silent victim is groundwater itself, as the aquifer system experiences a regular decline in water tables, often with a concomitant increase in salinity levels, and in the long run the pollution from more intensive farming systems. The second victims are water users that are either downstream, or – in our case – smallholders situated in the vicinity of larger over-pumping farms, raising issues of social justice and equity (Amichi et al., 2012).

Why continue to chase the red herring?

So, how come we have been chasing the red herring of dry water savings? A red herring is a logical fallacy, which is a 'seemingly plausible, though ultimately irrelevant, diversionary tactic' (Tindale, 2007). There have been quite a few informed debates over the past 20 years in international arenas criticizing the promotion of drip irrigation for water savings. So, why do people informed about the facts that groundwater is overexploited and that drip irrigation contributes to the agricultural boom responsible for groundwater overexploitation continue to sponsor drip irrigation projects? What are the reasons for which there is no water accountability of drip irrigation-in-use? In other words, did we follow the red herring by design? Perhaps partly so, as the stakes were high for many actors each taking a piece in the larger drip irrigation cake. These interests were financial, but also related to image or social capital both for the state and for farmers (Benouniche et al., 2016). Part of the answer also lies in the complexity of a long process of increased water use and progressive exclusion of certain users, which is relatively difficult to comprehend and visualise (Seckler, 1996). There was also a lack of evidence in this debate. On the one hand, much of the applauding research on the efficiency of drip irrigation was based on laboratory research, ignoring the more complex realities of drip irrigation in use (van der Kooij et al., 2013). On the other hand, most of the more critical analyses on the 'real' water savings of drip irrigation did not provide any grounded information either. This was probably because drip irrigation had left the technical sphere for a more political sphere, spurring policy analyses and focusing on the discourse surrounding drip irrigation.

Perhaps more importantly, many people also believed, or wanted to believe, in the saviour-nature of drip irrigation. A wide variety of actors were enrolled in the world of drip irrigation, and it was difficult for them to let go this fascinating technology. The beauty of drip irrigation was also the multiplicity of purposes that could be achieved through it. Once all the above-mentioned arguments have been accepted and it becomes common knowledge that drip irrigation does not lead to water savings, the debate generally focuses on some of the other perceived promises of drip irrigation to bring about desirable futures, for example increased crop production or water productivity (Venot et al., 2014). This was already the escape route proposed by Seckler (1996), although it is precisely increased water productivity that constitutes the engine of the galloping exploitation of groundwater resources.

Acknowledgments

The research for this chapter was conducted in the framework of the Groundwater Arena project (CEP S 09/11), financed by the French National Research Agency (ANR). We thank the two anonymous reviewers for their comments on an earlier draft.

References

Ababou, R. (1979). Les inconnues de l'irrigation au "goute a goutte. *Hommes, Terre, Eaux* 9: 3–41.

Akesbi, N. (2014). *Le Maghreb face aux nouveaux enjeux mondiaux: Les investissements verts dans l'agriculture au Maroc.* Paris: IFRI. Available at http://dev.ocppc.lnet.fr/sites/default/files/IFRI_noteifriocpnakesbi.pdf

Ameur, F., Kuper, M., and Hammani, A. (2016). Méthodes d'estimation et d'extrapolation des pompages des eaux souterraines par l'intégration des pratiques locales: Cas de la plaine du Saïss au Maroc. *Sciences Agronomiques et Vétérinaires Marocaines* 5(1) : 52–64.

Amichi, H., Bouarfa, S., Kuper, M., Ducourtieux, O., Imache, A., Fusillier, J., Bazin, G., Hartani, T., and Chehat, F. (2012). How does unequal access to groundwater contribute to marginalization of small farmers? The case of public lands in Algeria. *Irrigation and Drainage* 61(1): 34–44.

Barker, R., Scott, C., Fraiture, C. de, and Amarasinghe, U. (2000). Global water shortages and the challenge facing Mexico. *International Journal of Water Resources Development* 16(4): 525–542.

Baroud, A. (2002). Changements climatiques et gestion de l'irrigation dans la zone d'action de l'Ormva du Souss-Massa. *Hommes, Terre, Eaux* 124: 98–101.

Batchelor, C., Reddy, V., Linstead, C., Dhar, M., Roy, S., and May, R. (2014). Do water-saving technologies improve environmental flows? *Journal of Hydrology* 518: 140–149.

Benouniche, M., Errahj, M., and Kuper, M. (2016). The seductive power of an innovation: Enrolling non-conventional actors in a drip irrigation community in Morocco. *Journal of Agricultural Education and Extension* 22(1): 61–79.

Benouniche, M., Kuper, M., Hammani, A., and Boesveld, H. (2014). Making the user visible: Analysing irrigation practices and farmers' logic to explain actual drip irrigation performance. *Irrigation Science* 32(6): 405–420.

Berbel, J., Pedraza, V., and Giannoccaro, G. (2013). The trajectory towards basin closure of a European river: Guadalquivir. *International Journal of River Basin Management* 11(1): 111–119.

Boularbah, S. (2014). *Retour d'expériences des projets de reconversion individuelle et collective dans les secteurs N1, N5 et N9 dans le périmètre irrigué du Gharb*. MSc thesis, Institut Agronomique et Vétérinaire Hassan II, Rabat, Maroc.

Dumont, A., Mayor, B., and López-Gunn, E. (2013). Is the rebound effect or Jevons paradox a useful concept for better management of water resources? Insights from the Irrigation Modernisation Process in Spain. *Aquatic Procedia* 1: 64–76.

El Gueddari, A. (2004). Economie d'eau en irrigation au Maroc: Acquis et perspectives d'avenir. *Hommes, Terre, Eaux* 130: 4–7.

Elliott, K. (2013). Selective ignorance and agricultural research. *Science, Technology, & Human Values* 38(3): 328–350.

Gleick, P. (2003). Global freshwater resources: Soft-path solutions for the 21st century. *Science* 302(5650): 1524–1528.

Gross, M., and McGoey, L. (2015). Introduction. In M. Gross and L. McGoey (Eds.), *Routledge international handbook of ignorance studies*. Abingdon: Routledge, pp. 1–14.

Huffaker, R., and Whittlesey, N. (2000). The allocative efficiency and conservation potential of water laws encouraging investments in on-farm irrigation technology. *Agricultural Economics* 24(1): 47–60.

Kooij, S. van der, Zwarteveen, M., Boesveld, H., and Kuper, M. (2013). The efficiency of drip irrigation unpacked. *Agricultural Water Management* 123: 103–110.

Llamas, M., and Martínez-Santos, P. (2005). Intensive groundwater use: Silent revolution and potential source of social conflicts. *Journal of Water Resources Planning and Management* 131(5): 337–341.

MADRPM (Ministère de l'Agriculture, du Développement Rural et des Pêches Maritimes). 1999. Stratégie 2020 de développement rural. MADRPM: Casablanca, Morocco.

Margat, J., and Vallée, D. (1999). *Vision méditerranéenne sur l'eau, la population et l'environnement au XXIème siècle*. Sophia Antipolis: MEDTAC/Plan bleu.

Mitchell, M., Curtis, A., Sharp, E., and Mendham, E. (2012). Directions for social research to underpin improved groundwater management. *Journal of Hydrology* 448–449: 223–231.

Molden, D., Murray-Rust, H., Sakthivadivel, R., and Makin, I. (2003). A water-productivity framework for understanding and action. In J. Kijne, R. Barker, and D. Molden (Eds.), *Water productivity in agriculture: Limits and opportunities for improvement*. Wallingford: CAB International, pp. 1–18.

Peterson, J., and Ding, Y. (2005). Economic adjustments to groundwater depletion in the high plains: Do water-saving irrigation systems save water? *American Journal of Agricultural Economics* 87(1): 147–159.

PNEEI (Programme National d'Economie d'Eau en Irrigation). 2007. Ministère de l'Agriculture, du Développement Rural et des Pêches Maritimes (MADRPM): Rabat, Morocco.

PNEEI (Programme National d'Economie d'Eau en Irrigation). 2007. Ministère de l'Agriculture, du Développement Rural et des Pêches Maritimes (MADRPM): Rabat, Morocco.

PNEEI (Programme National d'Economie d'Eau en Irrigation). 2007.Ministère de l'Agriculture, du Développement Rural et des Pêches Maritimes (MADRPM): Casablanca, Morocco.

Popp, H. (1986). L'agriculture irriguée dans la vallée du Souss: Formes et conflits d'utilisation de l'eau. *Méditerranée* 59(4): 33–47.

Seckler, D. (1996). *The new era of water resources management: From 'dry' to 'wet' water savings.* Colombo: IWMI.

Shah, T. (2009). *Taming the anarchy: Groundwater governance in South Asia.* Washington, DC: Resources for the Future Press.

Tindale, C. (2007). *Fallacies and argument appraisal.* Cambridge: Cambridge University Press, pp. 28–33.

Venot, J., Zwarteveen, M., Kuper, M., Boesveld, H., Bossenbroek, L., Kooij, S. van der, . . . Verma, S. (2014). Beyond the promises of technology: A review of the discourses and actors who make drip irrigation. *Irrigation and Drainage* 63(2): 186–194.

Ward, F., and Pulido-Velazquez, M. (2008). Water conservation in irrigation can increase water use. *Proceedings of the National Academy of Sciences* 105(47): 18215–18220.

6 Sour grapes

Multiple groundwater enclosures in Morocco's Saïss region

Lisa Bossenbroek, Marcel Kuper and Margreet Zwarteveen

Introduction

While driving to the *douar*[1] Ait Ali in the agricultural plain of the Saïss in Morocco, we always passed by a fallow land plot of approximately 10 hectares. We passed this plot without noticing much, until the day we spotted a Syrian tube-well driller with a checked red scarf on his head and his drilling rig on the land. He attracted our attention and his presence, as well as the newly erected fence surrounding the plot, triggered many questions: Was this land recently bought? Who bought it? Why had the land been fenced? What would the Syrian tube-well digger do if he did not find any water? What if he would find water? We decided to pull over and to interview him. He told us that someone living in the nearby city of Meknes had recently bought the land. He showed us a little device with threads of different colours, which had helped him to locate the exact spot with the highest probability of finding water. When we asked him what would happen if he did not find any water he answered, 'the owner will resell this plot, and he will most likely buy another one and will give it another chance. But if there is water he will plant fruit trees and build a packing station'. Indeed, over the two years that followed we observed how the recently enclosed land gradually changed. Fruit trees were planted, a drip irrigation system was installed, and a packing station was built.

This story is emblematic of the changing realities of access to and use of groundwater and of land tenure relationships, and more generally, of agrarian transformations in the Saïss. The landscape is gradually changing as more and more fences are being built and the number of 'modern' farming projects is growing. These changes are promoted and subsidized by the Plan Maroc Vert (PMV, Green Morocco Plan), Morocco's ambitious agricultural policy, which goes hand-in hand with land tenure changes. Through various land policies, land is made available to various kinds of individuals. Yet, the buyers who are able to access land are mostly urban dwellers, as they can afford to pay the rising land prices.[2] What singles out these new farms is that most newcomers enclose the land with high wired fences, often immediately after having bought the land. Their farms also stand out because only high-value crops, such as grapes or fruit trees, are planted. The fencing of the land and changing agricultural practices are further accompanied by changes in water flows. Newcomers drill new deep tube-wells and use drip irrigation. Meanwhile, farmers who engage in a more peasant[3] way of farming, and who rely on shallow wells notice how over the past 30 years the groundwater level is dropping and how their wells are running dry during the hot summer months.

In this chapter, we seek to understand *how* and through *what* processes water flows are altered. We draw from both the concept of enclosure as well as from the critical water literature to analyse how this happens. Combining the two helps illustrate how changing groundwater access and use happens through, and is accompanied by, changing tenure relations, the use of new technologies, and a discourse of modernity. This is described in the first part of this chapter. We then illustrate how the groundwater enclosures are enacted by the gradual dissociation of groundwater from its sociocultural and territorial context; growing social inequities; and violent expropriations. Groundwater is increasingly flowing away from peasant families who relied on water for sustaining their livelihood, toward new 'entrepreneurs' who produce 'sour grapes'. This brings us to a strange paradox that is at the heart of the unfolding enclosure process: while the PMV has the ambition to create entrepreneurs, water flows are diverted to absentee 'investors' who all produce the same products and face problems selling them.

Case study area and methods

The research was carried out in the plain of the Saïss (Morocco), where a rapid groundwater-based process of agrarian change is taking place. The plain comprises 220,000 hectares, of which 49,677 hectares are irrigated, mainly with groundwater (Ministry of Agriculture, 2012). In the past, the region was known for rain-fed crops and only some crops were irrigated through small-scale irrigation schemes, depending on springs. Since the 1980s, due to severe droughts, and along with the structural adjustments programs liberalizing the agricultural sector, groundwater gradually became the lever of agricultural development and the extension of irrigated horticultural crops and fruit orchards in the region (Kuper et al., 2016).

The increase of groundwater access and use are also strongly interlinked with changing tenure relations. In 2004, the Moroccan state started to grant land formerly under state domain to private actors for a period of a maximum of 99 years (Mahdi, 2014). Only investors who manage to establish modern farming projects are eligible for these concessions. Additionally, in 2006, the state decided to privatize the land of the former socialist-inspired collective state cooperatives. These cooperatives were created in the early 1970s as part of a national land reform program. The lands that were redistributed had been gradually retrieved by the Moroccan state after independence from foreign settlers that had confiscated them during the French protectorate. The logic behind the land reforms of the 1970s was to create a class of peasant farmers who would contribute to agricultural development. These plots could not be sold until 2006, when former members could acquire a formal land title from the state and become landowners. Many families were subsequently forced to sell their land as they were either in debt or had inheritance problems, attracting urban investors and new 'entrepreneurs' to the region. The new owners, often from the city, do not live on the site and only come once in a while to visit their farming project. The new owner usually appoints a manager who is responsible for the daily management of the farming project.

For this study we focused on the *douar* Ait Ali, which consists of approximately 80 extended families. In this *douar*, the dissolving state cooperative Ait Ali has become the arena for a substantial number of land transactions and the arrival of new investors. In 2013, 8 out of 36 members had sold their land and 18 had bought their individual land rights. The peasant families who still live in the cooperative engage in mixed cropping systems and irrigate only 30 to 40% of their land. They typically have wells that are 30 to 50 meters deep, which generally can be used only part of the day due to a relatively low yield. Few of them started to dig deeper tube-wells, which guarantee a bigger discharge and a more continuous supply, or a second well. Seven families engage only in rain-fed agricultural and either do not have wells or their wells have dried up. In contrast, the newly arrived investors, subsidized by the PMV, have drilled new deep tube-wells (120–200 m) and installed drip irrigation to irrigate their recently planted fruit orchards.

Methods

The field research on which this chapter is based lasted one year, over a period stretching over 2011 and 2014. Additionally, in 2015, various field visits of several weeks were carried out. The field research was conducted by the first author together with her female Moroccan research assistant.

We chose for an ethnographic approach and concentrated on the experiences of individuals and how they make sense of and give meaning to the transformations happening around them. Here, experiences are considered as discursively constructed rather than to impute to it an incontestable authority (Scott, 1992). It implies that we studied identities as being 'socially constructed', rather than seeing them as primordial or given. We conducted life histories and interviews with different individuals (various members of peasant families; new investors; peasant families who sold their land; and representatives of the Ministry of Agriculture and Fisheries). Our systematic observations of the landscape further helped us to understand *what* is happening and *how* groundwater is gradually enclosed.

Linking the concept of 'enclosure' to water politics

We consider 'enclosure' to be a useful concept to rethink agrarian capitalist expansion. Enclosure offers critical insights to expose disruptive ecological and sociocultural dynamics. It is a concept, rooted in the process of communal land enclosure in England, which happened in the 16th century as described by Thomas More in his work *Utopia* (first published in 1516). The enclosure of communal land with fences excluded the community from their access to land and redefined the use of the land: from providing the daily livelihood for peasants, the lands were instead appointed to serve England's industrialization. Since then, the concept of enclosure has gained much attention, especially by Marxist scholars. Different scholars notably illustrate how this results in various processes of land and water grabbing, and of land and water privatization (White et al., 2012; Borras and Franco, 2012; Peluso and Lund, 2011). This body of literature highlights that processes of enclosure are inherent to the current neo-liberal era. Nevertheless, such processes have mostly been explained as emanating from outside forces, driven by the state and/or international private companies. Such a structural perspective focuses on the material consequences of processes of enclosure, yet reveals little about *how* enclosures take form and are set in motion. Moreover, such a view hampers the possibility to see the role and actions of subjects involved in, or concerned by, enclosures.

To understand *what* is happening and *how*, we therefore want to tell a slightly different enclosure story, which looks beyond structural explanations and material consequences. The work of various critical water scholars who shed light on how water distribution and access alter as part of capitalist development inspired us to do so. For example, Zwarteveen (2015: 15) illustrates how new water redistribution in irrigation 'happen through a combination of land transfers, new technologies, and the re-negotiation of farmers' relations with each other and with the government and private sector actors'. She states that where the water is flowing from, or to whom, cannot be simply read from policies and legislations. Instead, water (re-)distribution occurs through often messy, multi-layered and multiple negotiations, in which power, identity and politics are important (Zwarteveen, 2015: 15). Mehta et al. (2012) further illustrate how particular narratives are deployed (by states, private actors, bureaucrats, etc.) to justify processes of water (and land) grabbing. For example, an integral part of the modernization narrative is that existing uses of land and water resources (by farmers) are often portrayed as being 'inefficient', 'underutilized' and 'below potential'. Once this image is portrayed, it automatically follows that there is a place for 'modern' interventions. Mehta et al. (2012) further highlight an essential point related to the characteristics of water; it is fluid, it flows and has various manifestations, which makes it difficult to characterize the precise nature of the grabbing, appropriation and reallocation and their varied impacts across multiple scales and time frames (Mehta et al., 2012: 194). Kaika and Swyngedouw (2000) add an important point that marks the complexity of new water distributions. In their work on the commodification of urban water, they illustrate how, through the application of new technologies, water flows 'disappear' underground, locked in pipes, cables, conduits, tubes, passages and electronic waves. The locking up of water and its technological framing hides from certain subjects how water is re-allocated and 'renders occult the social relations and power dynamics that are scripted and enacted through these flows'

(2000: 121). Finally, the work of Ahlers (2010) adds an important aspect to the different scholarly literature cited above, by looking beyond the material consequences of contemporary enclosures. She illustrates, based on a case study on water privatization in a Mexican irrigation district, how water forges collective identities of communities who struggle to sustain their irrigated livelihoods and harbours knowledge about irrigation practices and soil management. She demonstrates how processes of water commodification disrupt these linkages and how collective identities become fragmented and knowledge is detached from its material and sociocultural environment.

The multiple groundwater enclosures in the Saïss region

By combining the concept of enclosure with the domain of critical water studies we have been able to untangle the different processes and mechanisms through which groundwater is enclosed. We identify and study 'enclosures' from three different angles: fenced water – physical enclosure through changes in tenure relations and new property demarcations ('enclosure 1'); hidden flows – enclosure through technology ('enclosure 2'); modern water – discursive enclosure of groundwater ('enclosure 3'). The enclosures are intrinsically connected and interrelated and difficult to disentangle.

1) Fenced water: physical enclosure through changes in tenure relations and new property demarcations

Groundwater enclosures happen in tandem with changing tenure relations and new property demarcations such as fences, cement poles with barbed wire, sometimes walls, guards and dogs. Tenure relations alter through the various land policies, which contribute to a sharp increase in land sales. These transactions are about land and about the groundwater below it. The importance of the access to groundwater is well reflected in land prices, as the following testimony of a *samsar*, an intermediary on the land market, illustrates:

> Lands where there is water are the most expensive, because the agricultural project will succeed. Lands with wells are around 550,000 Dirhams (c. €50,000). The lands next to the road of Azrou and Ifrane are only 50,000 Dirhams (c. €5,000) for a hectare, but the land is not sold because there is no water.

The lands situated in the Saïss region are thus ten times the price of lands situated outside of the region. This is due to the presence of the rich aquifer system and the relatively high chances of having access to groundwater. To materialize and make ownership visible, newcomers who recently bought a piece of land delineate the boundaries of their plot. These land demarcations are in stark contrast with the majority of the lands that have not been sold, which are usually not delimitated with fences. The following testimony of a new manager of a recently set up 'modern' farming project illustrates well how, as soon as the fences are installed, tube-wells are drilled: 'We started in 2010, ploughed the land and then installed the fences. There was only one well of 50 meters, which was used for drinking water. We dug a tube-well of 190 meters.' The fencing of the land and the groundwater below it, sometimes gathered in large water basins, serve communicative functions, signalling the creation of a 'close', that is, a space of exclusive use and entitlement (Blomley, 2007: 8). The owner owns the land and is eligible to dig tube-wells on his land. Consequently, through the fencing of the land, water is symbolically also fenced, but with less clear frontier delineations than for land.

2) Hidden flows: enclosure through technology

Groundwater is further enclosed through technology. Whereas water used to flow freely on the land, today it is hidden in drip lines and tubes. The water is pumped up by a submerged pump inside the tube-well and transported to the surface. It is then further transported through tubes and brought to a technological assembly of a drip irrigation system consisting of valves, filters and pressure regulators. The water is brought to the crops through drip irrigation lines, keeping the water hidden and eliminating any sound of running water. Through the use of drip irrigation and deep tube-wells, water disappears in black, blue or red plastic materials literally enclosing water. The following anecdote is emblematic for the technological enclosure of water. An elderly woman of 68 years, after coming back home from a trip, noticed that her son had installed drip irrigation on their plot. 'Where has the water gone?' she asked her son. Her son replied that the water was now brought through tubes to the plants. 'How will insects, birds and frogs benefit from the water now that it is not flowing anymore? How will they survive?' When observing the new projects and their irrigation systems, the confusion of the elderly lady who wonders where the water has gone is comprehensible. Her statement also reflects how in the 'old' days, water was used with care, in balance with the environment and living beings.

Moreover, the notion of efficiency closely associated to drip irrigation (for a discussion, see notably Chapter 4 of this volume) obscures the alteration of water flows. 'Modernizing' farmers and entrepreneurs increasingly install drip irrigation yet they not necessarily irrigate 'perfectly' in the sense hoped by engineers (Benouniche et al., 2014: 419). Some farmers explained that they would only close the valves once they could see that there was sufficient water at the roots of crops. They would not always worry about wasting water; their main concern being to make sure that their high-value crops would not suffer from water stress. Moreover, farmers would also often extend the irrigated area once they installed drip irrigation (see for instance, this volume, Chapter 5). They would argue that since they 'save water', they can irrigate more land, in the end using the same quantity of water if not more. Finally, farmers are not equal in front of technologies. As mentioned by a peasant farmer 'for the deep tube-well you need a powerful motor, not everyone can afford that'. Indeed, whoever has the most wealth, the social capital to obtain the authorization for installing a tube-well, and the deepest and strongest pumps can access water, which happens at the expense of others. Various peasant families of the former state cooperatives often encounter difficulties to dig new tube-wells, while their shallow wells are running dry in a context of declining water tables. In a recent study close to the douar Ait Ali, it was found that half of the wells (96 out of 193) of that area were no longer functional (Kuper et al., 2016).

3) Modern water: discursive enclosure of groundwater

Water is further enclosed through a discursive regime of modernity and the institutions, practices and procedures that are linked to it. In what follows we explain how the discursive regime functions and illustrate its various exclusionary mechanisms.

The modernity discourse is partly derived from and constructed around the PMV. The main objective of the PMV is to develop a 'modern' agricultural sector, which is professional, productive, intensive and competitive through the rational and efficient use of water (Ministry of Agriculture and Fisheries, 2014). The plan has two main components, called 'pillars'. The first pillar aims at developing a modern, intensive and competitive agricultural sector, based on high-value crops, new technologies and intensive use of inputs. This pillar has received the most attention and funding to date (for a critical analysis see Kadiri and El Farah, 2013; Akesbi, 2011) and is premised on the emergence and strengthening of private investors and entrepreneurial farmers. Although this pillar is based on the emergence of private entrepreneurs, there are many subsidies for irrigation technology, plantations, etc. The second pillar, with less financial commitments, aims at supporting 'solidarity agriculture'. More precisely, it assists small and medium scale farmers in marginal regions. Although the plan states that it includes all farmers and takes into account their diversity and socio-economic constraints (Ministry of Agriculture and Fisheries, 2014), it contains countless silences and ambiguities (Aloui, 2009; Benataya, 2008). For instance, key terms such as 'investor' and 'entrepreneur' remain undefined. This attracts a wide range of actors who benefit from state subsidy programs, but do not have any background in farming and rather invest in the sector to avoid taxes and to launder money or to obtain a farmhouse for the weekend.

Through the PMV, new farming and water imaginaries are invoked and current farm and irrigation practices are portrayed as being 'inefficient' and 'below potential'. New technologies and farm practices, such as drip irrigation systems and entrepreneurial farming projects, are legitimized through the use of a terminology highlighting they are 'efficient', 'rational' and 'productive'. Drip irrigation plays an important role in the formation of these new farming imaginaries. Government officials working at the department of agriculture would talk about drip irrigation in terms of 'increasing yields', 'water use efficiency', 'the percentage of water saved', and 'perfect control', sometimes together with formulas to explain how to calculate the irrigation efficiency and/or the crop water requirements. Consequently, the fully irrigated, 'intensive' and 'productive' vocation of recently bought lands where drip irrigation systems are installed, which used to be rain-fed or which were only partially irrigated, are highly praised. They are presented as new projects that contribute to the *real development* of the country.

The PMV and its various subsidy systems further contribute to the creation of new 'entrepreneurs'. They are discursively but also materially formed through the new farming projects, the drip irrigation technology, and the large four-wheel drive vehicles they use to visit their farming projects. Yet, the materially and discursively heralded entrepreneurs and their 'modern farms' are accompanied by the exclusion of existing farming identities, which do not conform to the dominant discourse. Managers of new farming projects and government officials generally refer to the peasant families as those 'who do not have the means to make real investments' and 'who are wasting water' and their farming practices as 'traditional', 'backward' and 'old-fashioned'. This framing draws a clear line of who contributes to the current (and desired) developments, and whose activities are therefore considered as legitimate, and those who do not contribute, delegitimizing and devaluating their activities.

The enactment of the multiple enclosures

Through the deployment of the concept of enclosure and our ethnographic work, we illustrated in the previous section *how* groundwater is enclosed through three distinctive yet closely entangled processes. The first enclosure refers to 'fencing off', 'closing' and the demarcation of a clear land, but more fuzzy water frontier. The second enclosure involves making invisible – through technology (tube-wells driven by powerful engines in combination with drip irrigation) – the water reallocations that happen through groundwater enclosure. Finally, the third enclosure involves modernity and 'making modern', which is highly exclusionary. In what follows we illustrate *what* is happening and *how* the groundwater enclosures come together and are enacted by the gradual dissociation of groundwater from its sociocultural and territorial context; by growing social inequities; and finally by violent expropriations. This finally brings us to meet the newly created subjects, the new 'entrepreneurs'.

The gradual dissociation of groundwater from its sociocultural and territorial context

In the Ait Ali state cooperative, groundwater had an important symbolic meaning in the past. After the end of the French protectorate, many of the cooperative members had the dream to become independent peasants. Instead, when the state cooperative was created in 1972 they fell under the responsibility of the Ministry of Agriculture, and all land was held in co-property: they became labourers coerced by the government (Pascon, 1977: 185; Bossenbroek and Zwarteveen, 2015). In the early 1990s, the state cooperative was split up into smaller state cooperatives, and the members received an individual plot of land with land use rights. From that moment onward, the first wells were dug and peasant families moved to the land and started to gradually realize their dream. By accessing groundwater, they could start to farm independently. Water became a symbol of liberating oneself from the state, engaging in more profitable crops, and of becoming an independent peasant farmer (Kuper et al., 2009). Today, for young male farmers, whose parents were members of the state cooperative, the access to groundwater opens up new horizons and possibilities to modernize the family farm (e.g. install drip irrigation). They wish to do so in continuity of the farming practices and values of their parents, while slightly adapting them (Bossenbroek and Kadiri, 2016; see also Box 6.1 on the gendered experience of drip irrigation among young farmers).

Box 6.1   Pride and shame of drip irrigation:
Gendered experiences of new farming practices

Drip irrigation allows and is part of changes in ways of farming. In the Saïss in Morocco, farming is increasingly becoming a professional and masculine activity. This can be partly explained by the growing number of 'modern' farming projects set up by farming entrepreneurs, consisting of high value crops and the use of new technologies, like drip irrigation. Young men dream of starting their own farming project and become 'modern' farmers according to their own values and practices instead of reproducing the more 'traditional' peasant identities of their parents. Drip irrigation plays a central role in the process of realizing their farming project. As this technology is discursively framed as modern, promising an opportunity for greater productivity, and is often combined with the cultivation of lucrative crops, it provides a modern air to their project. This is particularly important in order to convince their parents to be supportive of their project, and in particular their fathers who are often skeptical with regard to the farming plans of the younger generation. Young male farmers are further keen to work with drip irrigation, as it requires a particular know-how and creativity. For example, upon interviewing various young men they proudly stated that 'we do not call a technician whenever there is a leakage or when something breaks, we have the technical knowhow to fix almost any problem'. In addition, young men would also explain that the use of drip irrigation changes the labour division and irrigation activity as the following statement of a young farmer illustrate: 'Before my father irrigated with the sequia [small earthen channels through which water is conducted to crops in the fields]. It took a lot of time, sometimes four to five hours. His clothes were muddy and dirty. Today, I only have to open the valves and in the meantime I can do something else. I call my clients for example or overlook the labourers working on the field'.

All together, drip irrigation provides new exciting opportunities to some young men to combine the activity of farming with modern ideals of manhood. These young men picture their future selves as white-collar clean-shaven entrepreneurs, who operate their drip systems from their mobile phones while strolling in the soukh (market) or drinking a coffee with friends in the city (Plate 7 in the color plate shows this encounter between two generation of male farmers).

New opportunities for women to professionally distinguish themselves in farming are less obvious and rewarding, with tasks attributed to women often being least prestigious. At the same time, drip irrigation, new crops and land-tenure relations that accompany the professionalization of farming are gradually leading to stronger spatial and symbolic distinctions between the private-home-women's domain and the public-work-men's domain. This creates further distinctions and hierarchies between men and women in the countryside. Whereas new employment opportunities and farming identities offer sources of pride to men, young women are confined to a new home-bound traditionalism and actively seek new forms of identification within this confined space. Yet, other young women, often belonging to landless families are forced to engage in agricultural work to sustain their families, in which they find little pride and recognition as it reflects negatively on their virtues as women. They are hesitant to talk about it, and literally try to prevent others from seeing them doing it by covering their faces behind veils, leaving only their eyes to be revealed (for more on related topics, see Bossenbroek, 2016).

During the period of the state cooperative, a particular awareness of water scarcity existed. This became apparent through the well construction method and irrigation practices. The wells built during that period are between 50 and 80 meters and have several galleries, called *kifan,* which allow storing water. The peasants who have such wells explained that they irrigate according to the water availability in the well: 'If there is a lot of water we irrigate three to three and a half hectares. If there is less water available we irrigate two hectares. It depends on how much water there is in the well'. Peasant families in the dissolving state cooperative thus adapt their irrigation practices to the availability of water; none of the families irrigate the entire farm (between 7 and 13 hectares, depending on soil quality). Their awareness of water scarcity, reflected in their irrigation practices, is related to the limited availability and difficulty to access groundwater. Aquifer dynamics are complex (Kuper et al., 2016), and as stated by Ben Aicha, a former well digger, 'one can dig a well and easily not find any water, or find salty water'. To find the right spot to dig a well, 'gifted knowledge' is needed. Ben Aicha, for example, precisely knows where to dig a well and to find water. Farmers recruited him as a water diviner.

Ben Aicha continued to explain that, when he used to dig a well, the peasant family would organize a small celebration: 'Before people were happy when I dug a well. The women would do the joujou, and would distribute water to neighbours and invite them over to celebrate'. Digging a well took several months and up to a year (depending on the depth) of hard physical work because the wells were dug by hand. The required effort to actually dig a well further contributed to farmers' awareness of water scarcity. But today, the tide turned and Ben Aicha is a clandestine taxi driver in his old Renault 4: 'Before 2000, I used to dig four wells a year. From 2000 onward it became one or two wells a year. Since 2008 I completely stopped.' When we asked him if he deliberately chose to stop he replied; 'No, today people prefer the tube-wells'.

The reason that people do not want 'traditional' wells anymore is due to the lower capacity of wells and the time it takes to dig them. Today, new investors and farming families who have sufficient financial means want quick results and drill new deep tube-wells that are approximately 150–200 meters deep. With their deep tube-wells they pump directly from the confined aquifer and, in case they have the financial means to cover the pumping costs, they can irrigate their entire farm 24 hours a day. At that depth, water is abundant, at least in the short term, and consequently, the sense of scarcity and the awareness of water availability vanish and are replaced by an impression of water abundance. Also the traditional 'gifted know-how' of finding water is, according to Ben Aicha, substituted by 'a little machine [a geo-electric device], which looks like a pacemaker. It is a little box with threads in different colours'. The patience, the know-how and the hard work needed to obtain water are being replaced by the wish to obtain rapid results and finish the job quickly, further affecting notions of groundwater access, and even the way people 'see' water.

Growing social inequities in access to groundwater

The increase in the number of tube-wells contributes to growing social inequities, affecting livelihoods and farming identities. The wells that had been dug by the peasant families in the early 1990s and symbolized triumph and independence are now gradually falling dry, which forces some peasant families to dig deeper or to dig new wells. Not everyone has the financial means to do so or can provide for the pumping costs, which become higher as the depth of the well increases. The case of Aziza and Sliman, who own a plot of 10 hectares, illustrates well how water access is altered and how it affects livelihoods. Sliman had worked as a labourer in the state cooperative and recalled how he felt disappointed during that period, since little had changed since the French protectorate. He still did not own any land and was working under the auspices of the state. In the early 1990s, when the large state cooperative was divided into smaller cooperatives and the members received individual land plots, he was delighted and constructed a house close to the land and dug a well as soon as he had sufficient means to do so. But today his well is dry, and his feeling of euphoria and pride turned into worries with little perspective for the near future: 'We do not have any future plans. When we have a little bit of money we will deepen the well. Otherwise we will continue to cultivate rain-fed crops'. A couple of months after our first visit, we ran into Aziza. She explained that her husband was on the land with his herd and that she told him that they should think about selling the land. 'He is restless and I think it would be better to sell the land. But he does not want to think of selling it.' As the case of Aziza and Sliman suggests, in the near future the lack of water may become a reason why peasant families might be forced to sell their entire land or part of it. Having the possibility to dig deep and multiple tube-wells, as well as having the ability to pay for the pumping costs contribute to growing disparities between different actors in terms of access and use of groundwater.

Violent expropriations

The enclosure of water eventually forcefully and violently expels people from their land and from their water. The following case of Mona and Mohammed's family (seven members), who live in a barrack that used to be part of the state cooperative, shows how this happens. The grandfather of Mona (age 24) and Mohammed (age 19) owns a plot in the cooperative. Yet, due to family quarrels, Mohammed's family does not live with the extended family members. They rather wish to live in the barrack, which is state property. They have been living there for 25 years. In 2012, a new Moroccan entrepreneur living in France bought the land surrounding the barrack. A couple of months later, the president of the cooperative 'informally' sold the barrack to the new entrepreneur, although officially he was in no position to do so. On one of our visits, we ran into Mohammed. He was distressed. His brother had been arrested and was in jail: 'He did not want to leave the barrack. So the manager of the new farming project accused him of stealing some farm equipment and pesticides. He is in jail for a period of three months'. Since these events, Mohammed's mother had fallen ill and suffered from nerves and stomach issues and headaches. She stated that she did not know who exactly was expelling them. The family had never met the person who purchased the land, and the former president of the state cooperative acted as intermediary between them and the manager of the project. The problem remained unresolved. In the year that followed, Mohammed's father was arrested because 'he refused to leave' with nowhere else to go to.

After Mohammed's father had been arrested, we visited the barrack again. This time, we could not access it, as the last passageway had been fenced off, and barking dogs deterred us from approaching. A worker came to the fence and called the manager, Hichem. We asked him if we could visit the new farming project, and he gave us a tour. Hichem explained that as soon as the new owner bought the land he fenced it 'to demarcate the property and to be quiet'. He went on to explain that once fenced, the owner deepened the first well with a tube-well 170 meters deep and dug a second tube-well. He installed a drip irrigation system and planted four varieties of grapes (pergola, muscat, gretgrap, and victoria) on 5 out of the 7 hectares and peach trees on 0.5 hectare. The drip irrigation system, the poles used to grow the grapes, and the netting to protect the latter from being eaten by birds, were subsidized from 60 up to 100%. Hichem told us that the owner wanted to convert the barrack into a packing station. We stopped at the room where once we had taken a picture of Mona and her son showing us the wheat they had harvested from her grandfather's land (see Plate 6 in the color plate). The room was now filled with soluble pesticides and fertilizers, to be applied through the drip irrigation system. Despite having lived for over 25 years in the barrack and transformed it into a cosy house, the development of the new farming project by a new entrepreneur turned the family of Mohammed and Mona into 'illegal' squatters who were eventually forced to move out.

New 'entrepreneurs'

To understand *how* and through *what* processes water flows change, it was indispensable to actually talk to the new 'entrepreneurs'. It was difficult to meet them, as most of them do not live where the farming projects are located. The imagery that surrounds these entrepreneurs is one of efficiency, modernity and productivity. However, our various interviews with some of them revealed a slightly different image. The case of Driss and his wife, Hind, illustrates this clearly. We met him on a Sunday afternoon, while he was parking his Mercedes on the road close to his farming project planted with peaches and nectarines. Driss is an engineer, working in a public institution and lives with his wife, Hind, and two young children in Meknes. He started his farming project of 11 hectares three years ago. During our various interviews he explained how, after having bought the land, he first dug two tube-wells and planted 8 hectares of peaches and nectarines. On the rest of the land, he plans to build a packing station and a house with a swimming pool. He hired a consultancy company to make the study to design the drip irrigation system, which was subsidized up to 80%.

While giving us a tour on the farming project he explained how it developed. He stopped close to the house where he has a small vegetable garden, and while picking the tomatoes he said, 'This is why I love farming, the nature is so generous – Hamdullilah (thank God)'. He continued to explain that his father used to be a farmer and that, as a child, he always dreamed of becoming a farmer one day. He sees himself as 'amateur' when compared to the new investors who recently settled in the region and whom he describes as 'profitable', 'productive' and 'having money and making large investments'. After finishing the tour on the farm we drank a cup of mint tea and chatted with his wife Hind, who works at the Chamber of Agriculture. When Driss got up to check something with one of the labourers, Hind began to lament:

> Yes the farming project is great, but since Driss started it, he is hardly at home anymore. I have to take care of the children now. But I would also like to start my own project and set up a children's farm and organize

cooking classes. Children living in the city can come and see how rural life is like and how everything grows. Besides, I also want a swimming pool and to rebuild the house and redecorate the interior.

While staying in touch with her on Whatsapp, six months after our last interview, she sent us a picture of their newly built pool with the following text: 'Our new pool :–)! I hope you can come visit us soon'.

Whilst talking to Driss, we expected to meet an entrepreneur who was 'productive', 'efficient' and 'rational' and did not imagine meeting an entrepreneur who presented himself as a hobby farmer or 'amateur'. The other interviewed entrepreneurs presented the same hybridity. They all have similar agricultural projects (grapes, fruit trees) and do not live on site. Some of them used to be labourers working on French farms during the protectorate and left afterward with their employers to France. Today, they wish to retire 'au bled' (in their home area). With the money they earned in France and the various subsidies obtained, they bought a plot and installed a new modern farming project with fruit trees or grapes, which they visit during the weekends. Others, in the meantime, used new farming projects to launder money obtained in other activities.

The discrepancy that emerges between how new 'entrepreneurial' farmers are portrayed and their actual practices and self-presentation is interesting for two reasons. First, as the example of Driss and Hind illustrates, performing an entrepreneurial identity helps to access and secure water. In doing so, new 'entrepreneurs' actively contribute and enforce new enclosures. Second, it also reveals a strange paradox that is at the heart of the enclosure processes that are today unfolding in the Saïss. As most of the new entrepreneurs engaged in standardized highly subsidized modern farming projects consisting of grapes and fruit trees, they find difficulties in selling their products on the national market. As a manager of a new farming project mentions, 'The market is saturated, everyone is producing the same. I'm not sure we can make any benefits this year'. At the end of our interview, the manager gave us a bunch of grapes. They tasted sour, and once out of the view of the manager, we threw them away. In a way this incident illustrates the power of modernity and its disruptiveness and its contradictions. While the PMV has the ambition to create entrepreneurs, water flows are diverted to absentee 'investors' who all invest in the same normalized projects and face problems selling their products.

Conclusion: sour grapes

The modernity pathway underpinning the PMV encloses groundwater. We distinguished how this happens through three interrelated processes: (1) fenced water – physical enclosure through changes in tenure relations and new property demarcations; (2) hidden flows – enclosure through technology, notably drip irrigation; (3) modern water – discursive enclosure. We have illustrated how this development pathway is accompanied by the transition from a place-based culture-economy of water of sharing and living with what is available to a place-less culture of greed that, instead of promoting water conversation, is encouraging further extraction, leading to the depletion of aquifers and violently expropriating people.

New 'entrepreneurs' play an important role within these processes by performing an entrepreneurial identity through which they can claim and secure access to water. At the same time, they inhabit multiple subject positions, ranging from investing in a tax-free economic sector to hobby-farming or retirement plans. This multiplicity reflects on the one hand the power of the modernization discourse and its disruptiveness. It also illustrates that at the heart of the unfolding groundwater enclosure process lays an odd paradox. Whereas the PMV aims to create entrepreneurs, water flows are diverted to subsidized investors who all invest in the same normalized projects and face problems selling their products because similar projects pop up like mushrooms in the countryside. On the other hand, we would like to read these divergences as 'cracks' in the modernization discourse. The enclosure is not closed and it contains instabilities, which provide some space to probe and see alternative development trends that might be less disruptive.

When observing the farming practices of the peasant families who stayed, we might actually be able to observe such alternative development processes. Some peasant families reject the drilling of deeper tube-wells along with the fruit trees and grapes. They do not discredit the tube-wells and grapes because they cannot get it as the story of the fox goes (see 'Les Fables de la Fontaine: The Fox and the Grapes'), but rather wish to continue with their existing farming practices. Especially, the elder generations of peasant farmers uphold an image of farming consisting of mixed-cropping patterns and live with the notion of water scarcity. It is part of who they are, and of what works. Moreover, they also see the risks associated with intensive groundwater-based agriculture and have negative experiences with strongly fluctuating markets (Lejars and Courilleau, 2015). But for the younger generation of male farmers, the grapes are sour, and they slightly envy the new entrepreneurs who settle in their *douar*. As a young woman once told us, 'When he [young man] sees a newcomer exhibiting his car and his clothes, he does not appreciate it at all. He wonders and asks himself: what does he have that I do not have, to be able to own all this'. Indeed young men are confronted with the new farming projects and the new images that come along with the recently drilled deep tube-wells. They grow up in a situation where water abundance is for others, enclosed in a discourse of progress and development. They also want to attain this abundance, but in their own way. They wish to modernize the family farming project, but, at the same time, hold on to the same farming values as their parents. They see themselves as *fellah* (farmer) and *rajal aâmal* (businessman) at the same time (Bossenbroek et al., 2015). The emergence of these young farmers suggests hybrid forms of future entrepreneurs, who perhaps can combine modernity with tradition and have less disruptive farming practices. These hybrids need to be further looked into and brought to light to voice the possibilities and potentials of developing different visions of the future, to counter the grim future appearing on the horizon.

Acknowledgments

The authors would like to thank two anonymous reviewers for their valuable comments on earlier versions of this chapter, as well as the editor for his comments. We also wish to thank the people of Ait Ali for their willingness to share their stories with us.

Notes

1 The *douar* can officially be defined as an 'assemblage of households linked by real or fictitious kinship relations, which corresponds to a territorial unit' (Ministry of Interior, 1964).
2 Since the land privatization process the land prices have drastically increased. Land prices went up from around 120,000 Dirham (c. € 11,000) in 2006 to about 600,000 Dirham (c. € 55,000) in 2013 (see Bossenbroek and Zwarteveen, 2015).
3 As will be illustrated further in the chapter, this mode of farming is usually based on a sustained use of ecological capital and oriented toward defending and improving peasant livelihoods (van der Ploeg, 2008).

References

Ahlers, R. (2010). Fixing and nixing: The politics of water privatization. *Review of Radical Political Economics* 42(2): 213–230.

Akesbi, N. (2011). Le Plan Maroc Vert: une analyse critique. In Association marocaine de sciences économiques (Ed.), *Questions d'économie marocaine*. Rabat: Presses universitaires du Maroc, pp. 9–48.

Aloui, O. (2009). Silences et enjeux du plan maroc vert. *Economia* 7: 34–38.

Benataya, D. (2008). Les zones grises du Maroc. *Economia* 3: 6–11.

Benouniche, M., Kuper, M., Hammani, A., and Boesveld, H. (2014). Making the user visible: Analysing irrigation practices and farmers' logic to explain actual drip irrigation performance. *Irrigation Science* 32(6): 405–420.

Blomley, N. (2007). Making private property: Enclosure, common right and the work of hedges. *Rural History* 18(1): 1–21.

Borras, S.M., Jr., and Franco, J.C. (2012). A 'land sovereignty' alternative? Towards a peoples' counter-enclosure. *ISS Staff Group 4*, Rural Development, Environment and Population.

Bossenbroek, L. (2016). *Behind the veil of agricultural modernization: Gendered dynamics of rural change in the Saïss, Morocco*. PhD Thesis, Wageningen University. Available at http://library.wur.nl/WebQuery/wurpubs/fulltext/388046

Bossenbroek, L., and Kadiri, Z. (2016). Quête identitaire des jeunes et avenir du monde rural. *Economia* 27: 46–50.

Bossenbroek, L., Ploeg, J.D. van der, and Zwarteveen, M. (2015). Broken dreams? Youth experiences of agrarian change in Morocco's Saïss region. *Cahiers Agricultures* 24(6): 342–348.

Bossenbroek, L., and Zwarteveen, M. (2015). "One doesn't sell one's parents" Gendered experiences of shifting tenure regimes in the agricultural plain of the Sais in Morocco. In C.S. Archambault and A. Zoomers (Eds.), *Global trends in land tenure: Gender impacts*. London and New York: Routledge, pp. 152–169.

Kadiri, Z., and El Farah, F.Z. (2013). *L'agriculture et le rural au Maroc, entre inégalités territoriales et sociales*. Rabat: Blog scientifique Farzyates/Inégalités du Centre Jacques Berques.

Kaika, M., and Swyngedouw, E. (2000). Fetishizing the modern city: The phantasmagoria of urban technological networks. *International Journal of Urban and Regional Research* 24(1): 120–138.

Kuper, M., Dionnet, M., Hammani, A., Bekkar, Y., Garin, P., and Bluemling, B. (2009). Supporting the shift from "state" water to "community" water: Lessons from a social learning approach to design joint irrigation projects in Morocco. *Ecology & Society* 14(1). Available at www.ecologyandsociety.org/vol14/iss1/art19/.

Kuper, M., Faysse, N., Hammani, A., Hartani, T., Hamamouche, M.F., and Ameur, F. (2016). Liberation or anarchy? The Janus nature of groundwater on North Africa's new irrigation frontier. In T. Jakeman, O. Barreteau, R. Hunt, J.D. Rinaudo, and A. Ross (Eds.), *Integrated groundwater management*. Dordrecht, the Netherlands: Springer, pp. 583–615.

Lejars, C., and Courilleau, S. (2015). Impact du développement de l'accès à l'eau souterraine sur la dynamique d'une filière irriguée Le cas de l'oignon d'été dans le Saïs au Maroc. *Cahiers Agricultures* 24(1): 1–10.

Mahdi, M. (2014). Devenir du foncier agricole au Maroc. Un cas d'accaparement des terres. *New Medit* 4: 2–10.

Mehta, L., Veldwisch, G.J., and Franco, J. (2012). Introduction to the special issue: Water grabbing? Focus on the (re)appropriation of finite water resources. *Water Alternatives* 5(2): 193–207.

Ministry of Agriculture and Fisheries (Ministère de l'Agriculture et de la Pêche Maritime – MAPM). (2012). *Situation de l'agriculture marocaine*. Rabat: MAPM.

Ministry of Agriculture and Fisheries. (2014). *Official Website*. Available at www.agriculture. gov.ma/en/pages/strategy (Accessed June 11, 2016).

Ministry of Interior. (1964). Circulaire du ministère de l'Intérieur en date du 27 décembre 1964, relative à un projet de publication d'un recueil de circonscription administrative. In M. Rfass (Ed.), *L'organisation urbaine de la péninsule Tingitane*. Rabat Maroc: Université Mohamed V Publications de la Faculté des Lettres et des Sciences Humaines.

More, T. (1975 [1516]). *Utopia*. Middlesex, Harmondsworth: Penguin Classics.

Pascon, P. (1977). Interrogations autour de la réforme agraire. In N. Bouderbala, M. Chraïbi, and P. Pascon (Eds.), *La question agraire au Maroc 2*. Publication du bulletin économique et social du Maroc, Rabat, Morocco, pp. 183–200.

Peluso, N.L., and Lund, C. (2011). New frontiers of land control: Introduction. *The Journal of Peasant Studies* 38(4): 667–681.

Ploeg, J.D. van der. (2008). *The new peasantries: New struggles for autonomy and sustainability in an era of empire and globalization*. London: Earthscan.

Scott, J. (1992). Experience. In J. Butler and J. Scott (Eds.), *Feminists theorize the political*. London: Routledge, pp. 22–40.

White, B., Borras, S.M., Jr., Hall, R., Scoones, I., and Wolford, W. (2012). The new enclosures: Critical perspectives on corporate land deals. *The Journal of Peasant Studies* 39(3–4): 619–647.

Zwarteveen, M. (2015). *Regulating water, ordering society: Practices and politics of water governance*. University of Amsterdam: Inaugural lecture 529, Vossiuspers Amsterdam University, Amsterdam.

7 Creating small farm entrepreneurs or doing away with peasants? State-driven implementation of drip irrigation in Chile

Daniela Henriquez, Marcel Kuper, Manuel Escobar, Eduardo Chia and Claudio Vasquez

Introduction

The promotion of a neo-liberal economic model in Chile during the Pinochet dictatorship in the 1980s had a major impact on agricultural production and agrarian structures. At the time, the main regulatory mechanism was meant to be the market and the government liberalized land and water markets. Large companies, particularly agro-industrial complexes, appropriated these resources and currently concentrate land and water ownership. Pressurized drip irrigation systems allowed to expand irrigated areas, particularly in previously non-irrigated hills (Alvarez and Reyes, 2011), leading to an increase in water demand.

In the Coquimbo region, one of the main fruit producing areas for export, the critical water situation due to a 10-year drought (approximately since 2006) was exacerbated by the expansion of commercial farming and the ever-increasing use of water to irrigate vines and fruit trees. The shortage seriously affected peasants. In reaction to the water deficit faced by these small-scale family farmers, the state set up loans and credits and provided subsidies to enable them to install drip irrigation systems. These new systems were expected to help the small farmers produce more efficiently and reduce the use of water resources (*Instituto Nacional de Investigación Agropecuarias*, 2008). However, the last agricultural census conducted in 2007 (*Instituto Nacional de Estadísticas*, 2008) revealed that only 24% of the small farmers had installed micro-irrigation systems (drip or sprinkler) on their farms.

This article examines the discursive social construction of the introduction of drip irrigation for small-scale family farmers and the relationships between small farmers and the state. The analysis is based on a case study in the Limarí valley of Rio Grande, in the region of Coquimbo, Chile. Our objective was to analyse the state-led technology transfer process of drip irrigation for small-holders and the reactions of the latter to this program, in a context of water shortage and market regulation. To do so, we pose that innovation has to be understood as an interactive scheme (Busch, 2013; Coudel et al., 2013). It is the result of 'complex interactions, involving individuals and organizations acting in multiform networks which may lead to synergies or reveal oppositions' (Coudel et al., 2013: 27). As such, innovative activity is culturally rooted and socially produced and to understand it, one must study cultural meanings, institutions, as well as the structure of power and of social networks (Fernandez, 2012). Then, it is clear that the transfer of technology is not a neutral process, but rather a transmission belt of standards, norms and rules produced by the dominant model (Boelens, 1999). It 'embodies strong prescriptions on the directions of technical change to pursue' (Dosi, 1982: 152) and may clash with other alternatives ideas of what type of agriculture, for instance, might be desirable currently and in the future. Our results show that, for the state, drip irrigation is a way to promote a particular mode of farming – that of the agricultural entrepreneur. This involves the transformation of the meanings of and relationships of farmers to land and water.

Case study area and methods

Monte Patria, a place in the mountains

The municipality of Monte Patria is located in the Province of Limarí, Coquimbo Region in the central-north of Chile. It covers about 4,400 km² and has a population of about 30,000 inhabitants, 56% of whom live in rural areas (*Instituto Nacional de Estadísticas*, 2012).

Currently, extractive activities such as mining and agriculture are the main economic activities in the municipality. The total cultivated area is about 430,000 hectares (ha) of which about 13,000 ha are irrigated, 44% under gravity irrigation and 56% under drip or micro sprinkler irrigation (*Instituto Nacional de Estadísticas*, 2008). The ownership of land and irrigation water is heavily concentrated in the hands of large producers with 42 landowners owning 83% of the productive agricultural land with properties of 8,600 hectares on average (PAC, 2012). As far as water is concerned, 29 companies account for 45% of the consumptive water rights of the municipality of Monte Patria rivers (*Dirección General de Aguas*,[1] 2014). The remaining 54% of the water rights are distributed among 132 individual farmers.

Peasants, entrepreneurs and companies in Monte Patria

Peasants in Monte Patria are farmers who have inherited the activity of their families. They were often beneficiaries of the agrarian reform process in previous decades. Today, they are mainly elderly people who continue to cultivate small pieces of land. Generally their landholdings are small – i.e. from 0.5 to 3 hectares. Peasants usually live on their productive land or in a nearby town (still in a rural area). They mainly grow vegetables such as potatoes, tomatoes and cucumbers among other seasonal crops and they often adjust the size of the land they cultivate according to the availability of water. Some small farmers produce fruits but on small areas. These crops were traditionally irrigated by surface irrigation from canals. The peasants generally have one quarter or one half 'water share'; they may also 'rent' water from those who have water rights.[2] In addition to agriculture, peasants often exert a mining activity or have goats. They work the land by themselves, with families or neighbors.

Furthermore, small farmers usually work in group. Many of them belong to 'agricultural communities' which are collective landholdings. These farming communities bring together villagers, from only 5 to more than 100, on collective lands. This mode of land tenure means that all activities related to property must be decided together. Another important form of organization are agricultural cooperatives. The Coquimbo Region is famous for its production of pisco[3] and wine, so many peasants have chosen to produce grapes. As grape production is not sustainable if done on an individual basis due to the low prices, the risks of frost and the volatility of the demand, many small farmers are members of agricultural cooperatives, which provide agricultural support and may even produce their own pisco.

Medium-sized enterprises are usually owned by national investors who do not live on the land but in large nearby cities. They employ workers but fewer than larger companies and more erratically. They produce vegetables and fruits too, especially grapes, but usually for the domestic market. Farms are 100 hectares large on average. They have mixed irrigation systems with drip irrigation or other pressurized systems, but they also use traditional irrigation (furrow or flood irrigation). They own a 'water share' and have basins to store water.

Finally, the large companies in Monte Patria mainly produce fruits for export, in particular grapes, oranges and avocados. Their owners do not live in the region. The farm sizes of these companies are generally very large. 'Prohens', for example, an important agro-industrial company, has planted 795 hectares of grapes and oranges. As elsewhere in Coquimbo, these companies have a considerable number of water shares. They have storage basins and use drip irrigation.

Methodology

To achieve the objectives of the study, we chose a qualitative methodology. We opted for ethnographic methods involving comprehensive interviews and participatory observation of stakeholders (Valles, 1999; Hernández, 2006). We shared the daily life of eight small farmers and their families for approximately six months in the years 2012–2013 and conducted open in-depth interviews with them. These included small farmers with functioning drip irrigation systems, farmers with drip irrigation equipment that was not in use, and peasants who did not have drip irrigation but used gravity irrigation and furrows. The number of observations, interviews and the time allocated to fieldwork was set up according to the information saturation criterion, to the extent that we stopped collecting data once the information we obtained repeated itself. The information collected was recorded in a logbook along with field notes. In order to safeguard ethical principles, we used an oral or written informed consent procedure, thus maintaining confidentiality and participants' anonymity. We also participated in peasant organizations' events and peasant labour instances. Finally, we interviewed two public officials and were observers during (i) two meetings of the Local Development Council, (ii) one meeting of the Local Council of Health, (iii) one Water Sustainability Seminar organized by the government, (iv) one protest for water and (v) five meetings concerning rural development policy planning. In these instances, public officials, authorities and community members participated. We also interviewed three entrepreneurs and managers of agricultural corporations.

The research results were discussed with the community in a forum with community leaders. The main conclusions were shared, which was motivating for the audience as it was a sign researchers valued their identity and rural culture. In addition, to improve the reliability of our results, we compared the information we collected across the different strategies we adopted (observations and interviews), as well as with the interpretations of other researchers.

Neo-liberalism and peasant cultures in Chile

Chile is well known worldwide for the attempt of the dictatorship (1973–1989) to promote a neo-liberal political project through series of reforms toward a market-centric society (Jiménez, 2011). In so doing, the state lost its role of regulator to the benefit of markets, which were supposed to regulate social relations, resource allocation and financial systems in a process of globalisation of capital. According to Avendaño (2001), privatisation and other measures, promoted by supporters of economic liberalisation, facilitated the establishment of agribusinesses aimed at exports, at the expense of more traditional and peasant landowners who had benefited from the previous agrarian reforms (1962–1973). Economic liberalisation along with land market deregulation created the conditions for investment and territorial concentration by agribusinesses usually owned by multinational companies. As the beneficiaries of former land reforms were replaced by multinational exporters, the highly skewed structure of land distribution in place before the land reforms was restored (Avendaño, 2001).

According to Álvarez (2003), at the time of the dictatorship, water resources were considered as a productive factor that should be subjected to market logic and became a commodity that should be used efficiently. Accordingly, the 1981 Water Code (still in force) defined water resources as national property for public use, but 'rights of advantageous use' were attributed, with the aim of increasing private autonomy in water use and favouring free markets (Bauer, 1997). The state has limited means to control the way the holders of such rights exercise them, as it cannot impose penalties or cancel water rights for lack of use or 'misuse', thereby inversing the situation that had been set by the 1951 Water Code and the land reform law (Bauer, 1997).

Those who acquire water rights can rent, sell and transfer these rights through commercial transactions on the 'water market'. The logic behind offering the possibility of trading water (that an individual may have 'saved') was that it would boost the agricultural economy. To cite Hernan Büchi, Minister of Finance during the dictatorship, 'what advantage is there for a farmer to install drip irrigation if he is unable to sell to others the water saved with that system?' (in Álvarez, 2003: 153). In addition to separating water from land (i.e. water can be sold independently from the land), the 1981 Water Code also separates surface water and groundwater, enabling separate transactions. This promotion and support to an entrepreneurial – business – approach to agriculture did not happen in a vacuum but in a country in which the 'peasantry' has held, and still holds, a prominent position.

The peasant culture has been abundantly described in the literature, from Chayanov (1974) onward and most recently by van der Ploeg (2008). Such scholars often see the rural culture as being a non-capitalist economy where the labour force comes from within the household, which has control over the means of production and in which the worker and the employer are usually the same person. However, according to Chacón (1994), there is no such thing as a 'pure peasant'; being a 'peasant' is instead a dynamic identity, which depends on social relations grounded in time and place. Beyond the economic structure underlying the existence of the peasantry, the way the peasant identity gets shaped directs practices (Chacón, 1994). In the region of Coquimbo, there is a strong rural identity with its particular characteristics and, since 1998, peasant organizations have demanded respect for their identity, including during technology transfer programs. As such, it appears important to understand the tension between peasant identity and the state technology transfer programs that have been implemented in the study area.

Social construction of identities in a water shortage context

Drought as a political problem

The main transformation identified in peasants' discourse concerned landscape changes caused by new modes of agricultural production. The land has been increasingly colonized by perennial crops for export, mainly fruit, vineyards and citrus orchards: 'I have seen vineyards getting closer; [. . .] they are increasingly scratching the hills' (peasant).

This is linked to an extended use of drip irrigation by export companies, a technology which has allowed 'colonizing' new agricultural frontiers on the hills of the valley, called 'scratching' by peasants (see Plate 8 in the colour plate for an illustration of an agricultural 'frontier'). This form of capital-intensive production is not only inconsistent with peasants' logic but stresses the polit- ical nature of drought, as also stressed in the critical discourse of peasants themselves:

> Yes, previously there were dry years, but there were also fewer plantations; We see who is clearing the hills [. . .] It's rich people who want to earn

more money [. . .] while these people (peasants) are fighting for the few resources that are left over.

(peasant)

Construction of the peasant identity

Our interviews revealed that peasant identity attributes meanings to agricultural production and is defined through its relationship with the territory. This peasant identity is constructed in opposition to another one, that of the entrepreneur. Striking in the peasants' discourse is the strategies of legitimising or delegitimising relationships to the territory.

The peasants delegitimise the entrepreneur's identity. The peasants recognize differences between large companies and medium entrepreneurs. The latter would be more concerned about the territory and its people than large companies. However, both companies and entrepreneurs are considered to be different and, therefore, opposed to the identity of the peasant. The peasants consider the entrepreneur and the companies as profit-oriented with a capitalist rationality in which the territory is considered to be a means of obtaining income with no further significant meaning. According to the peasants' discourse, natural resources such as water and land are consistently considered as commercial inputs by the entrepreneur, who manages them in a way that enables him to make a profit. The 'imaginarium' of the entrepreneur/company is built on his alienation and non-connection to the environment. In the peasants' view, alienation from the land goes against the sense of traditional property, based on membership:

> the life of a peasant, of people in rural areas, is very different from the life of an entrepreneur. The entrepreneur may decide to sell the farm and buy another in the south where there is water. And we, if we sell it, where are we going to live?
>
> (peasant)

For peasants, the entrepreneur does not have a feeling of belonging, unlike the peasants and those living in the countryside. '[O]ver there is a company, and here we are small farmers. [. . .] they work the land as a company, we work like peasants' (peasant).

In contrast, for peasants, the construction of an agricultural rural identity is a goal in itself and heritage values are attached to agricultural work. Production is oriented toward subsistence and consumption.

> Here people live a subsistence life, because people plant their crops, work, are at home in their orchards and have to survive' (peasant). In that sense, peasants have no interest in maximizing earnings. 'I do not want more, what else do I need? I have enough [. . .] even if I won the lottery, I would not buy anything else.
>
> (peasant)

Peasants did not bring up the cost structure of their farms during interviews, especially concerning their own workforce as it is based on family relationships.

In contrast to a capitalist mode of farming, the peasants' culture is characterized by a multi-sector economic structure. Cash income comes from different sources including animal husbandry, trade, crafts, mining, and especially, temporary wage labour. 'Many people do temporary work from time to time, this is usually the type of work done here' (peasant). This is to meet new needs (e.g. schools, media communication) that have been brought by modernity. As far as cultivation is concerned, peasants mostly grow vegetables and cereals and food needs are mostly covered by on-farm production. This choice is justified by household needs and driven by water availability: the land cultivated is adjusted to the amount of irrigation water available, taking rainfall patterns into account: 'because we know, what we sow depends on the year, if we have enough water. We don't sow two hectares, we sow less, because we know there isn't going to be enough water' (peasant).

The peasant's symbolic relationship with water and land considers natural resources as parts of a whole. This worldview recognizes water and land as common property resources, where the actions of a few affect the others. This favours community discourse, which shapes group membership: 'we do care who is next to us, who is over there, who is on the other side' (peasant). Consistently, irrigation is viewed as a collective activity depending on other people's actions. Such a view conflicts with a commercial view of water and other natural resources, as represented in the dominant discourse referring to the entrepreneurs' viewpoint.

It is not surprising, then, if relationships between peasants and enterprises are tense according to the peasants themselves. On the one hand, business logic related to land is rejected; nevertheless, the economic dependence of many peasants on agricultural enterprises is recognized. 'Even though it is true that [agricultural enterprises] employ a lot of manpower, they are devastating something that is quite important, Nature' (peasant).

State actions to promote drip irrigation for small farmers

In our study area, drip irrigation was mainly introduced in smallholder agriculture by the state. This was notably made possible through the delivery of credits and subsidies. Normal financial institutions such as banks usually do not provide loans to small farmers, because they do not meet the institution's solvability requirements, such as having a stable permanent income. This is why the state uses the National Institute of Agricultural Development (INDAP) to provide financial support to small farmers in the form of soft loans or subsidies. During severe drought events, for instance, the state provides 'bonds' consisting of cash or baskets of necessities to families. INDAP also provides subsidies to small farmers to modernise their irrigation systems. This policy is consistent with the state's view of peasants as being 'vulnerable' and unable to compete on the market. There, the state plays a subsidiary role aimed to address small farmers' difficulties in a rather paternalistic way.

To benefit from loans or subsidies, small farmers have to become 'clients' of INDAP, which involves a bureaucratic process of registration and subsequent cumbersome and strongly controlled application to get any benefits. For instance, INDAP asks small farmers to use a private consultant to design and install the drip irrigation system to guarantee its quality. This is done by consulting companies and the peasants are not involved in the design or implementation process, which means they do not know how to operate the system once it is installed on their farm. Another important aspect is that applications for credit or subsidies must be individual. There are no collective instruments for obtaining credit or a subsidy, although peasant groups (e.g. cooperatives and 'farming communities') strongly structure agricultural production in the region studied and Chile as a whole.

Adopting an entrepreneurial discourse, state institutions justify the use of drip irrigation in farming by increased irrigation efficiency, reduced costs and improved product quality: 'We are on the same side as INDAP. We have both made people understand that this irrigation technology improves quality, optimizes water resources and saves a lot of energy' (local public official). In the dominant discourse, peasants must 'start up'; they need to become small entrepreneurs, always with the projection that they may someday become big businesses, which have come to embody the notion of 'success'. Technical innovation is meant to bring added value, especially concerning peasants' ability to produce new products and as a means through which they can be 'developed'. If peasants do not adopt drip irrigation, this is because they are ignorant, comfortable with present practices and resistant to change: 'The small farmers resist change. Innovation is harder for them' (local public official). In addition, as a 'client' of INDAP, the peasant is not seen as a source of innovation but rather as a passive receiver of black-boxed technology. Consistent with a dominant urban viewpoint, technical knowledge of consulting engineers is overvalued compared with peasant's knowledge, and technology is black boxed. Local adaptations made to artefacts ('bricolage') are dismissively referred to as 'amateur' tinkering:

> our technical advisor found some 'amateur' tinkering. When the farmer has used a lot of irrigation hoses to fix something, he will come here [the local government office], then the consultant who goes to his place to help will see that, basically, only 'amateur' fixing has taken place.
>
> (local public official)

Small farmers reactions to state efforts to promote drip irrigation

In the first place, in peasants' discourse, the state appears to be a financer: 'Of course, I was and I still am an INDAP customer. I still have a credit pending; I still have one instalment to pay' (peasant). In this sense, the state is emptied of political content and is not considered as a guardian of rights, but only as a moneylender. But getting access to the financer is not easy for peasants. State bureaucracy discourages applications. Peasants question the state's requirements to obtain access to such benefits.

> The National Agricultural Development Institute says: this farmer does not qualify for this reason, the other one does not either, another one is too small, another one is too big, and so on. So there's always a 'but'. Sometimes I've got all the necessary papers, I take them and then I am asked for others; eventually I go back again and all papers have been left forgotten some-where . . . eventually, no one is qualified to be granted benefits.
>
> (peasant)

Peasants do not only question the bureaucracy; the procedures are also incon-sistent with their culture. In the peasants' view, land is inherited, and there is no need for a legal document to prove it. Neither does a peasant understand the need to prove his farming activity with documents, tax, income, expenses, accounting documents and so on. Furthermore, the support provided by the state clashes with the peasants' sense of community. Drip irrigation projects are granted to individuals and farmers have to compete with each other for funds. Individualization hinders interaction among peasants, contrasting with the peasants' strong sense of community: 'Working together to find the best way of developing our people has to prevail over other interests' (peasant).

The peasants question the role of the state in the technology transfer, and they feel actions taken by the state do not match their needs: 'We do not want imposed life systems, irrigation system, reservoirs, that kind of thing, if they don't respect how we do things ourselves; the truth is that the state doesn't accept our way of life' (peasant). In addition, they question the sectorial way in which the state intervenes. INDAP, for example, provides drip irrigation but does not check whether the peasant family has access to drinking water. No institutions work on rural livelihoods with peasant families using an integrated approach. To the peasant, this does not make sense. 'To keep country people quiet, for them to continue living and to continue protecting the countryside, I think there must be a cross-cutting development, but not a half-baked one, it has to be a comprehensive one' (peasant). The peasants we interviewed fre-quently emphasised that the focus should not only be on developing the farm-ing sector; the sectoral logic of the state was surpassed by the multifunctional peasant producer.

More specifically, state officials justify drip irrigation by its efficiency and the fact it reduces the cost of agricultural production, but such arguments appear to have little relevance for peasants, who believe agriculture is a value in itself and appear little interested in increasing their production or income. Furthermore, the fact that farmers are prohibited to make any modifications to the equip-ment that is installed by external consultants as part of the state supported pro-grams and have to accept large signboards acknowledging the financial support of the state on their farm construes drip irrigation as an alien, untouchable and restricted access system. Generally, farmers are unable to volunteer their labour or to make suggestions, or ask questions when the external consultant installs the system. So the technology is emptied of substance. This happens even though evidence from other countries, such as Morocco, shows that this is exactly the ability small farmers have in tinkering with the drip irrigation technology that may be one of the main reasons for its widespread use (see Chapter 15, this volume). Denying them this opportunity is an indirect way to undermine the peasants' abilities and knowledge.

The 'clientelistic' relationships between the state and the peasants makes it seem that state projects are the only way to acquire new technology. 'Members of the cooperative [peasants] applied for drip irrigation provided by the Agriculture Development National Institute, as this process can only go through them [INDAP]' (peasant). This raises significant questions as peasants not only question the transparency of projects that are outsourced to private consultants but also express frustration as they feel 'excluded' from these projects. When stating that '[w]e are always obliged to use the services of a company, as INDAP uses external companies to carry out projects, the so-called "consultants". So, they design the project and money goes to consultants . . . so? . . . when the project is implemented there is no more money', a peasant clearly expresses the fact that drip irrigation is not his prime concern but also that the main beneficiaries of the subsidy schemes might not be the farmers but a network of intermediaries (consultants, engineers) who play the role of go-between them and the administration.

That peasants cannot engage with the promotion of drip irrigation by the state leads to dispossessing them from their own future through a transformation of the meaning of agricultural production

> small peasant family farms are lost because the national policy is to make small farmers entrepreneurial. What can we say? We want to remain farmers, we want to continue living in rural areas but we don't want to be entrepreneurs, because I believe that (even though the thought may be absurd) we are happy as we are.
>
> (peasant)

Conclusion

In the context of water shortage, the Chilean state was the main driver for introducing drip irrigation to peasants. The state justified the promotion of drip irrigation with an entrepreneurial discourse, whereby the technology would allow peasants to become competitive small businessmen thanks to productive and efficient water use.

In the programs we documented, the state is mostly an 'intermediary' between irrigation equipment companies, independent consultants and engineers, and the peasants, and its primary role is about financing the dissemination of drip irrigation through subsidies. By using a linear top-down approach to the transfer of technology, the state reproduces the dominant neo-liberal model but also disposes peasants of their agency because the latter are not considered actors of the innovation process in their own right. There, the top-down promotion of drip irrigation can lead to 'depeasantization' because the peasant is faced with many obstacles when trying to 'compete' in agricultural markets. Water rights abstraction, bureaucratic procedures and price volatility are as many constraints that peasants are ill equipped to deal with.

As a consequence, it is maybe not surprising that peasants link drip irrigation technology to the capacity of agro-exporting and big agricultural enterprises to expand their production area and the subsequent increased pressure on water resources. When linking the development of agro-entrepreneurship with water scarcity, peasants clearly highlight the socially and politically constructed nature of the latter. Strongly disagreeing with a capitalist approach to land production, peasants articulate an alternative discourse of belonging and relationships to their environment and natural resources as common property. In some instances, this alternative takes the form of local innovations – that are disparagingly referred to as 'bricolage' and underestimated by state officials and irrigation engineers. These local innovations (that are testimony to the fact that peasants can engage in their own terms with newly introduced technology) need to be recognized. They may indeed be important tool for communities in a context of a growing (yet socially constructed) water scarcity. Such recognition would also be a way to legitimize and recognize a rapidly changing 'peasant identity' which is currently undermined by state interventions. This means generating a constructive dialogue to enable the collective development of technological knowledge, without excluding the peasant's worldview, but integrating and recognising its potential. This will not only allow the technology to become a tool for sustainability and livelihoods enhancement in the face of growing pressure over water resources, but will also enable social integration by validating and recognizing the multiplicity of value and ideals about the rural world and farming.

Acknowledgments

The authors would like to thank the Social Development Program of the Central University of Chile for its support and the different partners who helped in conducting fieldwork. We acknowledge the contributions of Esteban, Noreen, Sergio Ríos, and Francisca Gonzalez who helped in translating the paper as well as the input of the editors in improving earlier versions of this chapter. We also appreciate the CEAZA support in the research. Special thanks to the peasants for welcoming us into their homes and sharing their experiences.

Notes

1 'General Directorate of Water Management'. This is the state institution that records water use rights exchanged on water markets.

2 Water in Chile is private property acquired through a right of use also called 'water share' that consists of a given flow rate per minute. The shares are traded in water markets. In times of drought, and in the study area a water share has a value of approximately $9,000.
3 Pisco is a traditional spirit from Chile and Peru, made by distilling grape wine.

References

Álvarez, P. (2003). Agua y sociedad chilena; Antecedentes del contexto histórico y jurídico. In *Dinámicas de los sistemas agrarios en Chile árido; La Región de Coquimbo*. Santiago: LOM Ediciones Ltda, pp. 121–157.

Alvarez, P., and Reyes, H. (2011). Le colonisation de la montagne aride par l'agriculture irriguée: un risque assumé. *Sécheresse* 22(4): 267–274.

Avendaño, O. (2001). *Diferenciación y conflicto en el empresariado agrícola chileno: Período 1975–1998*. Santiago: Programa de Estudios Desarrollo y Sociedad, Universidad de Chile.

Bauer, C.J. (1997). Bringing water markets down to earth: The political economy of water rigths in Chile, 19976–1995. *World Development* 25(5): 639–656.

Boelens, R. (1999). Gestión Colectiva y Construccion Social de Sistemas de Riego Campesino: Una introducción conceptual. In *Buscando la equidad; Concepciones sobre justicia y equidad en el riego campesino*. Los Países Bajos: Van Gorcum, pp. 87–106.

Busch, L. (2013). Thinking innovation differently. In E. Coudel, H. Devautour, C.T. Soulard, G. Faure and B. Huburt (Eds). *Renewing innovation system in agriculture and food: how to go towards more sustainability?* The Netherlands: Wageningen Academic Publishers, pp. 35–56.

Chacón, I. (1994). Elementos de análisis para la reproducción campesina. *Ciencias Sociales* 63: 101–108.

Chayanov, A. (1974). *La organización de la unidad económica campesina*. Buenos Aires: Nueva Visión.

Coudel, E., Devautour, H., Faure, G., and Soulard, C. (2013). Reconsidering innovation to adress sustainable development. In E. Coudel, H. Devautour, C.T. Soulard, G. Faure and B. Hubert (Eds). *Renewing innovation systems in agriculture and food*. The Netherlands: Wageningen Academic Publishers, pp. 17–33.

Dirección General de Aguas (DGA). (2014). *Ministerio de Obras Públicas* [Online]. Available at www.dga.cl/productosyservicios/derechos_historicos/Paginas/default.aspx (Accessed July 5, 2014).

Dosi, G. (1982). Technological paradigms and technological trajectories. *Research Policy* 11: 147–162.

Fernández, M. (2012). Hacia un programa de investigación en sociología de la innovación. *ARBOR Ciencia, Pensamiento y Cultura*. (188): 5–18.

Hernández, R. (2006). *Metodología de la Investigación*. Iztapalapa: McGraw-Hill Interamericana.

Instituto Nacional de Estadísticas (INE). (2008). *Censo Agropecuario 2007* [Online]. Available at www.ine.cl (Accessed April, 2012).

Instituto Nacional de Estadística (INE). (2012). *Censo*. Santiago: Gobierno de Chile.

Instituto Nacional de Investigación Agropecuarias (INIA). (2008). *Manejo productivo agropecuario en condiciones de escasez de lluvias*. La Serena: Intihuasi.

Jiménez, E. (2011). El agua en disputa; Gestionando el riego en territorios rurales semiáridos de Chile y Bolivia. *Revista América Latina; Revista del Doctorado en Procesos Sociales y Políticos en América Latina*, pp. 173–224.

PAC Consultores Ltda. (2012). *Plan de Desarrollo Comunal 2012–2018*. Santiago: Ilustre Municipalidad de Monte Patria.

Ploeg, J.D. van der. (2009). *The new peasantries: Struggles for autonomy and sustainability in an era of empire and globalization*. London: Earthscan.

Valles, M. (1999). *Técnicas Cualitativas de Investigación Social*. Madrid: Síntesis.

8 Conquering the desert

Drip irrigation in the Chavimochic system in Peru

Jeroen Vos and Anaïs Marshall

Introduction

Drip irrigation is a technology with much appeal to engineers, farmers, but also to a general public. It is strongly associated with a promise of 'modernity' and 'efficiency'. The last decade of intensive large-scale export-oriented agriculture in Latin America has seen a steady growth. In many parts, water availability is seen as one of the main constraints for further growth and prosperity. Especially the export sector enterprises are investing in irrigation technology. The export market demands constant volumes of high-quality products. Therefore, production that used to be mainly rain-fed (like export bananas in Ecuador) is now brought under irrigation. In addition, national government policies have promoted the expansion of pressurized irrigation. Legislation favouring drip and sprinkler irrigators and subsidies for constructing pressurized irrigation systems have greatly facilitated the expansion of the drip and sprinkler irrigated areas. Because of the increase in irrigated areas under drip irrigation the total volume of water extracted from surface and groundwater bodies has been increasing, despite the 'efficient' irrigation technologies being used.

This chapter will give a brief overview of the development of drip irrigation in Peru. In the remainder of the chapter, the Chavimochic irrigation system in the desert coast is described and analysed. This system is taken as an example of the irrigation modernization and the expanding of the agricultural frontier to illustrate the rationale and social and environmental effects of the use of drip irrigation in large-scale export agriculture in Peru.

The Chavimochic irrigation system takes its water from the Santa River, which is canalised to neighbouring valleys. The water in the Rio Santa is increasingly being appropriated by different groups of users in and outside the river basin, causing increased competition and tension. With the building of the Chavimochic system the water scarcity in the river basin has been displaced: water security for one group (the downstream users in Chavimochic) created water insecurity for another group (upstream users in the Callejón de Huaylas). This happens because many upstream users do not have legal concessions to use the water they use presently, let alone legally guaranteed allocations for future use. While the Chavimochic users have legal concessions not only for their present but also their future uses (Lynch, 2012).

The first part of the main canal of the Chavimochic canal was built between 1994 and 2000 and connects three valleys that had traditional run-of-the-river irrigation systems. These systems were used by many smallholders that cultivated mainly maize, beans and other subsistence crops. Now these 'old' valleys have a secure water supply and large agro-export companies cultivate some 20,000 ha of newly irrigated land in the areas between the 'old' irrigation systems. All the new land developed by the agribusiness companies is irrigated with drip irrigation. The drip irrigation is used to overcome the physical constraints in conquering the sandy soils of the desert. Without drip technology the sandy soils in the inter-valleys could not have been brought under production. The 'efficiency of drip' argument served to legitimize the large amounts of subsidy for the irrigation development for large-scale agriculture in Chavimochic. At the same time, the efficiency discourse brings the smallholders in discredit. The waterlogging, caused by bringing more water to the area, is attributed exclusively to the presumed inefficient irrigation practices of the smallholders. The smallholders received more, and more secure, irrigation water, but lost out in the economic competition with the agro-export companies.

Drip irrigation in Peru

The development of drip irrigation in Peru is concentrated on the desert coast (Plate 9 in the color plate illustrates how drip irrigation allows 'conquering the desert'). The extremely dry coast receives water from rivers with high seasonal fluctuations, or intermittent rivers, that flow from the Andean mountains. If irrigation water is available, the coast offers good conditions for agricultural production because of its stable and mild temperatures and optimal sunshine hours year round, making it possible to produce off-season fresh fruits and vegetables for the world market. The dry conditions and isolated valleys along the desert coast reduce the pressure of plant diseases and pests. Waterlogging and salinity form a threat to production on the coast. About 300,000 hectares in the coastal area (30% of the areas covered by irrigation infrastructure) have problems of salinization, drainage and/or waterlogging (World Bank, 2013). Groundwater is used at a minor scale, because it is either too deep to exploit or too saline. However, in some valleys, like the Ica valley, the groundwater is used intensively, in a rate beyond replenishment (Oré et al., 2012; Urteaga, 2014; Damonte-Valencia, 2015). The main commercial crops are sugar cane, rice, maize, avocados, grapes and asparagus (see Plate 10 in the color plate). Smallholders, many of whom got their land from *haciendas* during the land reform that started in 1969, produce the same crops at minor scale, but also subsistence crops like beans, cassava and sweet potato.

Each of the 50 valleys along the coast of Peru has one or more irrigation systems. In total about half a million hectares are irrigated in the coastal area (Oré et al., 2009). Of this area about 10% has drip irrigation. Most drip irrigation systems are installed on large plantations of sugar cane or export fruits and vegetables. River water is diverted by permanent or temporary weirs into large, often ancient, mostly unlined canals and distributed through a network of unlined canals in which the water is controlled by manually operated sliding gates. Increasingly drainage canals are being constructed. Some rivers are regulated by dams.

Large- and medium-sized agro-export companies invest in drip irrigation to expand the cultivated area or improve the productivity of land previously irrigated by surface irrigation (Plate 10 in the color plate shows the vast tracks of land that are irrigated with drip irrigation). The government supports the expansion of drip irrigation by direct and indirect subsidies to large-scale irrigation projects that provide pressured water for drip irrigation. The indirect subsidies were the unforeseen outcome of the low prices paid for the newly irrigated land, resulting in 'losses' for the government that could thus not recover the investments made with public money to develop an expensive irrigation infrastructure (Eguren, 2014). At a smaller scale, government programmes provide subsidies for drip irrigation equipment to medium and small farmers that use groundwater. The hot spots for drip irrigation development are located mainly on the desert coast of Peru in the valleys and interfluves of Ica, Villacuri and the newly developed irrigation areas of the Chavimochic irrigation system. Also parts of the large-scale gravity irrigation systems of Chancay-Lambayeque and Chira-Piura have some drip irrigation (particularly for sugar cane and export vegetables and fruits). The expansion of irrigated area in the region of Olmos will also be through drip irrigation.

Since the agricultural modernization policy of the Fujimori regime that started in 1990, the large infrastructure projects have been (and are being) built exclusively for large-scale export agriculture, projecting the use of sprinkler or drip irrigation to irrigate the desert coast. The most important and recent ones are the Olmos and Chavimochic projects in the north, while the Majes-Siguas II irrigation scheme in the south is under construction. Here, drip irrigation is not a technology for saving water, but rather an instrument to 'conquer the desert' with a model of large-scale export agriculture and hence use huge volumes of water transferred from other watersheds (in the case of Olmos, the water is transferred from the east side of the Andean Mountains by a large tunnel of 20 km). The water comes from rivers that flow into the Amazon basin or into the Pacific Ocean. As the water from the rivers becomes completely allocated, conflicts arise between different groups of water users (e.g. in Piura, Ica and the Rio Santa basin).

The large infrastructure projects are developed by the national state agencies – and tendered to large construction firms – and financed by international loans. The Olmos project is expected to irrigate more than 38,000 ha in the desert of Lambayeque. The total investments in the main irrigation infrastructure of this project is estimated at US$ 794 million dollars, of which 445 million are subsidies provided by the central and regional governments and 349 million are private investments by the Brazilian construction company Odebrecht that will own and operate the irrigation system for 25 years (Eguren, 2014). The Chavimochic project is planned to have 66,000 ha of newly irrigated area, all projected with drip irrigation, as will be explained in the next section.

Other major existing infrastructure systems in coastal desert areas are Majes (22,000 ha), Ica (25,000 ha), Tacna (18,000), Chinecas (4.200 ha), Jequetepeque-Zana (with a 5,700 ha newly irrigation area), with drip irrigation in the desert (asparagus and grapes). Also a large-scale infrastructure, but meant for smallholders, the Cachi project (14,000 ha) in the Andean mountains foresees sprinkler irrigation for livestock and subsistence crops of smallholders (Oré et al., 2009).

Government programmes have also supported the conversion to sprinkler or drip irrigation for individual or groups of small- and medium-sized farm holdings. The main programme in Peru is the World Bank and JICA supported Irrigation Subsector Project; PSI (acronym of the Spanish *Programa Subsectorial de Irrigaciones*). This program started in 1996 and is still ongoing. In the first phase (1996–2004) the project provided – amongst others – subsidies for groundwater development and drip irrigation in the dry coastal area of Peru. The groundwater is fed by the rivers flowing from the Andean mountains. During the second phase of PSI, subsidies were granted for sprinkler irrigation design and implementation for smallholders in the Andean mountains. Until 2015, in total approximately 26,000 ha were equipped with improved on-farm irrigation technologies (a relatively small part of this is drip and sprinkler), with a total cost of some US$ 15 million, benefiting some 5,500 families.

International irrigation technology companies benefited from the drip irrigation development in Peru. Netafim opened a dripline factory in the capital Lima in 2013 and the international president of Netafim reported that the growth of sales of his company worldwide were the highest in Peru.[1] Other international companies have distribution and selling points as well.

The luxury products like asparagus and avocado produced with the drip irrigation are consumed in the United States and Europe and do not add any significant value to the food security neither in the United States and Europe nor in Peru. Negative effects of the drip irrigation include unequal access to the technology; increased use of water; and shifting legislation and dominant discourses about 'efficiency', considering only the on-farm technical dimension of 'efficiency' but not the efficiency at watershed level nor the geophysical, social, cultural and environmental dimensions of 'efficiency'. A report on the asparagus production for export from some 10,000 ha of drip irrigation in Ica in the desert southern coast of Peru, showed the depletion of the aquifer. In 2008 total use of groundwater for asparagus in the Ica valley area (including Villacuri) was 500 Hm^3 with a recharge of only 244 Hm^3 (Progressio, 2010).

The Chavimochic project

Drip irrigation to expand the agricultural frontier

The Chavimochic project in the dry northern coast of Peru derives its name from the four valleys connected by the main canal: the Chao, Virú, Moche and Chicama valleys (see Figure 8.1). It is one of the largest irrigation projects in Peru and is projected to have a total of 144,385 hectares of improved and new irrigated land when fully constructed and operational.

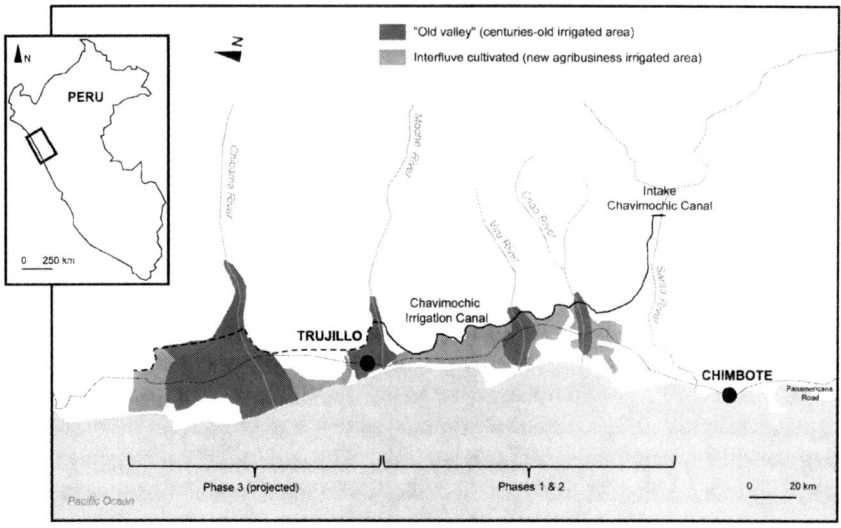

Figure 8.1 Map of the Chavimochic irrigation system.

Source: the authors; based on data from Escobedo (2012 and PECH (2013).

When the project was first conceived, the Chavimochic project was to support small farmers. In 1931, Victor Raul Haya de la Torre (founder of the APRA political party) proclaimed: 'The work of Chavimochic will become a reality and ensure the future of the new generations; a bright future, promising, waiting for the region La Libertad'. In 1990, Alan Garcia, then president of Peru and also of the APRA party, stated in the same fashion: 'This is CHAVI-MOCHIC, with a seal of the covenant with the people and witness that our future will be good and strong with this work; divert the waters of Santa is a privilege that destiny has brought me in this last years of my government' (Valle de Chao, July 20, 1990). But with the neo-liberal policies of the 1990s, the narratives and the project politics changed. The promotion of large-scale and thus 'efficient and modern' export agriculture became the main objective of the project.

The precipitation in the coast is near zero in normal years. Only in the El Niño years, occurring on average every seven years, significant rainfall is recorded. The water for the irrigation system is diverted from the Santa River south of the irrigated valleys. The Santa River flows from the Cordillera Blanca and Cordillera Negra mountain ranges in the Andean Mountains and is one of the rivers on the Pacific watershed of Peru with the highest yearly discharge. Nonetheless, concerns have risen on the downward trend in yearly discharge and base flow. The decrease in base flow is partly attributed to climate change (the melted glaciers), but mainly to the increase of water use in the upper part of the catchment. Future climate change and increased water use in the upper watershed are expected to severely impact the base flow (Pouyaud et al., 2003; Lynch, 2012). As explained above, the building of the Chavimochic and Chinecas irrigation systems taking water from the lower part of the river caused a relocating of water scarcity and increased water conflicts over the Rio Santa river water.

Before the development of Chavimochic project, the about 13,000 smallholder families in the three valleys irrigated their land with water from the unpredictable, highly fluctuating, intermittent rivers, groundwater and small springs. The agricultural areas have been established in the clay soils of the alluvial fan at the bottom of the valleys. In each valley a canal network, constructed from pre-colonial times onward, is in place to bring water from the river by gravity to each plot. The smallholders have on average some 2 to 3 hectares of irrigated land. They cultivate mainly maize, beans, avocado, vegetables and some sugar cane.

Since the Irrigation Management Transfer policy of 1989, a Water Users' Association (*Junta de Usuarios*) distributes the water in each valley; they determine the water allocations, following the requests of water users, and handle the payment by the water users. During low water in the river, some farmers used groundwater from wells (which operated on petrol) to continue to irrigate their crops.

Construction of the Chavimochic project (abbreviated to PECH, according to its Spanish acronym) started in 1984 in three phases. The two first phases of construction finished in 1996. The main canal has now a length of 155 km and connects three of the four 'Old Valleys' (*Valles Viejos*) mentioned above (see Figure 8.1). The existing irrigation systems in the Old Valleys were interconnected by means of the new Chavimochic canal. After the connection with the new canal, the smallholders received more, and more secure, irrigation water. Most smallholders stopped using groundwater because it was of lower quality and higher cost.

The Chavimochic canal brought water to the desert areas in between the Old Valleys. In these higher and sandier soils no irrigation was possible before the construction of the Chavimochic canal. This land was sold in big lots to agribusiness production and export companies. The sandy soils can only be irrigated by drip irrigation. The newly irrigated areas in the 'interfluves' desert lands constituted (in 2013) some 20,000 ha of the 55,000 sold by auction (see Table 8.1) to 15 agribusiness companies (Escobedo, 2012; PECH, 2013). Of this area some 8,000 ha are used for asparagus production, some 7,000 ha avocado and the rest of the area is cultivated with sugar cane, grapes and export vegetables (PECH, 2010b).

The third phase of the construction was started in 2013. It encompasses the construction of a storage reservoir called Palo Redondo (with a capacity of 400 million m^3, and a dam of 93 meters high) and the extension of the main canal to connect the fourth valley: Chicama.

The irrigation project has caused several ecological damages. Apart from waterlogging and salinization in the lower parts of the valleys (PECH, 2013),

Table 8.1 Potential and presently irrigated land in Chavimochic (2012)

Areas		Irrigated areas before the project (1981) (ha)	Area irrigated in 2012 (phase 1 and 2) (ha)	Totally projected irrigated land (phase 1, 2 and 3) (ha)
Old valleys	Santa	0	0	500
(mainly	Chao	984	5,331	5,331
smallholders, in	Virú	7,855	12,117	12,117
total 13,000)	Moche	8,382	10,315	10,315
	Chicama	50,000	50,000	50,047
Newly irrigated	Interfluves	0	20,022	66,075
areas	(mainly big		(55,000 sold)	
(interfluves),	landholdings)			
15 agribusiness				
companies				
	Total	67,221	97,785	144,385

Sources: The authors (adapted from PECH, 2010b: 16; Escobedo, 2012; Eguren, 2014).

the environment is likely to be affected by the contamination of soil and groundwater by agrochemicals. The transformation of desert zones into agricultural fields caused also a decrease of biodiversity. Especially as the extension of irrigated areas and the creation of new towns that host farm workers have reduced the dry forest areas. This is the case in the San José area where the *bosque seco* entirely disappeared between 1987 and 2005 (Marshall, 2014). Also the aquatic ecosystem of the Santa River is affected negatively by the abstraction of the major share of the water flow in dry periods (see for the effects of agricultural water extraction on biodiversity in another coastal watershed of Peru: Verona et al., 2012).

'Efficiency of drip irrigation' as an argument to legitimize subsidies to large-scale agricultural production

The construction of the Chavimochic irrigation system has been costly and the investments made by the government have not been recovered by the sales of the land to the agro-export companies. Since the neo-liberal agricultural policies of the 1990s the dominant discourse of 'efficient' and 'productive' drip irrigation has been used to legitimize the public investments and thus subsidise the large companies (see for a more general discussion on 'efficiency' discourses and irrigation development: Boelens and Vos, 2012).

The costs of the investments until 2012 have been some US$ 1,029 million (PECH, 2011; Eguren, 2014). In this project, as in the Olmos scheme, the newly irrigated land is auctioned to large agribusiness companies. When agribusiness companies buy land at the auction, they offer a prize consisting of two amounts: one corresponding to the price of the land, the other to the investment commitment to transform the desert soil into arable land. The companies obtained lots for a price between 45 and 3,430 dollars per hectare (with an average of 1,350 US$/ha). In 2012, 55,000 hectares had been sold. This resulted in a total recovery of 74 million dollars, which is only some 7% of the construction costs (Eguren, 2014). The investment commitment amounted to some 5,300 dollars per hectare. The low prices were justified by the fact that the companies needed to make investments to install the irrigation infrastructure on their lands. However, even when we take the $5,000 per hectare for investments into account, the total price is low for the agro-export companies. The price for the irrigated land is low compared to local land prices, international prices, and the potential profits that can be made with the land.

Every company that buys a lot in the Chavimochic area is granted a volume of 10,000 m^3 per hectare per year of water from the main canal. In most places the difference in height between the main canal and the irrigated land is sufficient to use gravity to pressurize the drip lines. This implies that the water from the main canal must be conducted by piped systems, but this also implies low energy costs for pumping.

The average landholding of the new large landholders is some 1,000 ha. Fifteen companies now own more than 94% of the newly cultivated land of which 19.6% (9,180 ha) belongs to Camposol SA group (owned by German and Norwegian funds)[2] (Escobedo, 2012). More than 90% is fed by drip irrigation and the rest with sprinklers.

Water prices are 0.010 US$/m^3 for smallholders and 0.020 US$/m^3 for the agribusiness companies (Marshall, 2014). The costs of irrigation water are relatively low compared with the total production costs, and also if compared with other export-oriented irrigation systems in the world. With a use of a total of 374 Hm^3 in 2010 the recovery is estimated to be some 5 million US$; of which three quarters go to the operation and maintenance (O&M), and a quarter to the payment of the investments. The budget requirement for the O&M in 2010 was some US$ 5 million.[3] Thus, the part of the fees that are recovered for O&M costs are lower than the budgeted O&M costs. In this way also the investment costs of the main irrigation infrastructure will not be recovered by the water fee payments.

In the 1990s, the dominant government discourse and policies in Peru focused on private investments and modernization of agriculture. In this discourse, drip irrigation was portrayed as being efficient and productive. The discourse on drip did not consider the increase of total water consumption, environmental effects or the distribution of the gains. New projects were formulated to stimulate drip irrigation. Moreover, many Peruvian water professionals were invited to travel to Israel by the Israeli Embassy to get to know the drip technologies in Israel on invitation, paid for by the Israeli Embassy.

The Chavimochic project became an emblematic project for the neo-liberal government of President Alberto Fujimori (1990–2000) and his successors. In 2009, for instance, the (then) president Alan Garcia stated that 'Chavimochic is a model of modern agriculture, an example for the rest of the world' in a conference for private investment.[4] In 2011, in a leaflet of the ruling party (APRA), the project was further described as:

> Conquering the desert . . . building the future, Chavimochic is the best example of efficiency and profitably, reaching milestones, recovering partly the investments while passing the valleys of La Libertad, bringing development, increasing living conditions of our population, showing the success of the APRA government in starting this majestic project, which is praised internationally for its social conception and good construction, a public project that generates much benefits for the people of La Libertad and Peru.[5]

At the international level, some critique was voiced by environmental NGOs and organizations that support smallholders (e.g. Eguren, 2014). A group of NGOs published a document on the negative effects on groundwater use of asparagus production for export to Europe from the Villacurí area (Progressio, 2010). This report got media attention in Europe, but in Peru the document did not get media attention, and the dominant discourse is still in favour of the expansion of the irrigation systems (worldwide, the use of drip irrigation is seen as an environmentally friendly practice, as illustrated by several water stewardship certification; see Box 8.1 as well as Chapter 1, this volume).

Box 8.1 Drip irrigation and water stewardship certification

Recent years have seen a rapid increase in the number of fresh agricultural products with private certifications sold in supermarkets in Europe and the United States. Retailers have three related reasons to require private certification: first, assure constant supply of high quality and traceability produce; second, prevent potential reputational damage by media attention for big social or ecological damages caused by the production; and third, access the niche market of 'conscious' consumers (Vos and Boelens, 2014). Most private certifications work with inspections by third party companies that do annual audits of the production companies. Examples of major private certification schemes are: GlobalGAP; the International Food Standard (IFS); the British Retail Consortium (BRC); and the Global Food Safety Initiative (GFSI). Furthermore, organic production and Fair Trade schemes have their own certification schemes. Recently, product-specific round tables on cotton, sugar cane, bio fuels and palm oil have also formulated sustainability criteria.

Water stewardship is part of most of these certification schemes. The water 'criteria' relates to the protection of water sources, both in terms of quantity (to not deplete the source) and quality and, sometimes, it makes reference to specific irrigation technologies that are meant guarantee efficient water application, such as drip irrigation. This is for example the case of the (emphasis added):

- GlobalGAP (V5): Control Point 5.2: *Water is a scarce natural resource and irrigation should be designed and planned by appropriate forecasting and/or by technical equipment allowing for the efficient use of irrigation water.* (5.2.2):) *a water management plan available that identifies water sources and measures to ensure the efficiency of application.*
- MPS-ABC Flower production[2]: Control Point 2.8.4: *Drip irrigation or recirculation* (requirement depends on region and type of production system).
- Rainforest Alliance – SAN Sustainable Agriculture Standard: Control Point 4.1: *the farm must have a water conservation program that ensures the rational use of water resources. The program activities must make use of the best available technology and resources. It must consider water re-circulation and reuse, maintenance of the water distribution network and the minimizing of water use.*

Another example is that of the EU (so not private, but in this case set by the private organization DEMETER) regulation on ecological production for olive production (DEMETER, LIFE08 INF/GR/000581) 4: *Drip irrigation must be applied where deemed necessary.*

Jeroen Vos

Waterlogging and the discourse about 'inefficient' smallholder water users

With the Rio Santa water being canalised to the Old Valleys (Chao, Virú, Moche) by the Chavimochic project, the lower parts of the Old Valleys started to suffer from waterlogging and salinization. Apart from the leaching of water from the old and new agricultural areas the waterlogging was aggravated because the smallholders and the drinking water company of the city of Trujillo reduced the pumping of groundwater (Marshall, 2011). The planned drainage system was only partly completed and proved insufficient to drain the affected areas. Notwithstanding the diverse reasons for the waterlogging, the Ministry of Agriculture and the PECH pointed to the 'inefficient' smallholder water users as the culprits for the waterlogging and proposed training, water measurement and more strict water allocation regulations to stop the overirrigation by the smallholders on the heavy clay soils. It was supposed that drip irrigation on the sandy soils did not contribute to the waterlogging.

In October 2013, around 38% of the area of the Virú valley (5,800 ha) was classified by PECH as affected or severely affected by waterlogging; showing groundwater levels of less than 1.5 meters below the surface (see also Figure 8.2). In the Chao and Moche valleys, this number was 4% (350 ha) and 7%

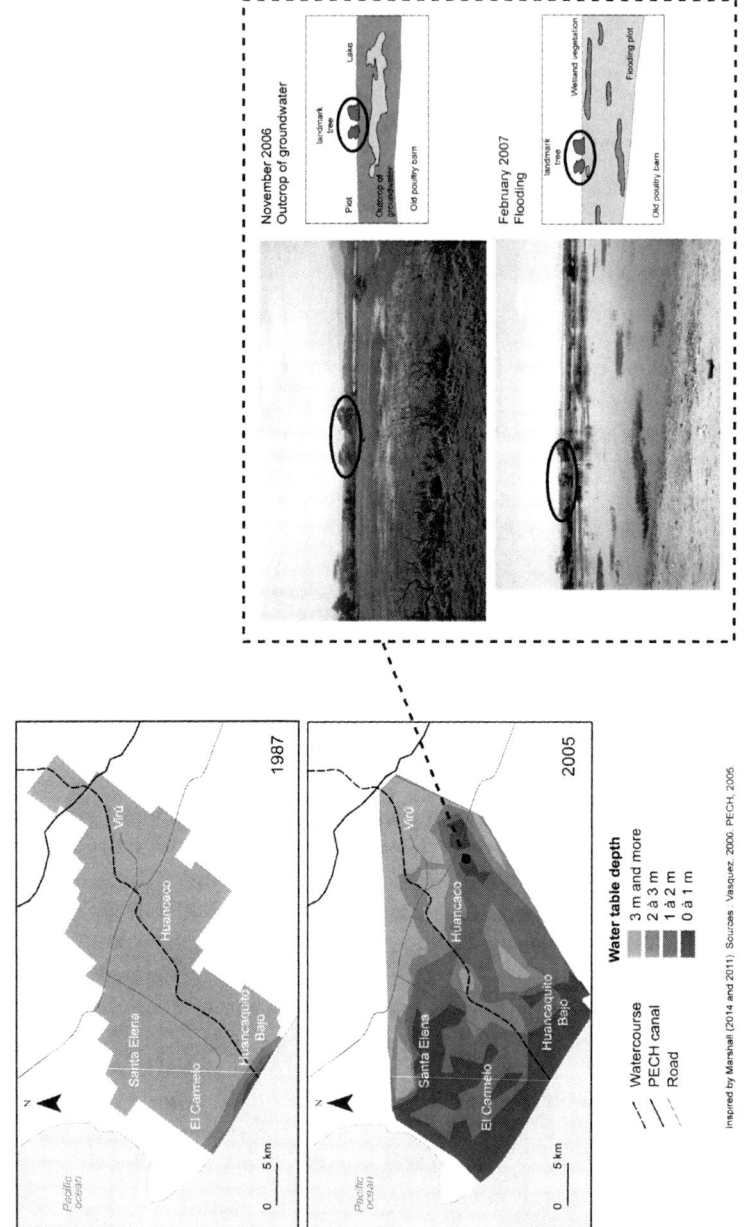

Figure 8.2 Impact of increasing groundwater level in the Virú Valley (1987–2007).

Source: the authors; adapted from Marshall, (2011).

(1,260 ha) respectively. More than 40% of the area in the valley of Chao had groundwater that was classified as highly affected with salinity due to waterlogging; with an electric conductivity (EC) of more than 2 mS/cm. Only 10% of the area was not affected by salinization, where soils had an EC of less than 1 mS/cm. In Virú, the same percentages could be identified for high affection, but in this valley 30% of the soil was not affected. In the Moche valley 14% of the area has saline groundwater, with 40% of the area showing low salinity of the groundwater (PECH, 2013: 68–70).

In a report published in 2010, the Ministry of Agriculture identified the need for an attitude change by the smallholders of the old valley of Moche to change their inefficient use of water and recommended to have another water management regulation to reduce their excessive water use:

> In the Moche valley there is a deficient water management that causes problems of water logging and salinity that affects all irrigation sectors. This inefficiency puts at risk the success of the irrigation, instead of providing a solution for the initial problems of water scarcity. If we also consider the high water demanding crops, the problems get worse. A strategy for a solution consists primarily of *a change in attitude* by the water users. The water users are requested to play a key role in the water distribution and use. The creation of blocks of water users that will have to fulfil a formal role in the allocation of the water will help to create this change in attitude towards an ordering in the distribution and management of the irrigation water, which shall be to the satisfaction of the users in terms of quantity, quality and timeliness.
>
> (Ministerio de Agricultura, 2010: 21, emphasis added)

> In the Moche valley there is excessive water use: in many cases *farmers apply double – or more –* the quantity recommended. And this in a context where it is known that water is a scarce resource. Also the lack of adequate infrastructure for water distribution and measurement and lack of systematic flow measurements, which has led to the lack of information about the real volumes being applied, generated conflicts among the water users and dissatisfaction about the service. Because of this, the programmed efficiencies for water distribution and use have not been attained, especially the field water application has been inefficient.
>
> (Ministerio de Agricultura, 2010: 24, emphasis added)

In the old Viru Valley, and more specifically in certain irrigation sectors (including Santa Elena), some PECH engineers and technicians had the task to help small farmers to switch their traditional irrigation system to a more efficient surface irrigation system. This would result in less water consumption (Marshall, 2014). This technical assistance to volunteer farmers was put into practice only occasionally and was not continued for a long time.

The dominant discourse in the Chavimochic area was that drip irrigation is efficient and productive. The small farmers were blamed for the waterlogging because their traditional way of irrigation is presumed to be inefficient. The

Chavimochic project planned an investigation on the irrigation efficiencies of six small farmers in the Chao valley in 2010 (PECH, 2010a). It was also planned to have training programmes for the smallholders in the 'Old Valleys' to teach them how to irrigate more efficiently. Meanwhile, the efficiencies of the water applications on the large plantations in the interfluve desert areas were not investigated.

Thus, while the ministry and PECH blamed the smallholders for the water-logging, attention was not turned to the percolations from the expanding fields of the new agro-export companies, while these fields also have their leaching requirements to prevent accumulation of salts. This shows the strong double discourse of 'efficient' yet unmeasured large-scale drip irrigation and 'inefficient' smallholders, while realities are much more complex.

Economic opportunities and constraints for the smallholder water users

The Chavimochic project has generated employment and increased access to irrigation water for the smallholders. However, the economic benefits of the project for poor people are small compared to the subsidies invested and the revenues generated by the agro-export companies.

The production does generate employment. Jobs are created for work in the plantations and in the post-harvest packing plants and service provision. Most workers come from poor families in the Andean mountains or Amazon region. The smallholders in the Old Valleys prefer to cultivate their own small plots or migrate to the coastal cities for work or study. Estimates on the number of permanent jobs created diverge widely. According to Mostacero (2013) some 65,900 permanent jobs were created. PECH (2015) mentions some 60,800 jobs created in total. PECH (2010b) mentions higher estimates. The daily wage is about US$ 7 per day. This income is in general not sufficient for a healthy diet, especially because prices for food, drinking water and housing have gone up, and health and education services are partly but increasingly privatized.

With an estimated 250 days of work annually this salary would generate some US$106 million in 2012 for workers in total. The value of the export crops was about US$437 million in 2012 (Mostacero, 2013). This implies that only about a quarter of the export value is benefiting the local population. The gains are mostly syphoned off by the export companies.

With more water available, many smallholders in the Old Valleys try to increase their income. Recently, contract farming of artichokes is practiced on the land of smallholders because of the adequate soil properties in the bottom of the valleys. The export companies bring in the inputs, knowledge of the cultivation of the crop and the market connections. The smallholders bring in the land, the water and sometimes the labour (Marshall, 2012). Direct sales of smallholder associations to export companies is also happening, but turns out to be difficult because of lack of expertise, contacts and credit.

From a landholding perspective, some major land ownership conflicts appeared with the delimitation of the area of the Chavimochic project. An example is the case of the Purpur association that saw its territory decreasing significantly from 350 to 50 ha, in favour of the companies (Marshall, 2014). The territory of the peasant community in San José has completely disappeared, when in 1991 it still had 596 ha (Velásquez, 2001; Marshall, 2014).

Conclusion

In Latin America, drip irrigation is generally associated with highly commercial farming (although with specific forms of government subsidy) and on a large scale replacing surface irrigation in large-scale sugar cane and banana plantations.

In Peru drip irrigation is mostly used on newly irrigated land. In this chapter we used the Chavimochic irrigation system to analyse the effects of drip irrigation. We presented two main findings: first, drip irrigation is used to 'conquer the desert' and not to save water. Thus, the irrigated area is increased resulting in more water use, instead of less. Second, the benefits and costs of drip irrigation are distributed unevenly over the stakeholders: the big export companies receive drip irrigation with subsidies from the national government (legitimised with the narrative of 'efficiency' and 'modernity') providing jobs to migrants from the jungle and mountain regions of Peru, while the smallholders that cultivated the lands in the 'Old Valleys' for centuries lose out economically and are confronted with salinization of part of their lands due to waterlogging.

In the desert coast of Peru, the 'agricultural frontier' is expanded by means of drip irrigation. Drip technology is the only way to irrigate the sandy soils of the interfluves. Here, drip irrigation is not a technology for saving water, but rather an instrument to expand the agricultural frontier. It is all based on private business, with the state giving large subsidies for drip irrigation expansion. The government financed large irrigation projects, in which land and water rights were auctioned with access only for big companies. Thus, national policies gave preference to large commercial farmers in the assignment of newly generated water availabilities and subsequent rights. This water comes mainly from rivers that discharge into the Pacific Ocean (or in the case of Ica and Olmos, partly or completely from water transferred from rivers that ultimately discharge in the Atlantic Ocean), and from groundwater. The increased use of water from the rivers in the coast for expanding the irrigated area has created new conflicts with other users (e.g. in Ica, Piura and the Santa River).

The Chavimochic project is an example of a project in which large subsidies for large commercial farmers expanded the irrigated area considerable by means of drip irrigation. The smallholders in the 'Old Valleys' benefitted also from the Chavimochic canal, as they received more, and more secure, water provision. However, it was quite difficult for the smallholders to engage in export crop production, and many farm out their land under contract farming.

The dominant discourse in Peru portrayed the Chavimochic project as a big achievement of turning the desert green, in benefit for the local people of the La Libertad region. The supposed efficiency of the drip irrigation was associated with development of the region. The smallholders in the Old Valleys were seen as inefficient, and in need of training. The newly irrigated land generated income for labourers; however the lion's share of the value of the exported fruits and vegetables is taken by the export companies.

The relatively little social progress and negative effects shine another light on the 'more crop per drop' and 'efficiency' discourses associated with drip irrigation. It is especially the private companies that reap the benefits of the heavily subsidized irrigation system. The food security of the local population might increase somewhat, but with relatively low income and high prices it can be questioned if overall the quality of life of the inhabitants and migrant workers increases more if compared with an alternative investment of the government in, for instance, subsidy for smallholder irrigation.

The use and depletion of the river affected (aquatic) ecosystems: the dry forest was affected and the environment polluted by agrochemicals. The increasing use of water in the higher parts of the Rio Santa watershed will decrease the water availability for the Chavimochic irrigation system in the lower part of the watershed. The leaching of irrigation water to the lowest parts caused waterlogging and salinity.

In this context, drip irrigation is not 'efficient' in the sense that it produces more output per available resources. In fact more resources (water, labour) are mobilized, and the benefits and costs are distributed quite unevenly over different social groups. The drip technology is used to expand the agricultural frontier not to save scarce resources. It generates also negative externalities because of the environmental impacts.

Notes

1   Source: www.freshplaza.com/article/109034/Peru-adopted-irrigation-systems-with-ease and Netafim Sustainability Report 2013: www.netafim.com/Data/Uploads/Sustainability Report-148_1.pdf.
2   The company's legal address is in Cyprus, and the Camposol Group has over 20 legal holdings, of which some are legally established in The Netherlands and Panama (countries with low tax rates for foreign holdings, also known as 'tax havens'). As from May 2008, the shares of the company are listed on the Oslo Axess Stock Exchange.
3   The annual budget for operation and maintenance of the main infrastructure was 21,590,000 Nuevos Soles (some US$5 million in 2010 according to the *Plan Operativo Institucional* of the Chavimochic project (PECH, 2010a).
4   Source: speech of president Alan Garcia (own translation), www.chavimochic.gob.pe/portal/ftp/informacion/boletines/b_octubre_2009.pdf.
5   Source: APRA, 2011, p. 7, (own translation), http://es.slideshare.net/susan1827/revista-chavimochic.

References

Boelens, R., and Vos, J. (2012). The danger of naturalizing water policy concepts: Water productivity and efficiency discourses from field irrigation to virtual water trade. *Agricultural Water Management* 108: 16–26.
Damonte-Valencia, G. (2015). Redefiniendo territorios hidrosociales: Control hídrico en el valle de Ica, Perú (1993–2013). *Cuadernos de Desarrollo Rural* 12(76): 109–133.

Eguren, L. (2014). Estimación de los subsidios en los principales proyectos de irrigación en la costa peruana. Perú: CEPES, Enero de 2014, 78p.

ELGO DEMETER (2011). Promoting sustainable production and consumption patterns: the example of olive oil. Greece: ELGO DEMETER.

Escobedo, J. (2012). Millonaria inversión del Estado en irrigaciones: pero en Chavimochic se cultiva menos de la mitad de las tierras. *Revista Agraria* n°138: 4–5.

Local GAP (2016). Foundation level- all farm base/crops base/fruit and vegetables check list. Available online at http://www.globalgap.org/.content/.galleries/documents/161028_lg_fl_cl_af_cb_fv_v2_0-1_protected_en.xlsx (accessed 8 March 2017).

Lynch, B. (2012). Vulnerabilities, competition and rights in a context of climate change toward equitable water governance in Peru's Rio Santa valley. *Global Environmental Change* 22: 364–373.

Marshall, A. (2011). Terres gagnées et terres perdues: Consequences environnementales de l'essor de l'agro-industrie dans un desert de piemont. Le cas de l'oasis de Viru, Perou. *Bulletin de l'Institut Francais d'Etudes Andines* 40(2): 375–396.

Marshall, A. (2012). Contrats agraires dans les oasis du piémont péruvien. Economies et sociétés. *Série "systèmes agroalimentaires"* n°10–11/2012. Isméa. Les Presses, pp. 1945–1968.

Marshall, A. (2014). *Apropiarse del desierto. Agricultura globalizada y dinámicas socioambientales en la costa peruana. El caso de los oasis de Virú e Ica – Villacuri.* Institut Français d'Études Andines – IFEA, IRD éditions, Tomo 321, 2014. 417p.

Ministerio de Agricultura (Ministry of Agriculture of Peru). (2010). *Proyecto "Obras de Control y Medición de Agua por Bloques de Riego en el Valle Moche" Estudio de Preinversión a nivel de Perfil*, Lima, Agosto 2010. Available at www.ana.gob.pe/sites/default/files/publication/files/informe_principal_moche_0.pdf

Mostacero, R. (2013). *Proyecto Especial Hidroenergético Chao-Virú-Moche-Chicama (Chavimochic)*. Presentation. Available at www.bcrp.gob.pe/docs/Proyeccion-Institucional/Encuentros-Regionales/2013/la-libertad/eer-la-libertad-2013-mostacero.pdf

Netafim (2013). *At the hearts of food, water and land nexus.* Netafim Sustainability report 2013. Israel, Netafim.

Oré, M.T., Bayer, D., Chiong, J., and Rendon, E. (2012). *Emergencia hídrica y explotación del acuífero en un valle de la costa peruana: el caso de Ica.* In R.H. Asensio, F. Eguren, and M. Ruiz (Eds.), *Péru: el problema agrario en debate.* Lima: Seminario permanente de investigación agraria, Sepia XIV, pp. 584–613.

Oré, M.T., Del Castillo, L., Van Orsel, S., and Vos, J. (2009). *El agua, ante nuevos disafíos: Actores e iniciativas en Ecuador, Perú y Bolivia.* Lima, Peru: Oxfam Internacional and IEP, 466 p.

PECH, Gobierno Regional La Libertad (2010a). *Plan Operativo Institucional 2010.* Trujillo, Peru: Proyecto Especial Chavimochic, p. 66.

PECH, Gobierno Regional La Libertad (2010b). *Chavimochic en Cifras 2000–2010.* Oficina de relaciones públicas, Proyecto Especial Chavimochic, 97p.

PECH, Gobierno Regional La Libertad. (2013). *Memoria 2013.* Gobierno Regional La Libertad, Perú. Available at www.chavimochic.gob.pe/portal/Ftp/Informacion/Memorias/M_2013.pdf

PECH and Ñique Alarcaon, E. (2015). *Proyecto Especial Chavimochic. Descripción general, beneficios y tercera etapa.* Presentacion en el Seminario Promoviendo el Desarrollo Económico y la Competitividad en el Peru. Available at http://fr.slideshare.net/ProGobernabilidadPer/1-chavimochic-pech

Pouyaud, B., Vignon, F., Zapata, M., Gomez, J., Tamayo, W., Yerren, J., Suarez, W., Vegas, F., and Rodriguez, A. (2003). *Glaciares y recursos hídricos en la cuenca del rio Santa, Huaraz Peru.* Rapport IRD, INRENA-UGRH, SENEMHI, EGENOR, p. 66.

Rainforest Alliance (2005). Sustainable Agriculture Standard with Indicators Sustainable Agriculture Network. Rainforest Alliance, San José, Costa Rica.

PROGRESSIO (2010). *Drop by drop, understanding the impacts of the UK's water footprint through a case study of Peruvian Asparagus.* Londres: Progressio, CEPES y WWI.

Urteaga, P. (2014). Creadores de paisajes hídricos: Abundancia de agua, discursos y el mercado en las cuencas de Ica y Pampas. In G. Damonte and M.T. Oré (Eds.), *Escasez de agua en la cuenca del río Ica.* Lima: PUCP, pp. 227–267.

Velásquez, O. (2001). *La communidad campesina en el Perú y los retos por la superviviencia.* Trujillo: Universidad nacional de Trujillo, 306p.

Verona, F., Bartl, K., Pfister, S., Jiménez, R., and Hellweg, S. (2012). Modeling the local biodiversity impacts of agricultural water use: Case study of a wetland in the coastal arid area of Peru. *Environmental Science & Technology* 46: 4966–4974.

Vos, J., and Boelens, R. (2014). Sustainability standards and the water question. *Development and Change* 45(2): 205–230.

World Bank (2013). *El Futuro del Riego en el Perú: desafíos y recomendaciones.* Washington: World Bank.

World Bank (2004). Peru. Irrigation subsector project. Implementation completion report. Washington: World Bank.

9 An elite technology? Drip irrigation, agro-export and agricultural policies in Guanajuato, Mexico

Jaime Hoogesteger

Introduction

In Mexico horticultural production for export, which is dominated by medium and large producers, packaging and export companies, started to increase exponentially since the second half of the 1980s due to several reasons such as Mexico's adhesion to the General Agreement on Tariffs and Trade (1986), signing of the North American Free Trade Agreement (NAFTA) (1994), the application of neo-liberal structural reforms, and the recurrent devaluation of the Mexican peso (Carton De Grammont and Lara Flores, 2010). In the last few years, and partly driven by the droughts in California, there has been a boom in the fresh-vegetable export sector (Ayala-Garay et al., 2012). One of the important production regions that took part in this boom is the state of Guanajuato in central Mexico (see Figure 9.1). The state has become the leading production area of broccoli, lettuce, cauliflower, asparagus and strawberries in the country (SDAyR, 2016).

According to official figures, in Mexico irrigated agriculture consumes 71% of total water withdrawals (CNA, 2010). Most of the irrigated area is concentrated in the centre and north of the country where water is scarce. Only 3% of the total irrigated area is found in the water abundant south (López-Morales and Duchin, 2011). According to the FAO-AQUASTAT (2009) database, 93% of irrigated land still uses flooding by gravity (mostly in furrows) as an irrigation technique; while the remaining 7% has switched to the use of sprinkler and drip irrigation. In the past five years, at least three large irrigation districts (DR 044 Jilotepec (4000 hectares), DR 001 Pabellón de Arteaga (6100 hectares), and Módulo La Purísima (4000 hectares) of DR 011 Alto Río Lerma) have been modernized to introduce the use of drip irrigation. Nonetheless, most large surface water irrigation systems that together compromise 3,498,164 irrigated hectares still function with furrow irrigation and great efforts are being made to pressurize these systems. The indicated figure for the surface area of drip irrigation nationwide is probably conservative especially as the use of sprinkler and drip irrigation is mostly concentrated in areas of intensive groundwater use in the centre and north of Mexico. Many of these areas are not accounted for in official statistics that focus mostly on state-built surface

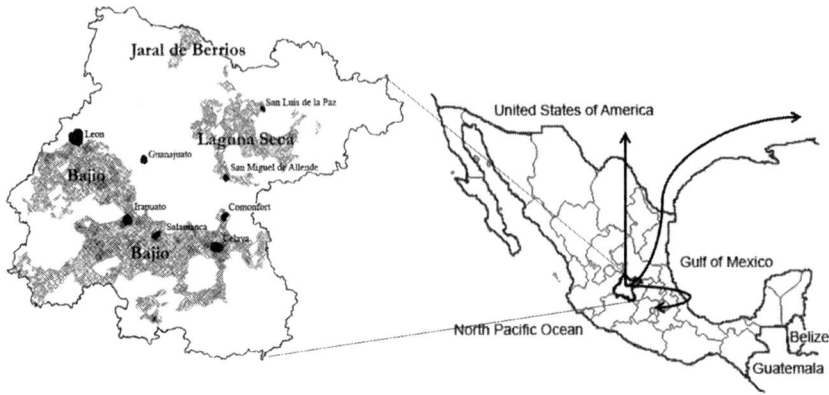

Figure 9.1 Areas of intensive groundwater use in Guanajuato and vegetable export flows.

Source: the author, adapted from Hoogesteger (2004).

water irrigation systems, many of which still focus on the production of basic grains. Nonetheless, in many areas of intensive groundwater use the production of more lucrative fresh fruits and vegetables, produced almost exclusively with drip, has become widespread.

In Guanajuato, amongst a wider set of policies aimed at supporting the agro-export sector and its international competitiveness, the state government of Guanajuato, through its Secretary of Agricultural and Rural Development (SDAyR) set up a subsidy program for the 'modernization' of individual and collective tube well-based groundwater irrigation systems. The program subsidized the installation of low pressure piped conduction systems for surge valve[1] systems and high pressurized systems for sprinkler (since 1996) and drip irrigation (since 2001). This program works based on the demands and co-investments by farmers, which means that farmers choose what kind of irrigation system they install, with which technical specifications and which installation company they hire for this. In the two decades that followed, the area under drip irrigation has expanded significantly. According to SDAyR, of the total 265,000 ha of groundwater irrigation in the state of Guanajuato, it has subsidized the installation of drip irrigation in around 30,000 hectares, sprinkler irrigation in about 18,000 ha and a whopping 146,000 ha with piped conduction systems at field level where furrow irrigation is still used (SDAyR, 2014). Added to this state subsidized modernization, there are a large number of farmers that have installed drip irrigation with other programs of the federal government or without subsidies, extending the drip irrigated area to a well-informed guess of at least 40,000 to 45,000 hectares, but official figures are missing.

As elsewhere in the world, the state subsidies for drip irrigation are mainly targeted at 1) intensification of production (more crop per cultivated land unit), 2) increases in irrigation efficiency (more crop per drop) as a means to 'save' water; 3) efficiency increases in the use of fertilizers and other inputs through fertigation; 4) reduction in labour costs. These 'efficiency' gains are incorporated as important social and environmental justifiers to support the development of 'international standards of competitiveness of agricultural activities within the framework of sustainable development' (SDA, 2000); echoing the discourse of international drip development, production and sale corporations, policymakers and practitioners (van der Kooij et al., 2013; Venot et al., 2014).

These discourses often portray the use of drip irrigation as 'the' way toward sustainable agricultural water use. In doing so, drip irrigation is ascribed with the status of an almost magical 'silver bullet' to solve water quantity problems at field, irrigation system and river basin levels. The underlying rationale behind these discourses rests on the idea that agriculture, which uses the bulk (70%) of the diverted water resources worldwide, is inefficient and 'wasteful' (Molle et al., 2010). Therefore, irrigation has to be modernized through drip irrigation, which according to a World Bank report 'uses 30–50% less water than surface irrigation, reduces salinization and waterlogging, and achieves up to 95% irrigation efficiency' (World Bank, 2006: 163). The efficiency gains are projected to liberate water from agriculture to the environment and other uses; while addressing the increasing world food and energy needs by producing more crop per drop and energy unit. Yet, this only applies from a specific scale perspective and as van der Kooij et al. (2013) point out 'that drip irrigation potentially allows using less water for a single plot without compromising (or even improving) yield . . . reveals little to nothing about the water saving potential of drip irrigation at the river basin or watershed level' (p. 108) (see also Seckler, 1996 and Chapter 5, this volume on the groundwater paradox).

Nonetheless, such facts find little resonance amongst policymakers, practitioners and drip production and installation companies that have by and large been able to effectively cast drip irrigation as an undisputed 'post-political' water management solution (see Swyngedouw, 2011) for solving local to global water challenges. According to Wilson and Swyngedouw (2014) some of the key markers of the 'post-political' are the apparent inevitability of market capitalism as the ordering principles to structure society; technocratic management that gets structured around specific fragmented problems (in this case irrigation efficiency and production), and emphasis on consensus building around possible solutions; albeit only within pre-established political parameters (stimulation of capital intensive market oriented agriculture). The emerging 'solutions', such as the use of drip irrigation, tend to be consistent with, and as such intrinsically advance, an often hidden neo-liberal agenda (such as the advancement of capitalized market oriented (large-scale agriculture), that relegates diverging discourses (such as food sovereignty) and actors (subsistence farmers) to the margins (Boelens and Vos, 2012; Zwarteveen and Boelens, 2014). In drip irrigation this has been achieved through the effective black-boxing of the deeply political assumptions and social relations that are ingrained in the pressurization of water for its transportation in plastic tubes and its application in the field by drippers. Some of these assumptions are: glocalization[2] of production, reliance on external/universal expert-knowledge systems, universal efficiency measures, globalized market integration of agricultural production, intensification of crop production, yield and profit maximization and the search for efficiency through technological modernization; which importantly includes the reduction of labour needs/costs.

With this as background, in this chapter, I argue that drip irrigation in the state of Guanajuato works as a technology first and foremost for capitalized farmers facilitating the accumulation of capital and resources; often at the cost of local labour and relegating small farmers as backward and inefficient resource users. To do so, I first present the development of capital intensive groundwater agriculture in the state of Guanajuato. Then I present actors and state programs (and their relations) that have stimulated the use of drip in the state. I present why and by whom drip irrigation is used or not. I conclude that in Guanajuato drip irrigation is and has been above all a technology that further advances capitalist modes of production that are dominated by political and economic elites. This chapter is based on field research conducted in the state of Guanajuato between September 2014 and January 2015 and January–May 2016; and builds on earlier work in the state of Guanajuato in 2003 and 2006–2007. Fieldwork consisted of interviews with state officials, irrigation technology retailers, commercial farmers, labourers and small producers (ejidatarios); as well as field visits to retailers and farms that depend on groundwater irrigation for their production.

Groundwater development and commercial agriculture in the state of Guanajuato

The state of Guanajuato is one of the most important agricultural regions of Mexico. Its fertile soils, favourable climate and extensive irrigation systems have made the state an intensively used agricultural region since colonial times (Wester, 2008). With the advent of tube-well technologies in the 1940s, little by little groundwater irrigation started to gain momentum marking a drastic increase in groundwater use. The number of tube-wells[3] rose from fewer than 100 in the 1940s to around 2,000 in the 1960s, 6,000 in the 1980s (CEAG, 2013) and over 17,300 in 2016 (CONAGUA, 2016). This translates in an increase of groundwater irrigated area from 24,000 ha in the 1960s to over 250,000 ha by the end of the 1990s (CEASG, 1999) and an estimated 265,000 at present.[4] The increase in groundwater irrigation went hand in hand with the expansion and intensification of the production of vegetables for agro-export and fodder for the national dairy industry; while also becoming more important in securing the production of basic grains such as maize and sorghum, wheat and oats. At present irrigated agriculture in the state contributes 94.2% of the total value of agricultural production while the remaining 5.8% comes from rain-fed agriculture (COTAS Rio Laja, 2013).

The rise in groundwater irrigation went hand in hand with the development of the commercial agricultural sector, which was spurred by the green revolution since the 1950s. According to Hewitt de Alcántara (1978), in the 1940s, agrarian policy in Mexico shifted from land reform (aimed at land distribution to rural communities) to emphasize commercial production aimed at the provision of cheap basic grains for national consumption and the production of high value export crops. In Guanajuato, this brought about a progressive but marked shift toward the production of vegetables; most of which are irrigated exclusively with groundwater. The first companies that started operations with fresh vegetables in Guanajuato at the end of the 1950s and early 1960s were the international food companies Gerber, Campbell's, Heinz and Del Monte (Steffen Riedeman and Echánove Huacuja, 2003). These companies contracted large local farmers to produce (under contract) the needed fresh vegetables such as sweet corn, carrots, peas and asparagus which were processed (mostly by canning) for the Mexican internal market (Bivings and Runsten, 1992). In 1967, the US company Birds Eye started operations for the production of frozen export vegetables introducing broccoli, okra and cauliflower in the region. In the years that followed some of the large farmers that had supplied the vegetables to these companies opened their own processing plants and industries.

As groundwater-based agriculture kept expanding in the region, new irrigated areas for the production of vegetables were opened, creating space for new agro-export industries. In 1983 Green Giant opened a processing complex in Irapuato and in the 1980s and 1990s several new Mexican-owned companies opened (Steffen Riedeman and Echánove Huacuja, 2003). What is remarkable is that the fresh vegetable production and processing sector developed almost exclusively through medium and large farmers (60 ha and up); while small farmers (mostly ejidatarios)[5] were by and large excluded because of the size of their landholdings and lack of capital to switch to vegetable production. By 2000, Steffen Riedeman and Echánove Huacuja (2003) identified 18 agro-export companies; a number that has kept increasing to over 70 broccoli export companies in 2014 (SDAyR, 2015). These companies greatly diverge in terms of size, export volume and vegetable processing capacity (from vegetables in bulk to individually packaged and from volumes that run in the tens of thousands to a few tonnes per year). Some of the companies that stand out because of their export volumes are Taylor Farms de Mexico, Covemex, Mr. Lucky, Mar Bran, Expor San Antonio, X tra Congelados Naturales, Congelados la Hacienda, Comercializadora GAB and Green Giant amongst others.

The bulk of the production of both agro-export vegetables as well as fodder crops for the dairy industry is in the hands of a few thousand large producers/landowners most of which operate as family businesses. These family businesses control between a hundred and sometimes up to a (few) thousand hectares[6] of land for their production under basically three modalities (or a mix of them) which are 1) contract farming, 2) production on own lands, and 3) production on rented lands. In the same period, a significant dairy sector developed in Mexico. Its growth spurred the demand for fodder crops, which in the groundwater dependent sector is mostly alfalfa that is increasingly irrigated with (subsurface) drip and sprinkler systems. In 2011, its production covered over 52,000 ha in the state (INEGI, 2015), a figure that by the end of 2014 had dropped to 45,000 ha (through its replacement by vegetables) (SDAyR, 2015). The groundwater 'boom' has brought about severe problems of over-exploitation in all of the aquifers of Guanajuato (CEAG, 2014). This brings the need to reposition and sink ever deeper wells and increasing pumping costs (Wester et al., 2011) as well as raising important questions with regard to groundwater governance, sustainability and equity (Hoogesteger and Wester, 2015).

Guanajuato's silver bullet for modern agricultural development

The state of Guanajuato is seen at the national level as an example in terms of its water and agricultural policies. The rise of its 'leading' role in these two fields is closely associated to the political power that the large agro-export producers have leveraged in the state government since the mid-1990s. In 1995, Vicente Fox Quesada, owner of amongst others the frozen vegetable exporting company Congelados Don José (established in 1985) became the governor of the state of Guanajuato. He appointed Javier Usabiaga Arroyo,[7] one of the largest garlic producers (and exporters) in the world, and whose family owns/operates the conglomerate of companies Grupo U (which consists amongst others of the largest packaged fresh vegetable company in Mexico (Mr. Lucky), COVE-MEX, Comercializadora GAB, Transportes GAB amongst others), as head of SDAyR. This duo set agro-export based agricultural development and water management high on its political agenda. This led to the full-fledged endorsement of the State Water Commission (CEAG), the creation of the Aquifer Management Councils (COTAS) (Wester et al., 2009) and the implementation of policies that supported the agro-export sector and its growth, including irrigation 'modernization' programmes (Hoogesteger, 2004).

   The latter were executed through SDAyR and with matching funds from the federal government through its Secretary of Agriculture, Livestock, Fisheries, Food and Rural Development (SAGARPA). This importantly included the program of 'modernization of groundwater irrigation in the state of Guanajuato'. The aim of this program as stated in its base documents is:

> To support increases in the productivity of the areas under irrigation based on projects that include the use of irrigation systems that permit an efficient management and use of water, reduce the costs of energy and fertilizers, and increase the yields, by granting resources for the installation of irrigation systems of high and low pressure, of which the sources is groundwater; and contribute to the stabilization of the aquifer levels in the state, by granting resources for the modernization of groundwater irrigation, which reduce present water withdrawals of these water bodies.
>
> (SDAyR, 2016: 5)

This program fits well into the overall strategy of SDAyR's policies which are mainly geared toward the support and stimulation of export-oriented agriculture by increasing its competitiveness in a context of free markets and globalized food production. The aim is to motivate producers to improve their irrigation systems through financial support for installing better and more efficient irrigation technology on the parcels. It is estimated that at the end of 2001 the area with 'improved' irrigation systems was 210,000 ha (FAO-SAGARPA, 2002), which accounted for 84% of the total groundwater irrigated area. It is claimed that the state program accounts for more than half of this area. Most of the improvements have consisted of low-pressure conveyance systems (surge valves) and an increasing number of sprinkler and drip irrigation systems.

The 'modernization of groundwater irrigation' program initially started with a subsidy scheme that subsidized up to 80% of the installation costs of the new irrigation equipment for smallholders; and 50% of the installation costs for projects larger than 50 ha with a fixed maximum subsidy value per hectare. The level of the subsidy as well as the differentiation according to the size of the project and the elements that are subsidized have varied through the years depending partly on the contributions and guidelines of operation of SAGARPA and SDAyR. These varied from 50–50% arrangements to an arrangement of 80% federal government and 20% state government as was the case in 2014. First, the differentiation between smallholder and larger projects was eliminated and the level of the subsidy reduced to 70%, later 60% of the whole irrigation system; and in 2014 only up to a maximum of 50% of the installation of the main conduction system was financed (program director modernization of groundwater irrigation of SDAyR, 2014, pers. comm.). In 2016 the subsidy was established at 50% of the investment costs with a maximum contribution per hectare of 10,000 pesos for low pressure surge valve irrigation systems, 15,000 pesos for sprinkler irrigation and 20,000 for drip irrigation; with a maximum allowable project budget of 750,000 pesos for individuals and 2,000,000 for companies or groups (SDAyR, 2016: 11). The aim of the program in 2016 is to support the 'improvement' of 3000 hectares restricted to budgetary constraints (idem).

Despite many applications and beneficiaries in its 2014 and 2015 exercise a part of the program's budget was not spent because of a lack of applications (pers. comm. 02/12/2014). Interviewed farmers stated that the hassle to apply for and obtain the subsidies is burdensome, which is attributed to the financial structures for payment of the subsidy (which usually takes a few months); the quality control standards; and the relatively low subsidy amount. As a result, many farmers choose to opt for more simple irrigation systems at lower costs. Another reason why there are fewer applications is that the state only finances an irrigation system for a user once and/or the upgrade from low pressure systems to drip or sprinkler.

Table 9.1 shows the investments and covered hectares as well as the estimated saved water volumes of the program between 1996 and 2013.

As this table shows, the program started with high numbers of beneficiaries as well as 'modernized' hectares with an average of just below 28,000 ha/year between 1996 and 1998. This was above all due to the massive conversion of both ejidatarios as well as medium and large producers to low pressure piped

Table 9.1 Results of the program 'Modernization of Groundwater Irrigation' in Guanajuato

Year	Area (ha)	Governmental investment (Mex Pesos)	Users investment (Mex Pesos)	Systems installed	Estimated water savings per year (Mm³)
1996	20484	40,353,000	28,727,000	561	41
1997	28351	58,190,000	65,699,000	771	60
1998	34623	95,710,000	106,796,000	970	73
1999	14794	61,572,000	31,064,000	412	33
2000	14294	61,300,000	45,305,000	402	31
2001	18301	85,156,000	61,063,000	601	42
2002	11190	52,500,000	35,747,000	327	26
2003	11790	62,251,000	69,098,000	394	28
2004	7193	42,026,000	42,760,000	287	17
2005	4749	30,678,000	34,024,000	183	12
2006	2919	23,579,000	25,544,000	123	7
2007	5153	44,251,000	40,350,000	190	13
2008	5041	32,341,000	55,703,000	219	13
2009	3371	43,208,000	33,632,000	161	8
2010	3779	40,927,000	47,360,000	198	9
2011	3368	37,756,000	37,509,000	262	8
2012	2960	31,957,000	42,738,000	224	7
2013	4965	55,039,000	66,021,000	341	12
Total	**197325**	**898,795,000**	**869,142,000**	**6626**	**440**

Source: the authors, adapted from SDAyR (2014).

conduction systems and surge valve irrigation. This number halved to around 14,000 ha/year between 1999 and 2003 ha/year, and again dropped to an average of 5,000 ha/year between 2004 and 2008; and 3,700 ha/year for the period 2009–2013. This drop in covered area is related to the fact that initially the subsidy scheme was much more interesting for producers and that through the years the area with a 'modernized' irrigation scheme has greatly increased and thus the eligible population and systems have been reduced considerably. Interviews with irrigation system installation companies, state officials, large producers and ejidatarios point to the fact that most beneficiaries, and especially the smaller producers and ejidatarios, have switched to low pressure irrigation systems, while by and large only the capitalized producers have adopted the use of sprinkler and drip irrigation.

The preferred option for most ejidos has been to modernize their irrigation system to low pressure piped irrigation systems fitted for surge valve irrigation. This technology basically reduces their water 'losses' from the well to the field through the piping of the whole system, while maintaining the same irrigation practice, crops and schedules of furrow irrigation. Investment costs are much lower than with drip and sprinkler irrigation and importantly the water allocation principles of the ejidos remain the same. While for ejidatarios energy costs are reduced, the investments as well as the changes to the production practices remain minimal (especially because most installed these systems with an 80% subsidy).

The total estimated 'saved' volume of this program is officially established at 440 million cubic meters (Mm³) per year. Nevertheless, important question marks can be established around these numbers. They depart from the notion that 'traditional' irrigation with earthen canals and furrows has an irrigation efficiency of 40%, which can be increased to 60% with surge valve irrigation, 80% with sprinkler irrigation and 90% with drip irrigation (program director modernization of groundwater irrigation of SDAyR, pers. comm. 10/10/2014). Yet as pointed out by van der Kooij et al. (2013), just as with most claims around water use efficiency with drip, these assumptions are based on specific boundary and scale assumptions. They do not take into account that 'percolation' losses might be an important source of recharge for the aquifers. Often with a change to drip irrigation production, water use intensifies, spurring the expansion of the irrigated area as put by a producer:

> If you have a well that gives 20 l/s theoretically it would suffice to irrigate 20 hectares with a surge valve system, maybe it would suffice for 1.2 ha/l/s with sprinkler, and being very efficient maybe 1.6, 1.8, sometimes 2 ha with drip irrigation. When you have all of your surface with drip and your 20 l/s suffice to irrigate 40 ha, you will irrigate 40 ha; you will not limit yourself to cultivate 20 ha. [...] So we have become much more efficient in the use of water but the total pumped volumes, the water we are extracting from the aquifers keeps on being the same.
>
> (April 2016)

Most of the funds have been channelled to medium and large producers[8] because of: 1) In sheer number of hectares, the ejidatarios represent a smaller portion of groundwater irrigated land than the private producers. 2) In terms of investment costs and related production systems, it is mostly the capitalized farmers that have been able to invest in drip and sprinkler irrigation. This is related to the type of production systems of ejidatarios that are generally characterized by low inputs, low risk taking and marginal profits as opposed to large-scale agriculture that is high input, high risk and often high profits or losses. Finally, 3) because of the legal requirements of the subsidy programs which include amongst others legalized wells, no electricity debts, land titles, installed flow meters and the means, know-how and time to apply, many of the small producers are not able to access the subsidies; and those that do comply with all these requisites have already benefitted from the program in the late 1990s and early 2000s. It is telling that in the last two years of the programme, no ejidatarios applied for subsidies (program director modernization of groundwater irrigation of SDAyR, 10–10–2014 pers. comm.) and that several of the interviewed irrigation retail companies could not remember installing an irrigation system for ejidatarios. At present SDAyR works with more than 50 installation companies from within Guanajuato and from other states of Mexico. Most of these companies not only design and sell the irrigation systems, they also offer advisory services which include the handling of the subsidy requests of the farmers to SDAyR. Some of the larger irrigation technologies retail and installation companies in the state of Guanajuato are also owned by the families involved in the agro-export business.

Drip for capitalized producers

The use of drip irrigation is widespread amongst capitalized vegetable and fodder producers in the state of Guanajuato. The use of drip irrigation fits well with the production of fresh vegetables and is increasingly used as subsurface drip irrigation for the production of alfalfa and has the following advantages for production:

- *Reduced pumping costs per irrigated unit of production*: As water losses due to percolation and evaporation are drastically reduced the pumped groundwater is used more efficiently in terms of crop production (more crop per pumped drop).
- *Reduced fertilizer and pesticide costs*: As most drip irrigation systems are installed with a fertigation system, fertilizers and some pesticides can be applied with more precision and directly through the water, reducing inputs and labour hours.
- *Reduced labour costs*: Drip irrigation reduces the costs of labour by eliminating the field irrigators. Although with most crops drip lines have to be placed and removed every cropping cycle, this is only a peak in incidental labour requirement and a lot of this work has been mechanized (see Plate 11). Other activities where labour hours are saved are: fertilization, fumigation and weeding. In alfalfas with subsurface drip, the quality of the harvested alfalfa is also better because of a lower germination rates of weeds in the dry season.
- *Higher yields especially in horticulture*: drip irrigation enables a higher plant density than furrow irrigation and a more targeted and precise administration of fertilizers leading to increased yields per cropped hectare.
- *Positive assessment from certifiers*: Finally many of the fresh vegetable producing companies need to comply with an increasing number of certification requirements to either enter the globalized markets and or to address specific consumer sectors such as organic. Most of these certification companies assess drip irrigation as a good water management practice.

Interviewed medium- and large-scale producers of fresh vegetables and alfalfa claim that investments in drip irrigation pay off. These vary between US$1250 and US$3000 per ha for a basic drip irrigation system; and the returning reposition costs of drip tape which is estimated at between US$800 and US$1000 per ha/year. Most users use drip lines that are calculated to last for one to two years of use (4–7 production cycles of vegetables whereby in every new cycle the lines have to be repositioned in the field) before they have to be replaced. The choice for these 'cheaper' options is linked to the fact that due to the process of installation and removal/repositioning in the field, most drip lines (regardless of their calibre) get mechanically damaged having to be replaced anyway every one or two years.

The production advantages of drip, as well as the rationale of its use, fit well with the highly mechanized commercial agriculture in Guanajuato (Plate 11 in the color plate clearly illustrates this mechanisation). This sector has become dependent for its technological input packages and know-how on commodified international technological knowledge systems. As stated by a large-scale farmer: 'in this globalized world Mexico imports almost all of its inputs, what do we put in [the production process], water and [cheap] labour' (April 2016).

The extent of 'modernized' irrigation use, and particularly drip, has further deepened the insertion of the commercial agricultural sector in the multiscalar networks of globalized agricultural production in which for instance most seeds for vegetables come from the Netherlands, fertilizers from Chile, Russia and Ukraine, tractors from the United States and Europe. In Guanajuato, as elsewhere in Mexico, all the international manufacturers of drip irrigation equipment such as TORO, NETAFIM, NaanDan-Jain, John Deere, Rain Bird, AZUD, Altamira (amongst others) are represented and selling their products through local retailers and irrigation installation companies.[9] Because most of these products, with the exception of the PVC pipes that are used to make the main conduction systems, are imported their retail prices are calculated in US dollars.

Drip irrigation and ejidatarios

Contrasting the widespread adoption of drip irrigation amongst capitalized farmers, ejidatarios have switched by and large to low pressure piped systems and surge valve irrigation and there is only a small segment of groundwater ejidatario-irrigators that have adopted the use of drip irrigation. The few ejidatarios that were identified during fieldwork using drip irrigation were ejidatarios that had engaged in the production of fresh vegetables such as tomatoes or strawberries on a small scale for the local market. All of these interviewed ejidatarios acknowledged that drip irrigation had benefits for their production systems such as increased yields, lower labour requirements, less weeding needs and water savings. Because of the size of the cultivated area, it was also the ejidatarios that had installed the irrigation system to tailor it to their specific needs and knowledge systems. For instance, an ejidatario that had drip in a small 600 square meter tomato greenhouse had installed a valve on every drip line to better regulate the water flows.

Interviewed ejidatarios all showed interest for drip irrigation but for many it seemed something that was far from their possibilities. A few of the interlinked factors that determine the low levels of drip use are outlined below:

- *Community organization*: Most groundwater irrigation in ejidos takes place through shared wells that irrigate from 5–40 hectares depending on the characteristics of the well. The shared wells are managed collectively by the ejidatarios that usually have access to between 1 and 5 hectares of land that are managed individually (with different crop choices, cropping calendars and production practices). Converting to drip would require the coordinated organizational and economic effort of all members using the well. This coordination is rarely achieved though exceptions exist where ejidatarios have built storage tanks to be able to use drip while maintaining the established water turns in the ejido (see Massink, 2016).

• *Production practices*: Most ejidatarios produce alfalfa or basic grains because of cultural reasons and/or because its production entails low economic risks and labour investments (cf. Eakin et al., 2015). In the case of alfalfa, its production also guarantees a monthly harvest and thus a more or less constant cash flow.

• *Financial capital*: As a result of the above, the returns of agricultural activities are usually low and there is little room for making new investments in the production system.

• *Knowledge*: Finally it is also a question of knowledge and practices. As drip entails a different way of irrigating and producing for many ejidatarios it is a black box that brings with it uncertainties and risks.

Because of these characteristics ejidatarios are usually blamed by state officials and large commercial farmers for being traditional, unorganized and 'inefficient' groundwater users that contribute little to nothing to the state's economy while stalling the development of the agricultural sector.

Conclusion: drip irrigation as technology for capitalized production

The case of Guanajuato shows that drip irrigation fits well with highly specialized commodity centred production systems that have become intrinsically globalised. The case of fresh vegetables is a special case in point as it greatly depends on global commodities in the form of knowledge, inputs and technologies (including drip) that are applied locally to use specific natural resources (in this case fertile soils, fossil groundwater and cheap labour) to produce a commodity that is fed back into the national and global markets. In the process a small and privileged elite benefits from these production systems as a means for capital accumulation; while leaving a few crumbs to the labourers that work the fields (around US$7 per labour day).

This agricultural commodity production process effectively externalizes knowledge, agricultural inputs and labour from the locales where production takes place to international companies that produce the technologies such as drip lines, pumps, filters, seeds, fertilizers, pesticides, herbicides, tractors and so forth and that depend on these local production systems. As such producers increasingly demand and become dependent on highly specialized commodified knowledge and technologies to increase their 'competitiveness' and monetary gains in international markets. These technologies such as drip irrigation, fertigation, hybrid seeds and pesticides portray an image of universality and many have effectively co-opted an environmentalist discourse which in the case of drip revolves around water use efficiency. Yet far from being a-political, these technologies are designed with specific characteristics making them suited for some forms of production and for some actors and not for others. At the same time, they inherently portray a rationale of production and use that effectively casts local know-how and 'unqualified' cheap labour, and with it local lives, traditional production systems and related livelihoods, into the realm of inefficient, backward and environmentally unfriendly. In this rationale local labour (and with it local lives and livelihoods) is put in an equation of production costs and benefits on which 'savings' and efficiency gains can be made. In the specific case of drip irrigation, local field irrigators, tractor drivers and other day labourers are effectively replaced by tubes and pipes. These subtle processes of commodification of the agricultural process through technological innovation, in which drip plays an important role, is re-patterning the flows of power, money, natural resources and labour. As such, drip irrigation helps to advance new capital intensive forms of agriculture while churning at the production systems of non-adopters.

The case of Guanajuato shows that this process of re-patterning is easily co-opted by economic and political elite groups. The introduction of neo-liberal policies in Mexico since the 1980s and the coming to power of agro-exporters in Guanajuato have marked a clear shift in the national and state agricultural policies. These have been increasingly geared toward the support of commercial production especially that of the vegetable and agro-export sector. With the neo-liberal dream of fully integrating Guanajuato's agricultural sector into global markets, their policy focus has become that of supporting and stimulating the agro-export sector in the state; a sector that is dominated by a few thousand well-off and capitalized medium to large producers. As such these elites have used governmental programs and subsidies to increase the international competitiveness of their own companies and through it the advancement of global capitalism in the sector. In parallel ejidatarios that depend on basic grains and alfalfa are now operating in a context of liberalized volatile markets in which basic grains have almost no value; a context in which reducing risks and finding other sources of income to sustain rural livelihoods has become normal (Massink, 2016). As a result, many ejidos have started to rent out and sell their lands and water to large producers. As such, there is a slow process of land re-concentration and accumulation, reverting the land and water redistribution efforts that were made in the country after the revolution of 1910. Though the time that the countryside was dominated by haciendas is part of history, new forms of accumulation and concentration are moving fast.

While environmental degradation keeps unabated, social inequity increases as little to nothing has been done to structurally support the ejido sector which is now expected to compete in an uneven battlefield with international knowledge systems, corporations and standards which have made drip irrigation one of their centrepiece technologies in groundwater irrigated production schemes.

Its discursively claimed 'universal' validity and applicability casts it into the post-political realm. There drip irrigation becomes a technological tool that further advances capitalist market modes of global agricultural production; while casting its non-users into the realms of backwardness and ignorance. As such, it can be argued that drip irrigation serves as a technology to advance new frontiers of capitalist agriculture and its contested socio-environmental effects.

Acknowledgments

I am grateful to the editors and the two reviewers as their support and suggestions greatly contributed to this chapter. This research was financed by The Netherlands Organization for Scientific Research division of Science for Global Development (NWO-WOTRO); grant number W 01.70.100.007. The research design, execution and publication are the initiative and responsibility of the author.

Notes

1 Surge valve is a surface irrigation technique that can be defined as the intermittent application of water to furrows through gated pipe systems through which the water applied can be regulated.
2 Glocalization is used here to define the use of globally marketed products and services in local markets and vice versa.
3 At present a typical tube-well in Guanajuato if perforated at a depth of 150–500 m (compared to 50 m in the 1960s), has a lift of 60–180 m and a 6 or 8' outlet pipe yielding 30–60 l/s.
4 Personal communication sub-secretary of SDAyR 18–03–2016.
5 Ejidatarios are official members of ejido, which is a communal land tenure system in which members are granted usufruct rights over agricultural parcels and/or common use areas (Assies, 2008).
6 Although legal restrictions exist on the total amount of land that one individual can own, the practice of 'name lenders' is widespread.
7 Between 2000 and 2005, he headed the national SAGARPA and in 2012 he was reappointed as head of SDAyR for the period 2012–2018.
8 Though acknowledged by the programme director, the data to back this statement was not provided.
9 All of these companies, as well as all other companies related to the agricultural sector, converge on a yearly basis at the 'Agro-expo Alimentaria' that is organized every year in Irapuato by one of the large agro-exporting families. This four-day event brings together hundreds of companies from around the world that aim to sell either their products or services to the thousands of visitors from Guanajuato and the rest of the country (see www.expoagrogto.com).

References

Assies, W. (2008). Land tenure and tenure regimes in Mexico: An overview. *Journal of Agrarian Change* 8(1): 33–63.
Ayala-Garay, A., Schwentesius Rindermann, R., and Carrera Chávez, B. (2012). Hortalizas en México: competitividad frente a EE.UU. y oportunidades de desarrollo. *Globalization, Competitiveness and Governability* 6(3): 70–88.

Boelens, R., and Vos, J. (2012). The danger of naturalizing water policy concepts: Water productivity and efficiency discourses from field irrigation to virtual water trade. *Agricultural Water Management* 108: 16–26.

Bivings, L., and Runsten, D. (1992). *Potential competitiveness of the Mexican processed vegetable and strawberry industries.* Report prepared for the Ministry of Agriculture, Fisheries and Food of British Columbia, Canada.

Carton De Grammont, H., and Lara Flores, S.M. (2010). Productive restructuring and 'standardization' in Mexican horticulture: Consequences for labour. *Journal of Agrarian Change* 10(2): 228–250.

CEAG (Comisión Estatal del Agua de Guanajuato). (2013). *El agua subterránea en Guanajuato (poster) Gobierno del Estado de Guanajuato.* Guanajuato: CEAG.

CEAG (Comisión Estatal del Agua de Guanajuato). (2014). *Balance de los Aquíferos de Guanajuato (PPT) Gobierno del Estado de Guanajuato.* Guanajuato: CEAG.

CEASG. (1999). *Plan Estatal Hidráulico de Guanajuato 2000–2025, Fase 1: Diagnóstico base de la situación hidráulica del Estado de Guanajuato.* Guanajuato City: CEASG.

CNA. (2010). *Estadísticas del agua en México 2010 Secretaría del Medio Ambiente y Recursos Naturales, México D.F COTAS Rio Laja, 2013. Síntesis del Acuífero Cuanca Alta del Rio Laja.* Dolores Hidalgo, Guanajuato: COTAS.

COTAS Rio Laja (2013). *Síntesis del Acuífero Cuanca Alta del Rio Laja.* Dolores Hidalgo, Guanajuato: COTAS.

Eakin, H., Appendini, K., Sweeney, S., and Perales, H. (2015). Correlates of maize land and livelihood change among maize farming households in Mexico. *World Development* 70: 78–91.

FAO-AQUASTAT. (2009). *Mexico: Summary fact sheet country fact sheets.* Electronic database. (Accessed June 23, 2011).

FAO-SAGARPA. (2002). *Evaluación de la Alianza para el Campo 2001, Informe de evaluación estatal tecnificación de riego.* SAGARPA, Guanajuato, Mexico, October.

Hewitt de Alcántara, C. (1978). *La modernización de la agricultura Mexicana, 1940–1970.* Mexico City: Siglo Veintiuno Editores.

Hoogesteger, J. (2004). *The underground: Understanding the failure of institutional responses to reduce groundwater exploitation in Guanajuato.* MSc Thesis, Wageningen University, Wageningen, The Netherlands.

Hoogesteger, J., and Wester, P. (2015). Intensive groundwater use and (in)equity: Processes and governance challenges. *Environmental Science and Policy* 51: 117–124.

INEGI. (2015). *Mexico en cifras: total estatal Guanajuato.* Available at www3.inegi.org.mx/sistemas/mexicocifras/default.aspx?e=11 (Accessed June 24, 2015).

Kooij, S. van der, Zwarteveen, M., Boesveld, H., and Kuper, M. (2013). The efficiency of drip irrigation unpacked. *Agricultural Water Management* 123: 103–110.

López-Morales, C., and Duchin, F. (2011). Policies and technologies for a sustainable use of water in Mexico: A scenario analysis. *Economic Systems Research* 23(4): 387–407.

Massink, G. (2016). *Accessing groundwater in a context of agrarian change: A case study on the infuence of Mexico's agrarian reforms on distributional patterns of groundwater among smallholder irrigators in an ejido in Central Mexico.* MSc Thesis Wageningen University, The Netherlands.

Molle, F., Venot, J.P., Lannerstad, M., and Hoogesteger, J. (2010). Villains or heroes? Farmers' adjustments to water scarcity. *Irrigation and Drainage* 59(4): 419–431.

SDA. (2000). *EVALUACION EXTERNA ALIANZA PARA EL CAMPO 2000: Desarrollo Productivo Sostenible en Zonas Rurales Marginadas.* Celaya, Mexico: SDA/SAGARPA.

SDAyR (Secretaría de Desarrollo Agrícola y Rural). (2014). *Presupuesto ejercido en el uso eficiente del agua en el estado de Guanajuato (excel sheet).* Celaya, Guanajuato.

SDAyR (Secretaría de Desarollo Agrícola y Rural). (2015). *Reglas de operación del Programa de Tecnificación del Riego con Agua Subterránea para el ejercicio 2016.* 31 de Diciembre 2015.

SDAyR (Secretaría de Desarollo Agrícola y Rural). (2016). *Exportaciones de Guanajuato 2014.* Excel sheet provided by SDAyR.

Seckler, D. (1996). *The new era of water resources management.* Colombo: International Irrigation Management Institute.

Steffen Riedeman, C., and Echánove Huacuja, F. (2003). *Efectos de las políticas de ajuste estructural en los productores de granos y hortalizas de Guanajuato.* Mexico City: Universidad Autónoma Metropolitana/Plaza y Valdes.

Swyngedouw, E. (2011). Interrogating post-democratization: Reclaiming egalitarian political spaces. *Political Geography* 30(7): 370–380.

Venot, J.P., Zwarteveen, M., Kuper, M., Boesveld, H., Bossenbroek, L., Kooij, S.V.D., Wanvoeke, J., Benouniche, M., Errahj, M., Fraiture, C.D., and Verma, S. (2014). Beyond the promises of technology: A review of the discourses and actors who make drip irrigation. *Irrigation and Drainage* 63(2): 186–194.

Wester, P. (2008). Shedding the Waters: Institutional Change and Water Control in the Lerma-Chapala Basin, Mexico. (PhD thesis). Wageningen University: Wageningen.

Wester, P., Hoogesteger, J., and Vincent, L. (2009). Local IWRM organizations for groundwater regulation: The experiences of the Aquifer Management Councils (COTAS) in Guanajuato, Mexico. *Natural Resources Forum* 33(1): 29–38.

Wester, P., Sandoval-Minero, R., and Hoogesteger, J. (2011). Assessment of the development of aquifer management councils (COTAS) for sustainable groundwater management in Guanajuato, Mexico. *Hydrogeology Journal* 19(4): 889–899.

Wilson, J., and Swyngedouw, E. (2014). Seeds of dystopia: Post-politics and the return of the political. In J. Wilson and E. Swyngedouw (Eds.), *The post-political and its discontents: spaces of depoliticisation, spectres of radical politics.* Edinburgh: University of Edinburgh Press, pp. 1–22.

World Bank. (2006). *Reengaging in agricultural water management: Challenges and options.* Washington, DC: World Bank.

Zwarteveen, M.Z., and Boelens, R. (2014). Defining, researching and struggling for water justice: Some conceptual building blocks for research and action. *Water International* 39(2): 143–158.

10 Collective drip irrigation projects between technological determinism and social construction

Some observations from Morocco

Mostafa Errahj and Jan Douwe van der Ploeg

Introduction

Agrarian dualism has dominated agricultural development discourse and policies in Morocco for more than a century, since the French protectorate first used it to distinguish 'modern', large-scale, often settler-based farms from the 'traditional' modes of agricultural of the Moroccan peasantry. Likewise in Latin America and Africa, 'agrarian dualism is an important feature of the agrarian political economy' (Pierri, 2013). This has remained the prominent discourse in Morocco despite the tremendous changes that have taken place: political independence, large-scale national projects like the 'Tillage programme' (Coz, 1961; Van Wersch, 1968), the agrarian reform of colonial farms (Bouderbala et al., 1977; Karrich, 1978), the development of large-scale irrigation schemes (Popp, 1978; Jouve, 1999; Errahj et al., 2006, 2009), and, more recently, the Green Morocco Plan (Akesbi, 2011, 2012). This latest plan distinguishes a modern export-oriented agriculture on the one hand and, on the other, 'social agriculture' intended to integrate the large peasant population. Nonetheless, the plan surprisingly announces that it 'eliminates the dualism and recognises the plurality of the Moroccan agricultures' (CGDA, 2014). Morocco is no exception when it comes to getting into and feeding a false debate on agrarian dualism: 'A false dualism lies at the heart of this debate. It sets smallholder and subsistence agriculture on one side against large-scale and commercial agriculture on the other' (Losch et al., 2012).

The different policies and programmes related to rural modernisation and agricultural development in Morocco after independence all focused on technical modernisation, and 'social treatment' as far as the 'traditional' agricultural sector was concerned (Le Coz, 1968; Kleich, 2001). Over this long period, the state organised these policies around impressive technical artefacts: the tractor in the 1957–1962 'tillage plan', fertilisers during 'Operation Fertilisers' in 1965–1973, and big dams and large-scale irrigation projects during the 'big dams policy' from 1967 to 1985. The state's aim was to legitimise its central/paternalist position and to control the redistribution of resources, while simultaneously maintaining strict control over the rural territories (Leveau, 1976). We argue that the recent programmes promoting water saving by subsidising

drip irrigation are a continuation of a dualistic approach to agricultural policies, organised around a technological artefact. The persistence of the dualistic
model in agricultural policies is closely linked to a deterministic view of development and change. This legitimises the focus on technological interventions,
thereby pushing the emancipatory needs and aspirations of the rural population
to the side.

In this chapter, our aim is to show that the dualistic paradigm hinders the
understanding of a far more complex rural society and simultaneously justifies
unequal distribution of resources. We critically discuss the dualistic discourse
and policies in Morocco as well as the technological determinism which often
accompanies such discourse. The chapter starts with a historical review of agricultural development in Morocco. It then briefly presents two case studies on
the diffusion and use of drip irrigation in collective projects (which promote
a collective shift from flood irrigation to water-saving irrigation systems) and
discusses findings regarding farmers' practices of drip irrigation in Morocco
more generally. The case studies and participatory observation in a diversity of
irrigated contexts show how peasants mastered the new irrigation technique,
among other techniques, through continuous learning processes. In so doing,
they were able to enhance their autonomy and resilience. Last, this chapter
critically discusses the diffusion of drip irrigation in more general terms by
placing it in its local and national contexts. This allows for a better understanding of the many social and the technical interactions which are decisive for the
diffusion, dynamics, impacts and hence for the success or failure of this socio-
technological package.

The historical construction of a fictional dualism

The book *European expansion and social change in Morocco* (Ennaji, 1996) is a
valuable reference in understanding the economic history of Morocco and
especially the transitions that are at the origins of its economic and political
dependence on Europe.

At the end of the 19th century, rural Morocco was weakened by the conjunc-
tion of three factors: (1) Subsistence crises resulting from cycles of drought and
human epidemics; (2) The penetration of credit relayed by traders connected
with merchant capitalism; (3) Abusive management of debts contracted by the
caïds[1] and their deviation for private use. According to the same author, the subsequent crisis in the rural areas resulted in land concentration which committed the country to a pre-capitalist production model. Lazarev (1976) described
this as a rent-seeking situation resulting from alliances between the commercial
bourgeoisie of Fez, as a political-economic elite, and the Makhzen.[2] From then
on, this historic turning point in Morocco would weigh on socio-economic
development policies.

Morocco became a French protectorate in 1912, and during the following
two decades, colonization focused on wheat production through the settle-
ment of colons on large farms. The 1929 economic crisis pushed the French

to reconsider this choice and deal with the lower agro-ecological potential of rain-fed farming in Morocco. In the 1930s, a second and more intensive colonization model inspired by the 'California Fruit Growers Exchange' agribusiness model was adopted and considerable investment in irrigation and fruit production was made by private colons (Swearingen, 1988). This type of agriculture was no longer referred to as colonial agriculture, but rather as the 'modern sector'.

In his paper on the history of the Moroccan rural modernisation, Pierre Marthelot (1961) used the term 'Modern Fellah' (*fellah* meaning cultivator) to describe the Moroccan peasant as a new category of farmer who would make rational use of the land and produce products for the market. According to the same author, this new category was 'created', because the previous selective and elitist funding politics had not succeeded in reaching the wider peasantry. Indeed, the capitalist colonial farming model was embedded as a strange entity, disconnected from the Moroccan peasantry. Apart from a few colonial family farms (especially with the development of irrigation in the 1930s), which can be considered as examples of an entrepreneurial farming model, the two worlds were antithetical.

The first attempts to reduce the gap between colonial and peasant agriculture were made in the 1940s and 1950s. On the one hand, through the implementation of the first large-scale irrigation scheme dedicated to indigenous farmers in the central Tadla region, where a 'centrally planned rural development structure accompanied farmers in "modernising" their farm holdings' (Kuper et al., 2012). On the other hand, an ambitious programme called Peasant Modernisation Sectors was implemented in the rain-fed areas. In these two projects, the support provided to farmers extended beyond agriculture to social and economic progress (education, health, community participation, etc.), following the famous slogan 'progress will be total or will not be' (Berque and Couleau, 1945, authors' translation). On the ground, the spirit of the slogan was expressed in two integrated actions (Swearingen, 1988):

- *Synthesis action*: modernisation should be linked to global, social and economic progress and the collectivization model was supposed to be the way to achieve community emancipation and restore equity and justice;
- *Shock action*: by introducing a technical artefact, in this case the tractor, and managing it collectively, the peasants would become empowered.

The colonial administration soon ended this emancipation process and 'brought down the level of peasantry ambitions to conform with colonial doctrine' (Oved, 1961). Resistance to the modernisation approach came from all protagonists, including the peasants, and was partly due to the design and implementation of the intervention itself: (1) the innovation diffusion hypothesis was not validated in practice; (2) the local communities involved in farming and, more broadly, located in the zone targeted by the development programme were not actively involved in the process, which decreased their trust in land restitution (and indeed, at the end of the protectorate, none of the land used in the Peasant Modernisation Sectors programme was returned to its owners; Swearingen, 1988), (3) the French settlers and the Moroccan rural nobility were scared by the (potentially) deep changes in the balance of power, the emergence of a 'subversive kolkhoz' movement and also of the risk of losing a low-cost labour force. Landlords helped put an end to these early attempts at general modernisation.

After independence, members of the rural nobility were co-opted by the 'Makhzen', and recruited by the territorial administration to control the rural areas (Leveau, 1976). This political choice blocked peasant's emancipation and converted rural communities into a stable support group of the monarchy. In so doing,

> the conformism of the rural areas was implemented at the cost of political conservatism and a renewed social control at the local level, which meant that all regional and national agricultural and rural development projects were evaluated from a security viewpoint, so that these projects would not question the social and agricultural stratification.
>
> (Desrues, 2006)

After independence, Morocco had to deal with serious food security issues which resulted from a reduction in agricultural production and the high rate of demographic growth. The central political power found itself at the crossroads between developing rain-fed agriculture, which applied to the largest area by far, or concentrating efforts on large-scale irrigation schemes, covering only a minor part of the area. Both options had already been experimented during the colonial period. To tilt the political power balance in its favour, the monarchy preferred immediate and startling results (Akesbi, 2006) and large-scale irrigation became the main priority and the strategic development option. The development of irrigation further widened the existing large gap between the modern and traditional sectors. The 'modern sector' continued to benefit from its privileged position in resource allocation (Belal and Agourram, 1971), and the real development questions related to the agrarian structures and rehabilitation of peasants were never put on the table: property rights and access to land, access to credit and appropriate extension services, and rural socio-educational infrastructure.

When the Moroccan nationalist movement, which had formed during the struggle for independence, called for a complete and profound agrarian reform, a political confrontation with the central power began. The monarchy focused on reinforcing its legitimacy by building alliances with the rural nobility (Leveau, 1976; Hammoudi, 1997; Desrues and Moyano, 2001). To protect the interests of the rural elite, the much-debated agrarian reform was abridged to a simple land distribution of 340,000 ha to 26,555 peasants (Bouderbala et al., 1977; Bouderbala, 2001; Jouve, 2002). The number of applicants was significantly higher than the number of beneficiaries who actually received land. For example, in the Gharb region, only one peasant out of five received land (Karrich, 1978), what is more, farmers who benefited from the land distribution did not obtain land titles. They only received land use rights and were clustered in state-led agrarian reform cooperatives. These cooperatives were given technical and financial support but little freedom to manage their own development pathways. The cooperatives were supposed to (1) ensure the full employment of family labour, (2) guarantee the income required to cover family needs, (3) generate agricultural development, and (4) help increase the national currency assets. Forty years after its implementation, the outcomes were ambivalent and in 2006 the state started a privatization process of the land used by the agrarian reform cooperatives to put an end to the collectivization experiment. Once more the general modernisation option failed, due to a technocratic and simplistic view of Moroccan agriculture.

In the 1990s, the structural adjustment process adopted by the Moroccan government led to a profound deregulation in the conventional role of the state. The lack of public funding combined with abrupt state disengagement produced an unprecedented situation. On the one hand, especially in the large-scale irrigation schemes, some farmers expressed their disappointment and their feeling of being abandoned by the state (Errahj et al., 2007). On the other hand, wealthy structured 'modern' farms took advantage of the situation to obtain access to land, water and credit, while building strong links to markets and value chains. More localised horizontal farmers' modes of coordination and successful experiments in collective action took place in many parts of the country, but most had no real connection with public policies (Errahj et al., 2009). The administration produced ambitious plans and prospective studies that were not implemented, thereby favouring a laissez-faire attitude.

In 2008, in the context of increasing prices for agricultural commodities at the international level, the Moroccan government adopted an ambitious plan called the Green Morocco Plan (*Plan Maroc Vert* in French – PMV) (Ait Kadi, 2012). With an expected public investment of US$15 billion up to 2020,[3] the plan aims to consolidate the role of the agricultural sector in the national economy and to act as a major catalyst for economic and social progress (Badraoui and Dahan, 2011). The plan suggests that by 2020, 1.15 million jobs will be created and the income of nearly 3 million people in rural areas will be tripled. The plan relies on two 'pillars'. The 'first pillar' centres on transforming the 'modern agricultural sector' into highly productive export-oriented agriculture specialised in high value products. Private investments supported by subsidies are supposed to play a key role here. The 'second pillar' aims to modernise traditional production. A positive social impact is a central objective. Public investments in social initiatives to combat rural poverty will play a key role.

A close examination of the philosophy behind the Green Morocco Plan, clearly reveals that public policy has not deviated from the view of agrarian dualism tainted with the same technological deterministic flavour that has prevailed thus far. It continues to focus on technical modernisation, massive investment and conspicuous preference for large-scale farms (Akesbi, 2012).

Early on, this recurring dualistic vision was criticised by academics (Belal and Agourram, 1971) as a simplistic view of Moroccan rural society which denies the complex social configuration of the rural population (Pascon, 1971). In our opinion, such a simplification runs counter to the multiple hybridisation of farming production models and their ongoing co-evolution (van der Ploeg, 2008, 2010). When politicians and technicians engage in such simplifications, the range of possible agricultural development pathways is inevitably narrowed down. In particular, the opportunities for the emancipation of the peasantry might get blocked.

The desire to act fast and to reach ambitious numerical targets once again ignores real local participation and empowerment. It is also characterised by massive procedural projects in which the modernisation discourse and a communication focused on 'marketing' the plan prevents raising any questions about rural realities, which generally appear to belong to a world apart. For instance, analyses of some projects implemented as part of the second pillar of the Green Morocco Plan revealed weak participation by farmers and a serious lack of learning (Mkadmi, 2011; Faysse et al., 2014).

Water saving is an important transversal part of the PMV through the PNNEI whose aim is to 'protect water resources and improve the living conditions of rural populations' through the conversion of surface irrigation to drip irrigation on nearly 550,000 ha toward 2020, with an investment of US$4.5 billion (ABD, 2009). In this context, drip irrigation is framed as a corollary to modern agriculture, and seen as a miraculous way to achieve prosperity. The programme (1) frames a particular technological solution (drip irrigation) as a comprehensive solution to the multifaceted rural problems, and highlights the need to (2) support small-scale farmers through public investment and (3) to promote agribusiness through public-private partnerships, which are, in turn, seen as levers for further development. Interestingly, the programme designed to support large-scale and small-scale farmers is again approached in a dualistic way. Large-scale farmers are provided with individual subsidies. Small-scale farmers are supposed to collectively engage in state-led drip irrigation projects, as they are believed not to have the necessary resources to equip their farms individually.

Moroccan agriculture between tradition and modernity: when rural development is blinded by technological artefacts

Modernisation and modernity are two terms used erroneously in most political plans in contemporary Moroccan political history. In Arabic, the words 'Asrana' or 'Tahdith' (modern can be also mean contemporary) are informally used to describe state-driven programmes: big dams, fertilisers, farm machinery, drip irrigation, etc. Balandier clearly expresses these plans constitute an endless flight driven by a massive economic process of capitalist accumulation (Balandier, 1985). Since the protectorate period, Morocco has not moved away from this

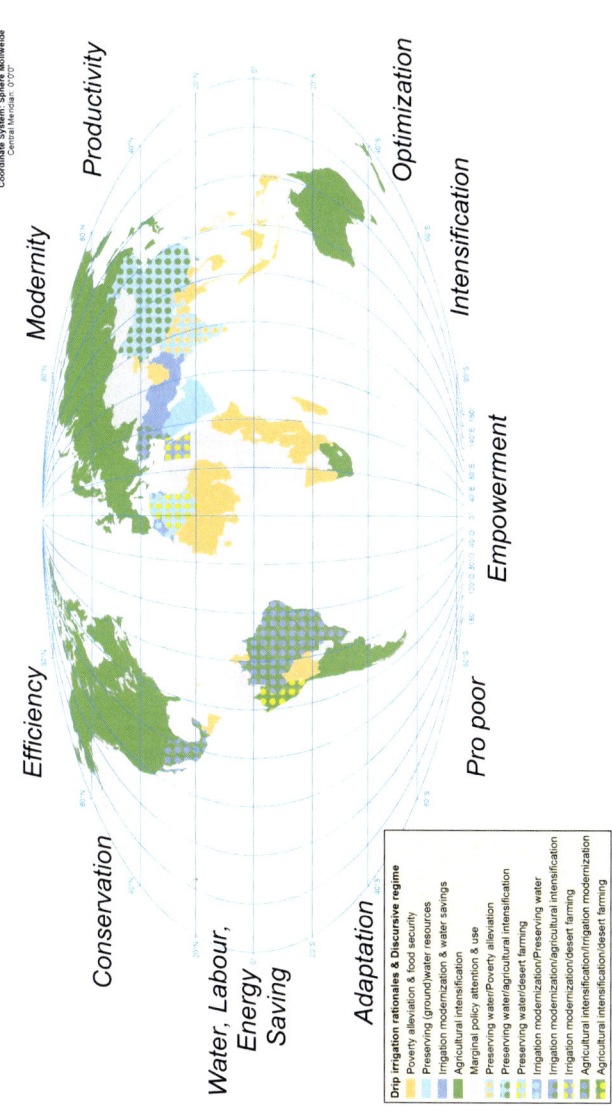

Plate 1 **A worldwide view of drip irrigation.** The map is a simplification of a complex reality; it only represents the main discursive framings of drip irrigation at country level. For each country only one or two main discourses are identified, even though other ideas of drip irrigation may also exist. Within a country, different discourses may play out differently across regions. Finally, the map may give the impression that different discursive regimes are neatly bounded; this is not the case as different discourses and imageries overlap and reinforce each others (see Chapter 1).

Source: Jean–Philippe Venot.

Plate 2 **Unclogging the drip lines.** When operating drip irrigation in the field, farmers often face practical problems. One of the most common ones is the clogging of drip tapes (see Chapter 3).

Source: Harm Boesveld, Morocco.

Plate 3 **Rolling the drip lines.** In annual crops like vegetables, there is a need for land preparation, planting, sowing, etc. The presence of a network of drip lines hampers these activities, while moving drip lines too often may damage them. Farmers have developed innovative simple devices to facilitate the frequent installation and removal of drip lines (see Chapter 3).

Source: Harm Boesveld, Morocco.

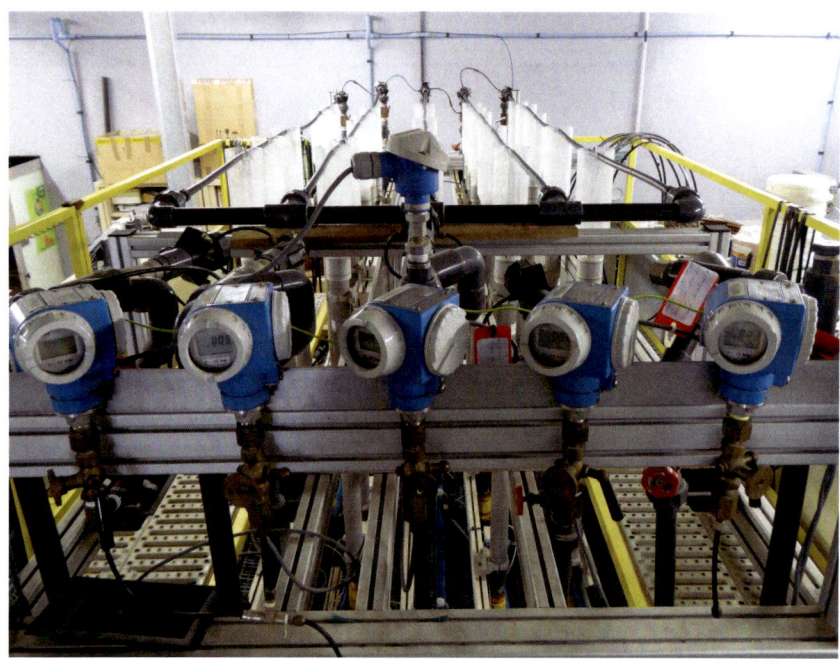

Plate 4 **Testing of drip lines in a certified laboratory.** Drip irrigation is continuously tested in certified laboratories to establish whether new equipment meets the specifications indicated by the manufacturers (see Chapters 1 and 2). There are, surprisingly, few measurements done in real-life conditions to measure how drip irrigation performs (see Chapter 5).

Source: Jean-Philippe Venot, France.

Plate 5 **A typical drip irrigation system linked to groundwater.** Farmers find ground-water attractive as a resource for drip irrigation. Obtaining access to groundwater enables farmers to escape a world of water scarcity and water sharing in collective irrigation systems, for a world of plenty with a perceived abundance of water, at least in the short run. Flexible use of groundwater allows intensification and diversification of cropping systems, and strengthens farmers' economic conditions. This explains why water consumption often increases under drip irrigation (see Chapter 5).

Source: Younes Bekkar, Morocco.

Plate 6 **Drip irrigation and Agrarian change.** This barrack once was the home of Mona and her son. After the land was sold to an entrepreneur, they were forced to leave it. Their former house became a shed to store soluble fertilizers and chemical products required for intensive agriculture under drip irrigation (see Chapter 6).

Source: Lisa Bossenbroek, Morocco.

Plate 7 **Different generations on a farm equipped with drip irrigation.** Drip irrigation is not only about water saving or increasing productivity. It also makes farming
more attractive, especially to young farmers interested in a 'modern' and 'cleaner'
agriculture (see Chapter 6).

Source: Saskia van der Kooij, Morocco.

Plate 8 **Expanding agricultural areas on hill slopes with the help of drip irrigation.** Pressurized drip irrigation makes it possible to farm on hill slopes, thus expanding the irrigated area (see Chapter 7).

Source: Daniela Henriquez, Chile.

Plate 9 **Conquering the desert.** Since the early experiences in Israel, drip irrigation is associated with ideals of 'making the desert bloom'. On the arid coast of Peru, drip irrigation did indeed make it possible to farm in the desert (see Chapter 8).

Source: Anaïs Marshall, Peru.

Plate 10 **Cultivation of asparagus on large-scale farms in Peru.** Vast stretches of land have been brought under cultivation by channelling water over long distances to commercial firms, growing fruits and vegetables, in this case asparagus (see Chapter 8).

Source: Anaïs Marshall, Peru.

Plate 11 **Mechanized drip line installation in Mexico.** Drip irrigation is in some places equated to an elite technology, for instance on the highly mechanized commercial farms in Guanajuato. The seasonal crops require the frequent installation and removal of drip lines (see Chapter 9).

Author: Jaime Hoogesteeger, Mexico.

Plate 12 **Smallholders prefer having access to system components of the drip irrigation system, Nepal.** Drip kits provide a quick and easy demonstration of how
a drip system works, but in the long run smallholders in India, Nepal and Myanmar
preferred access to drip line and system components over kits both to reduce costs
and to better fit their irregular field sizes and shapes (see Chapters 11 and 13).

Author: Bob Yoder, Nepal.

Plate 13 **Low-cost drip irrigation is not always in use.** In Burkina Faso, many development projects provide low-cost drip irrigation to groups of small-scale farmers,
men or women, as part of a larger package, including a well, seeds and training. Drip
irrigation is installed, but often not used (see Chapter 13).

Author: Jonas Wanvoeke, Burkina Faso

Plate 14 **Local Fabrication of a filter system.** In the rich agricultural plain of the Saïss in Morocco, local mechanics have started manufacturing filter systems for drip irrigation, copying and improving from imported versions. These are in high demand, as they are better adapted to local conditions (see Chapter 15).

Source: Maya Benouniche, Morocco.

Plate 15 **Drip irrigation in greenhouses in Algeria's Sahara.** In the district of Biskra, more than 100,000 greenhouses produce tomatoes and peppers for the national market. Every greenhouse is irrigated separately and has its own, locally fabricated, fertigation unit (see Chapter 16).

Source: Catherine Rollin, Algeria.

notion of development and change, which unfortunately does not really help the country to extract itself from the serious problems in which it is mired, rural poverty, vulnerability, and exclusion.

The dualist prism that has dominated agricultural development discourses entails technological determinism, just as it critically depends on this type of determinism. In this view, only one sector is assumed to meets the requirements of technological change: the modern sector. The other sector typically lacks what is required for a full-scale adoption of new technology packages. This is due to the small size of farms, the traditional attitude of the smallholders, their poverty and associated lack of savings, which preclude any substantial investment. In turn, this requires a dual policy.

This policy typically excludes dealing with sociotechnical and political questions. Instead, it focuses on technical artefacts far more than on people and processes. The norms and rules are supposed to be exclusively technical, and agrarian change is basically perceived as a technological transformation. As opposed to such a view, the paradigm in which technology is a social construction (Callon and Rip, 1992) clearly shows that sociotechnical norms are simultaneously negotiated between knowledge, social actors and procedures. In Moroccan agricultural history, these dynamic interactions, which inevitably produce controversies and can upset the established order, have always been ignored. Since the protectorate period, the administration has preferred fast track technical change. When in 1945, Berque (civil controller under the colonial administration) and Couleau (tax inspector) tried to promote in-depth and sustainable rural dynamics, they wanted to stop the tractor being seen only from a narrow technical point of view. Putting the 'local community on the tractor' would 'make it possible to bring about a psychological shock through the machine and to compensate the peasant for the ancestral upheaval by almost immediately increasing his power of production and his standard of living' (Berque and Couleau, 1945; authors' translation). This particular development project failed but its political and academic lessons are still valid today.

Other impressive state-driven technological projects took place: fertilizers, big dams, the agrarian reform, but none of them was explicitly managed in an integrated way. Neither the social actors' identities (in our case, the peasants) nor their knowledge were clearly valorised. Whilst the willingness of peasants to improve their livelihoods could have been a major driving force in and for agricultural growth and rural development, this potential was largely ignored. The same applies to their knowledge: it could have played an instrumental role in constructing well-functioning technologies, interventions and programmes that were adapted to local possibilities and constraints. But here the same applied: policy preparation and implementation rode rough shod over such a possibility. Consequently, for long periods, the Moroccan peasantry struggled with this incoherent approach to rural development and had to demonstrate, time and again, its ability to adapt to technical and institutional opportunities and constraints. This adaptation ability is demonstrated, for instance, in the recent National Irrigation Water Saving Program.

The centrality of social and the technical interactions

The National Irrigation Water Saving Program (PNEEI) which started in 2007 with the ambitious objective of converting 550,000 ha from surface irrigation to drip irrigation led to widespread dissemination of drip irrigation. Huge subsidies and massive interventions by the Moroccan administration and international programmes and agencies (FAO, JICA, USAID, IFAD) focused on individual and collective projects in different agro-ecological and institutional contexts with the leitmotif of 'modernising irrigation'. Collective drip irrigation projects, largely state led, were meant to benefit small famers and cover 220,000 ha (Box 10.1 describes an example of such project, in the Tadla irrigation scheme).

Box 10.1   Introducing drip irrigation in a conjunctive use environment

The 100,000 ha Tadla irrigation scheme in central Morocco was originally designed to be irrigated exclusively through surface water. Due to deep percolation of irrigation applications, water tables went up and became accessible to farmers who began tapping groundwater for agriculture in the 1980s in a context of prolonged droughts, increased irrigation water demand and gradual State withdrawal. Progressively, a conjunctive use environment emerged. By 2010, it was estimated that 70% of the irrigated area was served jointly by surface water from dams and groundwater from more than 8,300 private wells and tube-wells (Kuper et al., 2012). Groundwater provided autonomy from 'State' water and offered flexibility to farmers to deliver water to the crops whenever required. Farmers diversified and intensified cropping systems (fruit trees, horticulture, alfalfa) and invested in livestock. However, surface water remained important, in particular for the 54% of farmers (mainly smallholders) who did not have access to groundwater, but also to continuously recharge the phreatic aquifer that could then be used for irrigation.

This conjunctive use model illustrates perfectly David Seckler's distinction between 'dry' and 'wet' savings (Seckler, 1996). While at first glance surface irrigation can be labelled as 'wasteful' due to an estimated 50–60% irrigation efficiency, the use of groundwater from the phreatic aquifer recharged by the 'losses' of surface irrigation increases considerably the irrigation efficiency. An FAO report of 2014 indicated that, at the scale of the Oum-er-Rbia basin, which integrates the Tadla irrigation scheme, the overall water efficiency was 91% (FAO, 2014). This figure is remarkably close to the theoretical efficiency of drip irrigation. It was, therefore, somewhat of a surprise when the irrigation administration designed an ambitious 88,700 ha drip irrigation project in 2008 to convert a large part of the Tadla irrigation scheme from gravity to pressurized irrigation. The idea was to increase water productivity and to reduce the pressure on groundwater resources, as farmers had started to go beyond the phreatic aquifer to exploit the deeper confined aquifers. [Some] 39,700 ha were to be converted to drip irrigation by providing subsidies to individual farmers upon request (80–100% of the investment cost), and 49,000 ha through collective state-led projects mainly targeting smallholders who would not – it was thought – engage with drip irrigation individually. The first collective project, partly funded by international donors, is operational since early 2015 and 10,235 ha will be irrigated under (collective) drip irrigation shortly.

Different arguments plead in favour of this collective project. First, its main protagonists advance the water saving argument of drip irrigation, leading to increased irrigation efficiency. However, as shown above, this argument can be countered by the fact that the conjunctive use environment enabled comparable irrigation efficiencies. It is also expected that farmers will decrease or even cease groundwater use. Second, it is projected that drip irrigation will enable farmers to increase water productivity by at the same time increasing crop yields and revenues (shift to high-value crops) and decreasing the irrigation applications, due to more precise irrigation and improved irrigation service. This is favourable for agricultural development, but this may also increase the water demand in the long run due to more intensive agriculture (the so-called rebound effect). Third, the topography of the irrigation scheme means that water can be pressurized free of charge, without further energy requirements, at least for the first collective project.

On the other hand, the project dismantles a relatively robust conjunctive use environment, where irrigation responsibilities were shared between the irrigation administration distributing surface water, and farmers tapping groundwater whenever needed. The administration takes on more responsibilities in the new context, as all water is to be provided through the state-managed drip irrigation system. Also, possible breakdowns of the system need to be repaired more quickly than in the case of gravity irrigation, as farmers require water more frequently under drip irrigation. Second, the cost of the state-financed project with the help of international donors is substantial (more than 9,000 €/ha). Many farmers in the Tadla already converted to drip irrigation individually, encouraged by generous subsidies. It is not sure that the remaining farmers, who did not install drip irrigation individually, will be very motivated to 1) shift to drip irrigation, thus questioning the potential impact of this costly project, and 2) abandon their tube-well, as long as they are not sure that the irrigation service provided to them by the irrigation administration will be as good as the one they derive from their tube-well. Indeed, we observed that most farmers connected, on their own initiative, their tube-wells to the drip irrigation system, provided to them by the State. The risk is that farmers will continue to use both surface and groundwater, thus pursuing the conjunctive use model they are familiar with. However, this model is under threat, as the phreatic aquifer will be much less replenished by the piped drip system. This risk could be mitigated by improved irrigation service, inciting farmers to decrease groundwater use.

Marcel Kuper and Ali Hammani

To accelerate the dissemination process, arguments regarding the water savings (30% to 50% at the field level) and increased yields (10% to 100%) made possible by drip irrigation were added to a highly incentive subsidy rate that could even reach 100% of the cost of the equipment. By 2013, the areas equipped with 'modern techniques' for water saving (including drip irrigation) had jumped from the initial 11% to 24 % of the entire irrigated area (Akesbi, 2014). In 2014, according to the Ministry of Agriculture, drip irrigation accounted for 410,000 ha. These aggregated data conceal the distribution between individual and collective projects. Our own local observations and several case studies demonstrated persistent difficulties in implementing the collective projects (see, for instance, Zeine et al., 2015). In particular, scholars highlighted that several key elements were initially missing in state interventions aiming at a collective shift toward drip irrigation, which explained why such interventions faced persistent difficulties: engagement with specific user groups, an underestimation of the active role to be played by peasants themselves, and the absence of consideration given to local leaders, etc. Based on these observations, Zeine and Ouazzani (2015) categorized collective projects in three types: (1) farmer initiated projects; (2) projects initiated by local leaders, and (3) projects initiated by public administrations, international donor agencies, sometimes associating private agribusiness companies. Through their cross case analysis, these authors also identified three main factors required for the success of such projects. They qualified these factors as (1) essential: the existence, among the irrigators, of a competent and credible group who understands the potential benefits of the project and the need to construct it collectively; (2) stabilizing: the capacity for reinforcement and the potential for learning; and (3) accelerating: community organisation within a clear legal framework, with funding opportunities, and technical support. These 'success factors' highlight the centrality of the actors, notably the farmers: it is the social and the technical interaction (and mutual adaptation) that is decisive in processes of agricultural change. If this is not properly understood, frictions and difficulties will arise and the considerable potential for growth and development may remain untapped.

We build on these observations to describe two collective drip irrigation projects, one led by local leaders, the other led by an outside agency, implemented

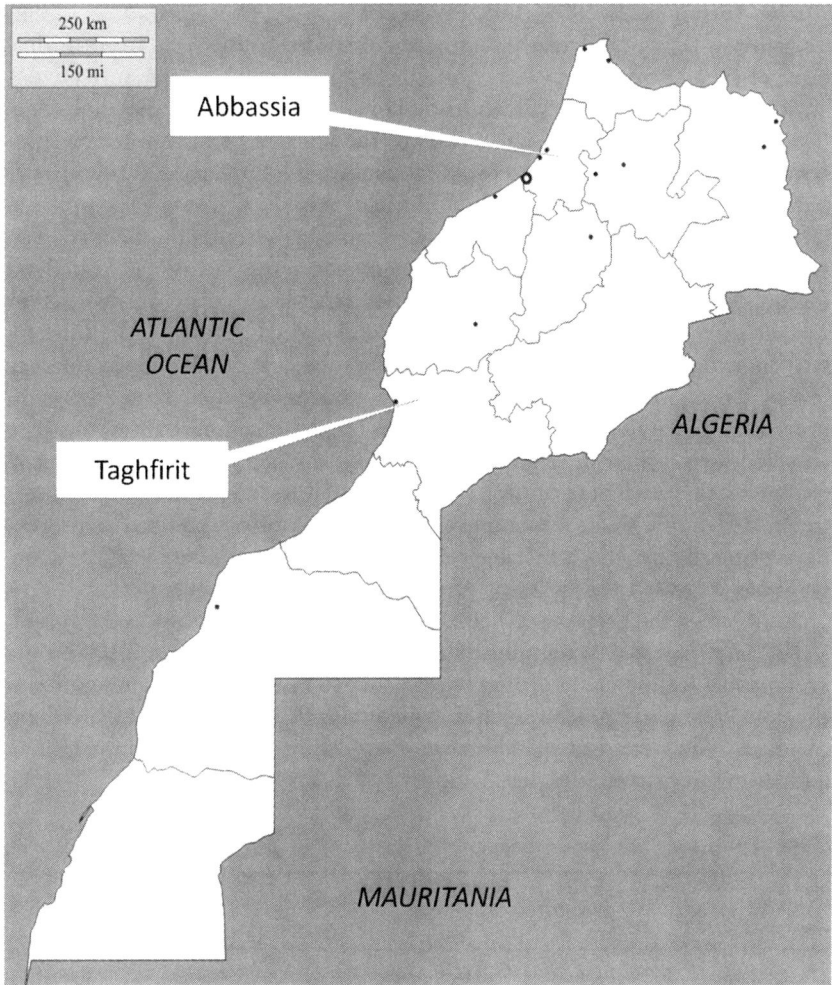

Figure 10.1 Map showing the location of the case studies.

in two contrasting regions of Morocco (see Figure 10.1). We also draw from firsthand experience of engaging with farmers in different irrigation contexts in the Moroccan countries The case studies and our participatory observations illustrate (1) the diversity in the use of drip irrigation and (2) the many-sided interactions between the state and the irrigators.

Abbassia in the Gharb: technocracy and state domination

The state-managed Gharb irrigation scheme, located in northwest Morocco, is the largest scheme in the country and has recurrent water service crises. The irrigation service has long been undermined by technical and institutional difficulties (Errahj et al., 2007, 2009; Faysse et al., 2010). The area under sprinkler irrigation is facing particularly acute difficulties due to the need for coordination between farmers, high water cost due to energy requirements and the degradation of the equipment, which has led to a spectacular free riding behaviour (Errahj et al., 2006) and in some cases to the collapse of the entire irrigation system. The Abbassia agrarian reform cooperative is located in an area with such difficulties, associated with a high level of farmers' debt toward a sugar manufacturing company (formerly state-owned, now privatized) and the water service administration. To release the cooperative from this vicious circle, in 2003, the public administration tried to convince farmers to collectively shift from sprinkler irrigation to drip irrigation (Benouniche et al., 2011). At the time, the farmers would have had to cover 70% of the total cost of the conversion; their fear of getting more deeply into debt did not encourage them to adhere to the project. In 2005, the subsidy was raised to 60%; the administration's proposal attracted the business interests of a sugar company, and a project including 25 farmers with a total area of 68 ha was submitted to the farmers' organization (see table below). What was commonly called a 'win-win contract' appears to have primarily benefited the sugar company: thanks to the project, the company has increased the quantity of sugarcane it processes. Farmers were 'put forward' when the project was being designed so as to get access to the subsidy (60%). They also had to reimburse the bank, and the farmers' organization was accountable for the credit farmers contracted to pay for the remaining 40% of the costs. The sugar-processing industry ensured that most of the financial and technical risks were shouldered by the state (drip irrigation subsidies), farmers and farmers' organisations' (See Table 10.1).

Table 10.1 Sharing of responsibilities for the Gharb drip irrigation project

Main actor responsible for:		*Farmers and their organisations*	*State*	*Sugar processing company*
Designing	Project specifications, design, monitoring and final technical acceptation of the project;	X	X	
Funding and implementation	Obtaining the 60% subsidy	X	X	
	Technical supervision			X
	Seeking additional credits from banks;	X	X	
Running the project	Supply of inputs for the sugar cane			X
	Technical support for sugar cane production and irrigation;	X	X	X
	Growing sugar cane on 60% of land;	X		

Main actor responsible for:		Farmers and their organisations	State	Sugar processing company
Reimbursement	Transferring the entire state grant to the Bank;	X		
	Recovering credits on revenues for four years	X		X
	Guaranteeing the credit	X		

Source: The authors (adapted from Benouniche et al., 2011).

The new modern irrigation system was implemented in 2007, according to international level standards. In the first irrigation season, the farmers adapted the system to meet their needs and local conditions. In particular, they adjusted the irrigation practices, as the built-in water turns did not suit them. Progressive adaptation through iterative adjustments and knowledge transfer can be a response to economic and technical difficulties or seeking way to improve one's social status (Benouniche et al., 2016), but in this case the farmers were obliged to pay for expensive options they never used. For four years, the sugar company collected the loans repayments directly from the farmers for the banks as well as water fees for the public irrigation agency. Although the water bill decreased significantly (from US$700/ha to US$300/ha), farmers' debt increased to US$2000/farmer due to the fact they shouldered 40% of the investment cost (Boularbah, 2014). The farmers' efforts to adapt the functioning of the irrigation system did not lead to sustained appropriation of the project for two reasons: 1) the farmers did not master this top-down turnkey project which had been implemented by an irrigation company that did not report to them, but instead to the agricultural administration and the sugar-processing company, 2) the adaptations made by farmers did not reduce the initial high cost of the project which had been implemented in one go. As a result, the project was abandoned by the farmers who simply stopped irrigating.

Taghfirit in the Souss (collective action and community development)

The Souss region is one of the most important agricultural regions in Morocco. The Souss is located in southern Morocco, where intensification through irrigation has taken place for a long time. After an early water crisis in the 1970s (Popp, 1986), farmers became highly sensitive to groundwater availability and degradation (Bekkar et al., 2009) and also showed significant aptitude for coordination and collective work (Errahj et al., 2009).

Taghfirit and Ouled Issa are two villages where, in the past, when surface water from the River Souss was easily available, small farmers grew cereals, legumes and citrus. By the 1980s, water scarcity had increased dramatically, and in the early 1990s, the state set up a large-scale irrigation project with two tube-wells, collectively managed by famers. The valuable traditional irrigation heritage (Popp, 1986) and previous experience in collective action simplified project implementation and the water users associations (WUA) played a key role in managing the water. In contrast to other WUAs created elsewhere in Morocco which were mostly inactive, the WUA in the two villages also started other projects aiming at local rural development (drinking water facilities, dairy cooperative, etc.).

At the start of the project, the area to be irrigated by the tube-wells (the 'project area') had been identified by farmers themselves. The real irrigated area in the villages was larger. Indeed, the famers, who were used to avoiding state control and statistics, actually underestimated the area on purpose. In this context, the tube-wells were under-designed and, for years, the Water Users Association (WUA) suffered from structural water scarcity, which increased further with the droughts of the early 2000s. In the 2000s, the water flow was nearly 50% of what it had been in the early 1990s (from 210 to 120 l/s according to the president of the WUA) and the farmers decided to collectively shift to water saving techniques. They chose drip irrigation because they thought it would (1) make it possible to increase the extent of their irrigated land using the same water flow; (2) increase agricultural productivity; (3) reduce work discomfort; (4) reduce management costs, and (5) save fertilizers.

To achieve these goals, the WUA designed a drip irrigation project, which they initially intended to implement in one shot. They negotiated three agreements separately and progressively, one with the river basin authority, one with the Ministry of Agriculture, and one with a drip installation company. First, the farmers participated by purchasing the land where collective equipment would be installed and obtained a financial contribution from the river basin authority, which covered 20% of the total cost of the project. They built a storage basin and the head station. When subsequently negotiating with the Ministry of Agriculture for public subsidies from the water saving programme, and evaluating their financial situation, the farmers realised that they would not be able to implement the whole project at one go as originally planned, because they would have to advance the financial costs of the investments. They consequently decided to divide the project into three stages and to implement them one by one. For each new stage they implemented, the investment cost per ha was lower because of the collective technical and managerial knowledge acquired in the meantime and also because the cost of the drip irrigation equipment had gone down. During this process, the famers were actively involved in technical choices and also in shaping the new water management rules. Moreover, and in contrast to the collective Abbassia project, they managed the process themselves, slowing it down or accelerating it as required, in particular when they decided to carry out the project in three stages. This clearly demonstrates the importance of putting the peasants on the central stage when it comes to designing a particular project and its implementation. Peasants' local knowledge is often decisive so as to have an appropriate design and their motivation to move ahead becomes a major driving force for projects that might otherwise fall flat.

A new panorama

Drip irrigation is now widely practiced in Morocco. Together with the rapid expansion of new export crops, and the use of deep tube-wells which exploit aquifers as deep as 200 meters below the surface, drip irrigation has changed the appearance of the Moroccan countryside. However, the large-scale use of drip irrigation cannot be simply understood as the outcome of a linear diffusion process. It is not the simple result of push-factors (subsidies, extension, etc.) supported by the state. In this process, the peasants and other local actors played an active if not decisive role. While the state-led technology promotion inspired the Taghfirit WUA, it was only when the community took charge of the process that peasants in the area adopted drip irrigation. The widespread adoption of drip irrigation is a process of social-technical change in which the interaction between the social and the technical has been decisive.

Beyond the two case studies described above, drip irrigation is now widespread in the Moroccan countryside. This has happened because Moroccan peasants found many innovative ways, often differing from state supported practices, to engage with the technology. Notably, they quickly found ways to run the pumps required for drip irrigation using butane, which is widely available in rural Morocco but is primarily intended for domestic use (mainly cooking). Another important technical adaptation was the development of far cheaper filters than the imported one that were partly subsidised by the state. The filters were developed in collaboration with local craftsmen and small enterprises (Benouniche et al., 2016). The newly constructed filters and the use of butane resulted in a sharp reduction in costs: the price of a drip irrigation system went down from around 2,000 Euros per hectare to only 500 Euros per hectare. At the same time, new maintenance and repair services were established. Without these, the spread of drip irrigation would probably have been far less impressive, especially for peasants for whom it would have remained out of (financial) reach (see also Chapter 17).

Important 'contextual' and 'technical' adaptations were also made at farm level. Examples of the latter are fine-tuning and making drip irrigation 'mobile'. Fine-tuning means that instead of being constant, the water flow is continually increased or decreased according to the needs of the growing plants. This is made possible by using small valves which open and close part of the irrigation system network. More generally, 'mobile' drip irrigation (widely used in onion cultivation in the Saïss region; Benouniche et al., 2016) enables a combination of surface irrigation (at the start of the season) and drip irrigation (during the vegetative stage), while coping with 'insecure' land tenancy for farmers who rent land and may have to move plots from one year to another. 'Contextual' adaptations were mainly in the form of changes in cropping systems with a shift toward high-yielding agricultural products whose value compensates for the high investment costs. However, this also increased the farmers' vulnerability to price volatility and possible loss of income.

Equally, if not even more important are the social changes that go together with widespread use of drip irrigation. As subtle as they might appear at first sight, their consequences are far-reaching. For instance, Bossenbroek (2016), describes how young men are looking for new identities to distinguish themselves from the previous generation without being too threatening toward their fathers. They are doing so by actively using drip irrigation (see the box on the topic in Chapter 6, this volume). In such instances, drip irrigation is no longer simply a device to rationalize agricultural production; instead, and above all, it is a device which helps to subtly re-order social hierarchies. In this case, the search for emancipation (as limited as it may appear) is a major social driver of technological change.

In terms of social interaction, an interesting and telling pattern characterised the diffusion of drip irrigation. It started in large-scale commercial farms, where drip irrigation was incorporated for three reasons: (1) it helped private entrepreneurs demonstrate their superiority; (2) it reduced their dependence on day labourers (and thus reduced the danger of strikes at critical periods); (3) the large state subsidies that were provided. Next, drip irrigation started to be adopted by and adapted for small-scale peasant agriculture, which led to new technical designs, which, in turn were picked up by large-scale farms. The widespread adoption of drip irrigation in the small-scale sector occurred mainly without subsidies.

The diffusion of drip irrigation has been successful – at least so far. But will it prove to be successful in the longer term? Here again the social and technical interactions will be decisive – be it at a different scale. Drip irrigation is being justified – at national scale – as a water saving technology (see Chapter 1 and 5, this volume). In the longer run, this might turn out to be only partly true or not true at all. Whilst it certainly does save water during transport (no leakages or evaporation) and limits water use to the immediate surroundings of the plant, it is quite possible that the cultivation of water demanding crops will expand. This might result in the use of even more water. The fact that in practice, the application and further diffusion of drip irrigation involves creating more access points to new, deep aquifers is, in this respect highly informative. As a matter of fact, the spread of drip irrigation and the associated construction of many deep tube-wells is starting an invisible, but no less real, subterranean war for water. In this 'war', peasants with standard shallow wells, which are only around 50 deep, are ending up as losers, whilst those who dispose of tube-wells will, in the end, be 'winners'. In such a scenario, drip irrigation appears to be one of the 'weapons'. Instead of being a device to save water, it is a tool being used in the multi-actor scramble for water.

Conclusion

Our results reveal two contrasting examples of technological change and rural development. The Gharb project illustrates a deterministic understanding of technical change in agriculture, and reflects a dualistic view and policy of agricultural development. It clearly failed to provide opportunities for small farmers. On the contrary, the collective project in the Souss illustrates the complex realities of Moroccan farming where modernity and traditional ways are closely intertwined. The dynamics at play in this case are eye opening; they allow moving away from the largely artificial dual understanding of the Moroccan agriculture that has dominated the debate to date and, as such, offer a way of envisioning empowering and sustainable agricultural development.

To better understand the processes of innovation in (drip) irrigation, and to create the conditions for peasant emancipation, it is not only the design and implementation of current drip irrigation projects that should be called into question. What requires attention is the view and policies in which these processes are embedded. The latter are grounded in a technological deterministic paradigm and have framed Moroccan agriculture as being dual, whereby modernity and tradition are opposed; the former being equated with technological change and the later with the impossibility or refusal to evolve.

We argue that the reality of rural Morocco is more complex, and agrarian structures better understood as a result of complex interactions between social and technical dimensions. Drip irrigation, for instance, is not an isolated artefact. It is part of, and tightly interwoven with, a wider constellation of actors and objects and it is through those interactions that it reveals its nature and impacts, which might very well be different from those that are attributed to it by promoters who still adhere to a technological deterministic (hence, bureaucratic) paradigm.

To predict possible interactions, to make use of the positive ones and to correct the negative ones, a paradigm that understands technology as a social construct is needed. Such paradigm puts forth that both 'the path of innovation and the consequences of technology for humans are strongly if not entirely shaped by society itself' (Botha, 2013). Consequently, it theoretically opens the possibility of a multi-track process of development that has no previously established hierarchy. Many real development opportunities that remain concealed by a dual vision of the Moroccan countryside (and the underlying simplistic understanding of technological change) would, then, be opened up. This is urgently needed but this diversity cannot be captured through national typology and characterization indicators. Only territorial and local analyses can help politicians and extension agents to best target and accompany agricultural modernisation.

Acknowledgments

The authors would like to acknowledge the comments of the external reviewers, which helped improving the flow of their argument.

Notes

1   District officer.
2   'The administrative structure, legal framework and military manpower to extend Moroccan sultans' authority over self-governing tribes' (Maghraoui, 2001).
3   Ministry of Finance and economy (www.finances.gov.ma/fr/pages/strat%C3%A9gies/strat%C3%A9gie-de-d%C3%A9veloppement-agricole--le-plan-maroc-vert.aspx?m=Investisseur&m2=Investissement).

References

African Development Bank (ADB). (2009). *National irrigation water saving program support.* Project Appraisal Report. Tunis: ADB.

Ait Kadi, M. (2012). *Agriculture 2030: A future for Morocco.* The Futures of Agriculture. Rome: Global Forum on Agricultural Research (GFAR).

Akesbi, N. (2006). *Evolution et perspectives de l'agriculture marocaine.* Rapport Thématique, Cinquante ans de Développement Humain au Maroc, pp. 85–198.

Akesbi, N. (2011). La nouvelle stratégie agricole du Maroc annonce-t-elle l'insécurité alimentaire du pays? *Confluences Méditerranée* 78(3): 93–105.

Akesbi, N. (2012). Une nouvelle stratégie pour l'agriculture marocaine: le Plan Maroc Vert. *New Medit* 11(2): 12–23.

Akesbi, N. (2014). *Le Maghreb face aux nouveaux enjeux mondiaux. Les investissements verts dans l'agriculture au Maroc.* Paris: Note de l'Ifri. Ifri.

Badraoui, M., and Dahan, R. (2011). The Green Morocco Plan in relation to food security and climate change. In M. Solh and M.C. Saxena (Eds.), *Food security and climate change in dry areas. Proceedings of an international conference,* February 1–4, 2010, Amman, Jordan. Amman: ICARDA, pp. 61–70.

Balandier, G. (1985). *Le détour: pouvoir et modernité.* Paris: Fayard.

Bekkar, Y., Kuper, M., Errahj, M., Faysse, N., and Gafsi, M. (2009). On the difficulties of managing an invisible resource: Farmers' strategies and perceptions on groundwater use, field evidence from Morroco. *Irrigation and Drainage* 58: S252–S263.

Belal, A.A., and Agourram, A.J. (1971). Les problèmes posés par la politique agricole dans une économie 'dualiste': les leçons d'une expérience: le cas marocain. *Bulletin économique et social du Maroc* 33(122): 1–36.

Benouniche, M., Errahj, M., and Kuper, M. (2016). The seductive power of an innovation: Enrolling non-conventional actors in a drip irrigation community in Morocco. *The Journal of Agricultural Education and Extension* 22(1): 61–79.

Benouniche, M., Kuper, M., Poncet, J., Hartani, T., and Hammani, A. (2011). Quand les petites exploitations adoptent le goutte-à-goutte: Initiatives locales et programmes étatiques dans le Gharb (Maroc). *Cahiers Agricultures* 20(1): 40–47.

Berque, C., and Couleau, J. (1945). Vers la modernisation du fellah marocain. *Bulletin Économique et Social du Maroc* 26: 1–12.

Bossenbroek (2016). *Behind the veil of agricultural modernization: Gendered dynamics of rural change in the Saïss, Morroc.* (PhD thesis). Wageningen: Wageningen University.

Botha, D.F. (2013). Africa rural communities as knowledge prospecting domains for emerging e-business models. *European Journal of Business and Social Sciences* 2(3): 37–47.

Bouderbala, N. (2001). La lutte contre le morcellement: un thème idéologique. La stabilité des structures foncières marocaines. In A.M. Jouve (Ed.), *Terres méditerranéennes: le morcellement, richesse ou danger?* Karthala: Paris, pp. 41–56.

Bouderbala, N., Chraïbi, M., and Pascon, P. (1977). *La question agraire au Maroc.* Rabat: Société marocaine des éditeurs réunis.

Boularbah, S. (2014). *Retour d'expériences des projets de reconversion individuelle et collective dans les secteurs N1, N5 et N9 dans le périmètre irrigué du Gharb.* Thèse d'ingénieur d'Etat, Institut Agronomique et Vétérinaire Hassan II.

Callon, M., and Rip, A. (1992). Humains, non-humains: morale d'une coexistence. In J. Theys and B. Kalaora (Eds.), *La Terre outragée : Les experts sont formel!* Paris : Editions Autrement, pp. 140–156.

CGDA. (2014). *L'agriculture familiale en Méditerranée et en Afrique de l'Ouest: de nouvelles dynamiques entrepreneuriales et territoriales.* Paper presented at the "Séminaire Eau et Sécurité Alimentaire en Méditerranée", SESAME 2, Meknès: Morocco.

Coz, J.L. (1961). L'opération-labour au Maroc: tracteur et sous-développement. *Méditerranée* 2(3): 3–34.

Desrues, T. (2006). Le corporatisme agrarien au Maroc La trajectoire de l'Union marocaine de l'agriculture. *Revue des mondes musulmans et de la Méditerranée* 40(111–112): 197–217.

Desrues, T., and Moyano, E. (2001). Social change and political transition in Morocco. *Mediterranean Politics* 6(1): 21–47.

Ennaji, M. (1996). *Expansion européenne et changement social au Maroc.* XVIe-XIXe siècles. Casablanca: Eddif.

Errahj, K., Abdellaoui, E., Mahdi, M., and Kemmoun, H. (2006). *Les adaptations de l'agriculture familiale en grande hydraulique: quelques enseignements de la plaine du Gharb, Maroc.* Paper presented at the conference L'avenir de l'agriculture irriguée en Méditerranée. Nouveaux arrangements institutionnels pour une gestion de la demande en eau, 2006, Cahors, France.

Errahj, M., Kuper, M., and Caron, P. (2007). *L'action collective entre le rationalisme économique et les motivations psychosociales.* Actes du séminaire Euro Méditerranéen Les instruments économiques et la modernisation des périmètres irrigués, 21–22 novembre 2007, Sousse-Tunisie.

Errahj, M., Kuper, M., Faysse, N., and Djebbara, M. (2009). Finding a way to legality, local coordination modes and public policies in large-scale irrigation schemes in Algeria and Morocco. *Irrigation and Drainage* 58(S3): S358–S369.

Faysse, N., El Amrani, M., Errahj, M., Addou, H., Slaoui, Z., Thomas, L., and Mkadmi, S. (2014). Des hommes et des arbres: relation entre acteurs dans les projets du Pilier II du Plan Maroc Vert. *Alternatives Rurales* (1): 75–83.

Faysse, N., Errahj, M., Kuper, M., and Mahdi, M. (2010). Learning to voice? The evolving roles of family farmers in the coordination of large-scale irrigation schemes in Morocco. *Water Alternatives* 3(1): 48–67.

Food and Agricultural organization (FAO). (2014). *Initiative régionale pour faire face à la pénurie d'eau dans la région du Proche Orient et Afrique du Nord, Evaluation Nationale Maroc.* Rome: FAO, 72 pp.

Hammoudi, A. (1997). *Master and disciple: The cultural foundations of Moroccan authoritarianism.* Chicago: University of Chicago Press.

Jouve, A.M. (2002). Cinquante ans d'agriculture marocaine. In P. Blanc (Ed.), *Du Maghreb au Proche-Orient, les défis de l'agriculture.* Paris: L'Harmattan, pp. 51–71.

Jouve, P. (1999). Un modèle d'aménagement hydro-agricole à l'épreuve du temps et de l'évolution des systèmes de production. Le cas des grands périmètres irrigués marocains. *Cahiers de la Recherche-Développement* 17(45): 122–131.

Karrich, J. (1978). La distribution des terres et les coopératives de la réforme agraire dans le Gharb. *Hommes, Terre et Eaux Revue marocaine des sciences Agronomiques et Vétérinaires* 26: S38–S46.

Kleich, M. (2001). Aux origines du concept de développement. Quand l'irrigation devient enjeu de réforme agricole: nouvelle mise en ordre du paysage rural marocain dans l'entre-deux-guerres. *Hesperis Tamuda* 39(2): 175–194.

Kuper, M., Hammani, A., Chohin, A., Garin, P., and Saaf, M. (2012). When groundwater takes over: Linking 40 years of agricultural and groundwater dynamics in a large scale irrigation scheme in Morocco. *Irrigation and Drainage* 61(S1): 45–53.

Lazarev, G. (1976). Aspects du capitalisme agraire au Maroc avant le protectorat. *Annuaire de l'Afrique du Nord* 14: 57–90.

Le Coz, J. (1968). Le troisième âge agraire du Maroc. *Annales de Géographie* 77(422): 385–413.

Leveau, R. (1976). *Le fellah marocain défenseur du trône*. Paris: Presses de Sciences Po.

Losch, B., Fréguin-Gresh, S., and White, E.T. (2012). *Structural transformation and rural change revisited: Challenges for late developing countries in a globalizing world*. Washington, DC: The World Bank.

Maghraoui, A.M. (2001). Monarchy and political reform in Morocco. *Journal of Democracy* 12(1): 73–86.

Marthelot, P. (1961). Histoire et réalité de la modernisation du monde rural au Maroc. *Revue Tiers-Monde* 2(6): 137–168.

Mkadmi, S.E. (2011). *Analyse du modèle d'accompagnement technique des projets du deuxième pilier du Plan Maroc Vert (Cas du projet olivier dans la région de Meknès)*, Mémoire d'étude, Ecole Nationale d'Agriculture Meknès, Morocco.

Oved, G. (1961). Problèmes du développement économique au Maroc. *Revue Tiers-Monde* 2(7): 355–398.

Pascon, P. (1971). La formation de la société marocaine. *Bulletin économique et social du Maroc* 33(120): 1–25.

Pierri, F.M. (2013). How Brazil's Agrarian dynamics shape development cooperation in Africa. *IDS Bulletin* 44(4): 69–79.

Ploeg, J.D. van der. (2008). *The new peasantries: Struggles for autonomy and sustainability in an era of empire and globalization*. London and Sterling: Earthscan.

Ploeg, J.D. van der. (2010). The peasantries of the twenty-first century: The commoditisation debate revisited. *The Journal of Peasant Studies* 37(1): 1–30.

Popp, H. (1978). Les périmètres irrigués du Gharb : Actes de Durham. *Bulletin Economique et Social du Maroc* 138–139: 157–177.

Popp, H. (1986). L'agriculture irriguée dans la vallée du Souss : Formes et conflits d'utilisation de l'eau. *Méditerranée* 59(4): 33–47.

Seckler, D.W. (1996). *The new era of water resources management: From 'dry' to 'wet' water savings* (Vol. 1). Colombo: International Irrigation Management Institute.

Swearingen, W.D. (1988). *Moroccan mirages: Agrarian dreams and deceptions, 1912–1986*. Princeton, NJ: Princeton University Press.

Van Wersch, H.J. (1968). Rural development in Morocco: "Opération labour". *Economic Development and Cultural Change* 17(1): 33–49.

Zeine, M., Faysse, N., Errahj, M., Bekkari, L., and El Amrani, M. (2015). Grille d'analyse de la maturation de projets collectifs de conversion à l'irrigation localisée: application dans des oasis du Maroc. *Canadian Journal of Development Studies* 36(4):1–15.

11 Historical perspective on low-cost drip irrigation design and promotion

Robert Yoder and Brent Rowell

Drip irrigation: benefits and challenges

Crops are most productive when soil moisture closely matches the plants' transpiration rates. Unfortunately, the timing and quantity of water supplied by rainfall is often unable to keep soil moisture optimal. For millennia, human ingenuity and effort have moved water from surface and groundwater reservoirs to address this water deficit.

A continuing challenge is to reduce the cost of moving and applying irrigation water. Diverting surface streams or lifting surface and groundwater is the first component of irrigation cost. Additional costs include conveying water to the field and applying irrigation water within the field so that it becomes accessible to plants. In many cases, excess water must be moved away from plants and fields, making drainage another cost.

Water is heavy, expensive to move, and in many locations in limited supply. Implementing precise control of water delivery by applying a measured amount of water to only the soil encompassing the crop's roots, greatly increases the overall efficiency and effectiveness of the applied water by: 1) reducing the area of wetted soil surface from which evaporation takes place and, 2) reducing the amount of water that infiltrates below or moves laterally by capillary action beyond the crop's root zone. Drip irrigation[1] provides this precise water application by leading the water through pipes, called drip lines,[2] and delivering it through emitters.[3]

Precision application minimizes the amount of water that needs to be withdrawn from surface or ground reservoirs and therefore minimizes the energy required for pumping. For a given crop and field size, drip irrigation typically requires 15 to 25% less volume of water than a comparable pressurized sprinkle irrigation system and 40 to 60% less than surface irrigation application by flood or furrow.[4] Since irrigation water does not wet the plants above the surface, drip irrigation keeps foliage dry and root zones less saturated thereby reducing certain fungal and bacterial diseases. Soluble fertilizer can be conveniently added to the irrigation water and applied via the drip irrigation system. This allows better timing of nutrient applications to meet plant needs with cost savings due to a lower application rate with precise placement. Other advantages include reduced soil crusting, better weed management, ability to work the crop while irrigating, labour savings, less sensitivity to water salinity, etc., while the overwhelming advantages in most contexts are increased crop yield and quality.

The precise localized application of water requires a pipe network and drip lines. For many farmers, particularly smallholder farmers, the cost of this infrastructure is a significant barrier to adopting drip irrigation. The current capital investment required for installing drip irrigation in the United States is in the order of $1100 to $2500 per hectare excluding the well and pump (Simonne et al., 2015; Burt, 2011). Part of this investment is for components that will last for three or more years but some parts, such as the drip lines and fittings, may require more frequent replacement, adding future annual costs in the range of 10–30% of the initial cost.

The drip irrigation system

As pumping to provide irrigation water and conveyance by pipe to fields became more common over the past six decades, drip irrigation evolved rapidly. At the point of delivery, the 'emitter' dissipates the water pressure and discharges a small uniform flow of water to the soil. The drip irrigation design challenge has been to develop the pipe and emitter systems that deliver the same amount of water from each emitter in the system (Box 11.1; see Chapter 3).

Box 11.1   Historical perspective on pipes and emitters

Clay pots filled with water within a planting area were an early way of watering a plant's root zone. In the 1920s, growers in Germany began using perforated pipe to irrigate. With the development of plastics, an Australian inventor, Hannis Thill, configured a small diameter tube to evenly distribute water to crops. In 1959, Simcha Blass and Kibbutz Hatzerim developed and patented the first practical drip irrigation emitter that could be embedded inside the drip line as it was being manufactured. This was a small piece of moulded plastic with a long passageway to decrease the water pressure before it exited through the drip line wall. The development of microtube and plastic emitters enabled precise metering of water, achieving the concept developed by Dr. Daniel Hillel to use drip irrigation to shift from low frequency, high-volume (flood) irrigation to high-frequency, low-volume irrigation.

Irrigation efficiency is a concept widely used in the design and management of irrigation systems. It is determined by three components: 1) losses due to evaporation and deep percolation; and, 2) losses in the conveyance and distribution of water in the system, and 3) application uniformity in the field. Losses due to evaporation and deep percolation are minimized in drip systems by placing the drip line next to the row of plants and controlling the application rate and duration so that most of the water is retained within the soil profile in close proximity to the roots.

The uniformity of water application via a drip line depends on the combined effect of the water supply pressure, elevation differences throughout the irrigated area, friction loss in the pipe network and drip line, and the discharge characteristics of the emitter, as well as manufacturing variation among emitters. A measure of uniformity used for smallholder drip irrigation systems is the 'coefficient of variation uniformity' (CvU). By using a series of 'catch cans' to measure the discharge of emitters at different locations along the drip line the CvU can be easily computed with a spreadsheet or hand calculator to confirm that performance is acceptable. Commercial drip irrigation systems typically achieve CvU > 90%. For low-cost drip serving small plots CvU > 88% is considered excellent and CvU > 80% good, especially in developing country contexts (Keller and Keller, 2005).

Designing a drip system that properly matches crop water requirement[5] requires information that is site and crop specific. Frequently, especially when drip irrigation is being introduced, site and crop specific information is not available and estimates must be based on more generalized information. A starting point for pump and irrigation delivery pipe specifications is to estimate evapotranspiration of the field or irrigation zone based on a full canopy crop. This gives the expected maximum water delivery and fully accommodates crops with row spacing requiring less water than a full canopy crop. Design involves adjusting the drip installation to match plant and row spacing of the farmer's selected crop that also best fits the shape and topography of the field. Drip line manufacturers provide a variety of emitter flow rate options to be matched with crop water requirements. The water pressure, length and diameter of the drip line, emitter flow rate, spacing between drip lines and between emitters together determine the flow rate into the drip line. These parameters, along with emitter characteristics provided by the manufacturer, are then used to determine if irrigation delivery to the far end of the drip line is sufficient to achieve the desired application uniformity. Where drip irrigation is newly introduced there may not be a wide selection of drip line options with varied emitter spacing and flow rates. Selection limitations can be overcome by adjusting the row length and zone[6] size to match the flow characteristics of the drip line.

Most farmers need considerable assistance in planning an appropriate layout the first time they install a drip system. Many drip irrigation manufacturing and supply companies provide system design assistance to farmers, even in developed countries. In some states in India, for example, where capital investment in drip irrigation systems is subsidized by the government, a system design plan is required as part of the subsidy application. However, for smaller farms the cost of professionally planned and designed drip irrigation is often prohibitive. Design complexity (for further information, see Chapter 3, this volume) as well as cost spurred development of 'low cost' drip for smallholders. This development envisioned cutting through the need for on-site design by using prefabricated packages that would meet most of the real-world needs of a smallholder without further engineering analysis.

History of low-cost drip irrigation development

Reducing complexity and lowering costs: the drip kit

Large farmers purchase their drip products in bulk from a manufacturer or retailer. Often their initial investment is based on a design prepared by an irrigation professional that is tailored to their fields and crops. Such a design includes layout drawings and a detailed list of all the necessary components and their specifications. As farmers gain experience with the technology, they modify and adapt the design to fit changes in their cropping system. They keep extra components at hand to easily make necessary adjustments during installation or to replace defective parts. Smallholders have difficulty arranging capital to purchase a drip system and, particularly for small fields, are unable to engage a design service to provide technical assistance.

The late Richard Chapin, founder of Chapin Watermatics (now owned by Jain Irrigation Systems Ltd.), was the first large commercial drip manufacturer in the United States. He was perhaps the first to conceive and develop the idea of a 'drip kit' as a means to simplify access to drip irrigation by smallholders. This came about when Mr. Chapin, at the request of Catholic Relief Services, travelled to Senegal in 1974 to investigate ways for farmers to use drip irrigation for household vegetable production. Without a pump to pressurize the water, he concluded that an elevated tank supplied by buckets of water could take its place. His experience led to the formation of Chapin Living Waters Foundation and promotion of 'bucket kits' for irrigation. By preselecting the water supply pressure (height of the supply tank), system size, layout shape, row spacing, emitter spacing and emitter flow rate, he could compute the delivery and manifold pipe dimensions and select the proper filter and fittings to ensure appropriate hydraulic performance (high CvU). All of these components could be conveniently packaged along with assembly and use instructions for delivery to farmers.

A farmer would simply open the drip kit package, assemble the parts in his or her field and plant the crop along the drip lines as the kit designer had taken care of all critical design decisions. Drip kits greatly reduced the need and cost for collecting site-specific information. A farmer only needs to determine what size kit would best fit his/her available land and has reasonable assurance that the kit will perform reliably and efficiently for almost any row crop. Assembly line packing assures that drip system components of the correct type, size and quality will reach the farmer at the lowest possible cost. While convenient for mass dissemination, the one size (now multiple sizes) kit frequently does not fit the irregular shapes and sizes of smallholders' fields. More critically, the supply chains necessary to sustain a 'kit' approach to introducing drip irrigation have seldom evolved to support replacement and expansion. These issues will be explored more fully after reviewing the evolution of the kit.

Evolution of low-cost drip irrigation in India

Commercial drip irrigation was introduced in India via government support programs in the 1980s. Manufacturing companies were required to obtain Bureau of Indian Standards 'ISI certification' for their products to qualify for the subsidy program. By the 1990s, a growing number of smallholder farmers who did not qualify for or found it too costly to access the drip irrigation subsidy program were contacting small plastic manufacturers with requests for drip irrigation equipment. The manufacturers responded by adding drip tape[7] and fittings to their product lines. Most chose not to apply for ISI certification, enabling them to sell their products at a lower price, often called the 'grey' market.

In the mid-1990s, iDE (formerly International Development Enterprises) encountered drip irrigation in India and searched for ways to make it more accessible to its smallholder farmer clients. iDE adopted the microtube emitter-based drip technology already widely used at that time in western India.[8]

Initially, iDE field staff served as advisors, providing farmers guidance in purchasing components and installing drip systems. This provided opportunity to experiment with the range of products available in the market and meet with manufacturers to communicate the opportunity and requirements for expanding drip irrigation to smallholders in the water scarce areas of western India. Drip kits seemed an appropriate approach to rapidly introduce smallholders to irrigation. Drawing upon Mr. Chapin's experience in Africa, iDE began packaging drip irrigation systems as kits to facilitate marketing to new farmers.

Kits were designed as stand-alone irrigation systems for a range of field sizes from 20 m^2 to 1000 m^2. iDE's initial kits used a tank (often a discarded oil drum) supported about 1–2 m above the ground. An outlet tap connected the tank to a 16 mm outside diameter plastic tube with 1 mm wall thickness to a small screen filter and continued on to 'sub-main' or manifold tubing of the same size. A series of 12 mm tubes were connected to the manifold as drip lines. The emitters consisted of 3 mm diameter microtubes with 1.2 mm inside diameter flow path cut to 20–30 cm length. The farmer inserted the microtube emitters into the drip line in the field. The spacing between emitters was generally about 25 cm, or one emitter for each plant if their spacing was more than 25 cm. The high flow rate of the microtube emitters limited the drip line length to about 30 m. All of the components (except the tank sourced by the farmer) were packaged in a box making it easy to establish local distributors and carry out mass marketing (see Figure 11.1).

The modular design of drip kits was expected to make it easy for farmers to expand the area irrigated by simply adding additional kits. As packaged, the cost of a 100 m^2 system was about \$20 without the tank. This was slightly less than the subsidized custom designed name-brand systems of similar design. Because

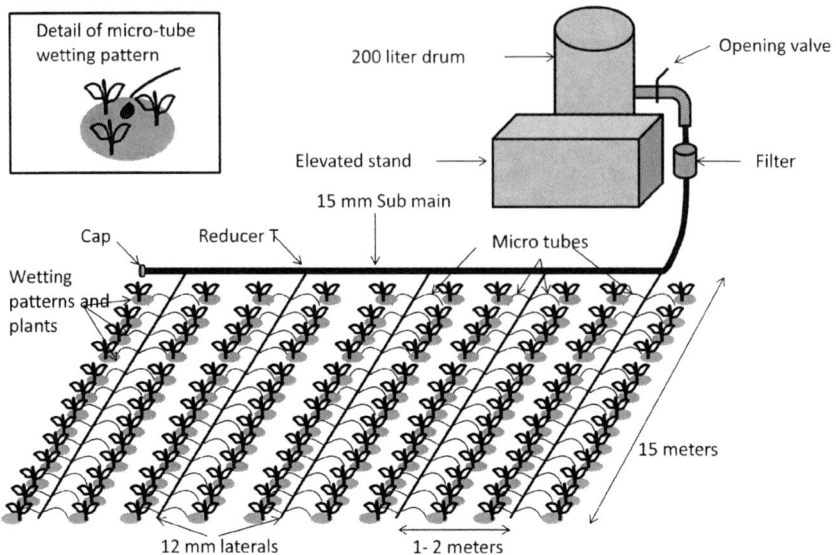

Figure 11.1 An iDE low-cost drip kit.

Source: drawing by Jean-Philippe Venot, adapted from iDE global communication material with their authorization.

of the price, sales of these earliest drip kits never reached the numbers expected and highlighted the need for a lower-cost alternative.

Influence of pepsee drip

Several farmers with sizable landholdings in a cotton growing area of Madhya Pradesh used the government subsidy program to install drip irrigation. Smallholders without access to subsidies saw the increased production opportunity drip irrigation provided for planting cotton prior to the rainy season and were receptive when iDE introduced the microtube-based drip kits. However, rather than purchasing additional kits to expand their drip irrigated area, farmers turned to the grey market and purchased components and built their own microtube-based systems to better match the size and shape of their fields.

In about 1999, an innovation called Pepsee drip appeared in the area (Verma et al., 2004). The name comes from the 'ice candy' packaged and sold in short, clear plastic tubes. It is not known exactly where or how rolls of this thin-walled (65–135 micron) plastic tubing first came to be used as drip lines for irrigation.[9] Someone recognized that the uncut rolls used to package the Pepsee ice candy could be used to convey water and soon it became common in the area to use Pepsee tape as drip line for cotton. After installing the Pepsee drip line along the cotton row, a series of small thorn pricks allowed the water to flow directly to the plants.

Even though Pepsee drip had a field life of less than a full growing season, if installed at planting time before the rainy season, it resulted in higher germination rates and lower incidence of pest attack than furrow irrigation. The initial capital cost of $85 per acre ($0.021/m^2) for Pepsee drip was very attractive to smallholders. The local cost for the same 1-acre installation using grey market microtube drip at the time was $144/acre ($0.036/m^2) and the cost of Pepsee drip was less than one fourth the cost of the cheapest commercial drip systems ($386/acre or $0.095/m^2 (Verma et al., 2004). The result was that the Pepsee innovation spread rapidly in spite of its limitations – easily punctured and rapid algae growth inside the transparent tubing that reduced its usefulness to two to three months. Pepsee drip provided an alternative to furrow irrigation for pre-rainy season planting of cotton at an affordable price adding considerable value to the crop.

With assistance from the late Professor Jack Keller, iDE took a lesson from the appeal of Pepsee drip and worked on ways to improve on the concept. This led to revised specifications for iDE India drip products in 2001. By experimentation, they found a mix of polyethylene grades that enabled a dramatic reduction in wall thickness (hence cost) of the main, sub-main and manifold pipes and drip line components. This was introduced to the low-cost plastic manufacturing industry along with quality control standards. The microtube emitter-based drip kit was reduced from 1000-micron to 125-micron wall thickness drip lines (250-micron for kits if longer life cycle was desired) and still had strong resistance to stress and cracking. Sunlight was blocked from entering the tube to resist algae growth by adding carbon black. UV stabilizers were added to slow the sunlight induced deterioration. With these changes low-cost drip could be used for multiple seasons and cost was reduced to between $0.03 and $0.05/m^2 depending on drip line spacing. The 100 m^2 drip kit price was reduced from $20 to $11. The new drip system was initially called EasyDrip.[10] In addition to its low cost, iDE's drip system had the following features (Keller et al., 2001):

- Operation at low inlet pressure (1–2 m water) allowing lightweight tubing and inexpensive fittings to be used and leaks to be easily repaired
- Main, sub-main and drip lines of lightweight tubing assembled with simple fittings
- Lateral and sub-main tubing packaged in tight rolls reducing handling and transportation costs
- Dimensional tolerances of tubing and fittings that allowed inexpensive manufacturing
- Microtube emitters
  - Minimal clogging so a small screen filter is adequate
  - Achieve flow rates ranging from 2 to 7 liters/hour, ideal for small vegetable plots
  - Application uniformity in level fields comparable to conventional drip
  - Few leaks that are easily repaired

- Manufactured using manually controlled extruders and moulds, resulting in a low entry cost for manufacturers.
- Easily assembled without sophisticated tools: a farmer can install a drip system with microtube emitters in a 1–2,000 m^2 field in one day.

Factors influencing adoption of low-cost drip irrigation

Local availability of grey market manufactured products

The robust plastic extrusion and moulding industry in western India welcomed the growing interest in drip irrigation in the 1990s. The same 'blow extrusion' machines used for making plastic bags can be used to make drip lines. A multitude of small plastic moulding industries added drip system accessories (filters, off-take valves, drip line splicers, etc.) to their product lines. The supply chain from manufacturers to retailers spread rapidly in western Indian states making drip equipment widely available in local markets. Competition became strong and this brought prices down.

Though iDE designed its drip kits for the western India market, their usefulness there was short-lived and primarily served as a marketing tool to smallholders. Often farmers purchased a single kit to test its usefulness, but instead of buying additional kits to expand the system to the full extent of their water supply or available land, they simply went to the local market to purchase parts needed to tailor their drip system to fit their field(s) at a lower price than purchasing another kit.

The downside to the competitive grey market manufacturing was compromised quality. This was not a serious handicap for farmers living a bicycle ride away from a shop selling drip line and fittings. To these farmers, lowest cost was more important than quality since defective components are easily and cheaply replaced. However, for iDE this became a difficult problem, as demand for export to other countries grew but defective parts could not be easily replaced. Considerable cost and effort was expended trying to select manufacturers willing to introduce and enforce quality control measures with only partial success.

Purchase cost more important than water application uniformity

In sugar cane and cotton growing areas of India, there has been a growing demand for ultra low-cost drip line that forgoes high application uniformity and water saving for lower costs. iDE assisted in establishing a manufacturing process to make precision orifices in 125- and 250-micron wall thickness drip line. Blank drip line is passed through a machine that uses a hot, temperature-controlled needle or laser to burn or cut a precise hole at made-to-order spacing. These orifice emitters have a higher flow rate and lower application uniformity than microtube or internal emitters. While the high flow rate limits the drip line to about 35 m, making it less suitable for large fields, this is not a serious handicap for smallholders. These 'direct punched' drip lines are similar to their predecessor 'Pepsee' drip but with a useful life of several years instead of half a growing season. For a one-acre field installation the cost of direct punched 'black Pepsee drip' in 2010 was about $100 per acre ($0.025/m²).[11]

Installation cost of internal and external emitters

As labour prices increased in India, it was observed that farmers shifted their search from lowest off-the-shelf purchase price to lowest installation cost. To install microtube emitters for a one-acre cotton field takes one person two to three days. Rolling out and connecting drip lines with hot-punched or laser-cut holes as emitters or using drip line with internal emitters for a one-acre cotton field can be done in less than half a day. If labour cannot be hired easily, the opportunity cost for the farmer's time pushed adoption of the labour saving approach. Lower installation cost, potential for longer drip lines resulting in fewer zones and easier irrigation delivery management together with lower prices for in-line emitter drip tape have greatly reduced the demand for microtube emitter drip systems in India in the past few years.

Water supply availability and adequacy

Fundamental to the long-term success of irrigated agriculture development is confirmation that the water resource is adequate to support the potential irrigation demand. Shifting from flood to drip application can reduce the water required per unit area irrigated, but where additional un-irrigated land is available it will more likely result in using all of the water to increase the irrigated area (see Chapters 5, 8, and 16 in this volume). Particularly in water scarce environments, a constant concern is that successful introduction of drip irrigation does not increase consumptive water use beyond the sustainable resource limit.

A recent solar pump test in western Gujarat, India, used water application by drip as a way to minimize the pumping energy requirement. Farmers in the area use large diameter dug wells to access groundwater to complement the monsoon rainfall, especially toward the end of the rainy season. After their monsoon crop is harvested, groundwater was traditionally reserved for animal watering and a small area of flood irrigated fodder. Farmers were skeptical about depending on dry season groundwater to grow vegetables and were pleasantly surprised when they saw they could grow 3000–4000 m² of drip-irrigated vegetables with a single well. All agreed that they wanted to continue using drip irrigation while concluding that the solar pumps were too expensive, reverting to diesel pumps the following year. Although the test to use high-flow rate internal emitters operated at 1–2 m pressure with solar pumping was highly successful, it also identified a larger water resource issue if drip technology was to be widely adopted as this area with finite groundwater. As long as farmers only use their annually recharged upper aquifer all will have relatively equal access to the water. However, if they tap a deeper aquifer with their well, the need for assessing sustainability will be essential.

When drip kits were first promoted in the hills of Nepal in the mid-1990s, they appeared to be an ideal solution for fields with springs and streams in close proximity and whose elevation made them inaccessible for surface gravity channels used extensively in the area for irrigation. The overflow water piped from distant springs for domestic needs proved to be an excellent water source. However, design specifications for village drinking water systems only supported domestic needs with little water to spare for irrigation. This ultimately led to what became known as the MUS (multiple use services) program in Nepal where village water systems were designed to use one water source for both domestic and productive water needs (Polak et al., 2002; Plate 12 in the color plates shows a system coupling MUS and low-cost drip irrigation as installed in Nepal). This required development of an entirely new approach including modified water supply regulations safeguarding domestic water needs and regulations for water allocation among farming households for irrigation (Mikhail and Yoder, 2008).

The Lake Ziway area in the Rift Valley of Ethiopia has fresh water and is like an oasis in an arid climate. Without irrigation, crops in this area can be grown only during the rainy season. Farmers with access to lake water or shallow groundwater use engine pumps to irrigate. Many smallholder farmers a bit further from the lake constructed hand-dug wells for domestic water needs. To enable the smallholders to grow vegetables, iDE promoted drip kits of 20 m^2 to 200 m^2. The difficulty in lifting groundwater with manual pumps from 7 to 15 m depths together with the lack of locally accessible replacement components for drip kits resulted in small irrigated areas and the abandonment of drip systems by many farmers.

Challenges in promoting low-cost drip irrigation

Manufacturing quality

Manufacturing issues, identified both through field observations and laboratory analysis, were primarily related to the quality control of production (Thompson, 2009). The low-cost extrusion process used to manufacture the drip line requires careful monitoring and continual adjustment by the operator to control the drip line wall thickness and diameter. Necessary adjustments are made manually. However, temperature and feed rate of the extrusion material varies with the power supply voltage making quality control a difficult challenge. Given the large number of low-cost extruders manufacturing drip lines in India, it was difficult to monitor and control product quality. For example, drip kit packaging uses the weight of the drip line roll to estimate its length. Many drip kits were found to have drip line length up to 15% shorter than stated in the packing list. This was due to manufacturing errors where the wall thickness on one side of the tape was thinner than specified and the other side thicker. Instead of making extruder adjustments to correct this problem, wall thickness was increased so that the thin side reached the specified thickness. This added to the weight of the thicker side resulting in increased weight per length of drip line and a shorter than stated drip line.

Off-take valves not properly moulded did not achieve acceptable hydraulic performance. The internal passage diameter was often smaller than the specification and the passageway frequently had fabrication burrs that reduced the flow rate. Because of price competition the manufacturers were reluctant to spend extra money to improve quality.

Supply chain for spare parts

In western India, pushback on poor quality has been minimal. There has been little incentive or ability for the grey market manufacturers to establish higher quality products than demanded by local consumers, resulting in serious quality issues with exported drip kits.

Extending the largely positive drip irrigation experience from western India to other parts of the world has exposed a supply chain dilemma. A good installation will operate with little or no maintenance for a cropping season but if it must be removed and stored between crops, inevitable damage will require replacement of some components when reinstalled. A rule of thumb is that about 10–30% of the drip line and fittings need to be replaced annually. Care in handling and storage can reduce but not eliminate the need for spare parts. It is also difficult to anticipate which parts will need to be replaced and at what frequency when a drip kit is being assembled. Spare components are seldom included in a kit with the expectation that vendor will stock them. A retailer that has only complete unsold kits is reluctant to use them to supply replacement parts and lose the opportunity for a sale of the kit. As a result, kits installed in new locations tend to be abandoned as soon as a critical component breaks. The dilemma is that a vendor cannot afford to stock parts until there is sufficient demand, and demand only builds if enough farmers install drip systems and see value in continuing to keep them operating. This dilemma has surfaced in many programs trying to promote drip kits.

System layout and installation

The water supply for low-cost drip is sometimes pumped manually or delivered by bucket to an elevated tank. This limits the operating pressure to 1 to 2 m of water instead of the typical 7 to 10 m used in commercial drip systems. To maintain acceptable application uniformity at low pressure, the drip line length is generally limited to less than 35 m. This works well with small fields but may require multiple zones in larger fields requiring additional delivery pipe. Designing the layout is often an iterative process to determining the orientation that minimizes the cost of the delivery pipe, drip lines and fittings. Most critical is that drip lines not be longer than the combination of water pressure, emitter spacing and discharge allow for the desired application uniformity. It is also important that zone valves or individual drip line off-take valves be installed and used to maintain the desired pressure in the delivery system. Farmers new to drip irrigation require training and assistance in design, layout and operation of their first installation.

System operation: clogging

Even though microtube emitters have large flow paths, they can clog when using only a simple screen filter. There are two common causes for clogging. One is from a dirty (usually surface) water supply that is not adequately filtered. Instructions suggest that cloth supplied with the kit be used to filter the water as it is pumped or poured into the header tank. Then an 'in-line' screen filter at the tank outlet provides a second level of filtering to guard against dirt entering the drip line. In many cases the cloth for filtering the water soon disappears. The small screen filter at the tank outlet is sometimes not adequate for dirty surface water and more sophisticated media or disc filters may be required.

The second and often more severe clogging problem is encountered when the microtube outlet end touches the soil surface. After irrigation is completed for the day the drip line will drain through lower elevation emitters drawing air into those at higher elevation. If the outlet end of a higher elevation microtube extends to the wet soil surface it can draw dirty water into the microtube. When this dries it plugs the emitter. While it is relatively easy to clean microtube emitters by flicking them with a finger or removing them and blowing out the dirt, this is a time consuming activity for the irrigator. Furthermore, frequent removal of the microtube for cleaning caused the drip line hole to stretch and leak. The nuisance of emitter clogging was observed to be serious enough to discourage drip irrigation adoption in some situations.

Post-harvest drip line removal

An important consideration for using direct-punched as well as internal emitter drip line instead of drip line with microtube emitters is the ease with which it can be removed from the field after the growing season for reuse in following seasons. Post-harvest removal of drip lines so that soil can be tilled and prepared for the next crop is a laborious task where microtubes must first be removed before trying to pull the drip line out of crop residue. The microtubes are difficult to locate so some are missed. Microtubes that are not removed catch on the crop residue and damage the drip line or get lost, requiring repair and replacement before the drip line can be reused. Direct punched drip line and most drip lines with internal emitters have a smooth exterior that more easily slides from under the crop residue with less danger of being damaged.

Off-season storage

Storage of drip kits during the rainy season frequently results in damage. Rodents are the primary culprits but lack of appropriate storage facilities often exposes the drip line and fittings to mechanical damage requiring replacement parts. Lack of access to spare parts when drip equipment is first being introduced discourages famers from continuing to use drip systems. This problem has not been addressed with the same effort as upfront cost and continues to be a factor in drip irrigation's low appeal in new settings.

Rethinking the approach for low-cost drip irrigation promotion

The nature and depth of problems observed when first-time irrigators attempt to use drip irrigation, especially drip kits, illustrates the critical importance of understanding the full range of technical, social and economic issues the farmer faces. Too often, when irrigation water is scarce, drip irrigation is proposed as a remedy without first confirming that the farmer has access to a reliable water supply and a means to move it to the field. Drip is promoted before determining if the farmer has access to seed, fertilizer and knowledge for growing, post-harvest handling, and marketing of a succesful crop. If a complement of support services is not already in place, then establishing it may be the first required activity. These take many forms, from strengthening traditional government-supported extension services to field-based farmer advisors.

In Nepal, iDE's irrigation promotion program was reconfigured to identify and assist communities with potential for both improving their domestic water supply and in the process establishing a reliable water supply for small plot irrigation. This required staff with skills in facilitating community mobilization to identify their water resources, develop construction plans and mobilize funds from local and district governments. It required technical support for implementing construction. It also required on-going research and testing to find ways to lower the costs and improve performance of both the domestic water and irrigation systems. Agronomic support for vegetable production and training for marketing produce also had to be introduced. Within the overall activity, introduction and use of drip kits for irrigation was an important but minor component of the overall MUS promotion activity.

In new locations such as the Lake Ziway area of Ethiopia, where smallholder farmers had seen but not experienced irrigated crops, drip kits for smallholders perhaps should not have been the entry point for promoting irrigated agriculture. Access to water was costly and difficult. Hands-on experience in growing high value crops was limited as was post-harvest handling and marketing; all these steps must be mastered if income generation is the goal. Perhaps a better approach would have been to start with the small commercial entrepreneurs who leased land near the lake and furrow irrigated vegetables using engine pumps. They already had knowledge and experience necessary for growing and marketing vegetables.

Assisting them to test and compare drip irrigation to furrow irrigation would have required preparation of a site specific drip design rather than supplying a drip kit, but if dozens of these commercial farmers found drip irrigation to be beneficial it would have produced enough demand for an equipment supplier to set up a sales depot in the area. Demonstration of success on the larger farms would perhaps have then made it possible to introduce drip irrigation to nearby smallholders. The suggestion here is that introduction of drip technology should start with more experienced farmers who can better afford the upfront cost and demonstrate greater profit in the long term as a way to establish and support the necessary supply chain. It is then possible, as in western India, to find ways to lower the cost barrier for smallholder farmers so that they can benefit from the advantages drip irrigation provides in water-scarce environments.

Introduction and extension of drip irrigation in Myanmar

Experimentation with drip irrigation in Myanmar began in late 2006 with dual goals: 1) develop and promote affordable, easy-to-use drip irrigation for small vegetable, fruit and flower growers on commercial-scale plots of about 400 m^2, and 2) establish a local private manufacturing capacity. Local manufacturing was essential due to import restrictions at the time. While the Myanmar program started as part of an iDE country program (with significant investment in research and development), it became an independent non-profit organization, Proximity Designs, in 2011.

It was assumed that introducing drip would be relatively quick and easy and that drip kits imported from India could be demonstrated while local manufacturing was set up. It soon became apparent, however, that the iDE-India drip kits were not well suited to Myanmar's horticultural crop production systems. The pre-packaged kits were either too small or too large, and drip line-to-manifold connectors were preinstalled at spacing not used in Myanmar. There were also no fittings available to connect the system to the flexible delivery pipe readily available in most local markets. The components of the iDE-India kits were thus used to custom fit on-farm installations but were quickly replaced as locally made components were manufactured.

Before committing to a particular type of system, simple tests were carried out with six drip line/emitter systems in farmers' vegetable plots in Yangon. Low pressures from elevated header tanks were used with surface water and only simple iDE screen filters. Labyrinth path lower flow internal emitters from several name-brand drip manufacturers provided better uniformity and higher crop yield than microtube emitters; however, clogging of these emitters occurred within 90 days and they had to be abandoned. Microtube emitters also clogged at times but were easily cleaned. More importantly, microtube emitters could be easily manufactured locally.

Further testing of microtube emitter systems in farmers' fields resulted in complaints of insufficient water from emitters farthest from the manifold. Extensive experimentation was carried out to determine a suitable operating pressure, delivery pipe diameter and maximum drip line length (and number of emitters) that would still give acceptable application uniformity. New designs were eventually prepared for manufacturing delivery pipe, drip line, microtubes, filter, connectors and fittings. A local manufacturer was selected and quality control procedures established to test samples confirming that design specifications were being met. Over a three-year period, a complete system (including Proximity's elevated plastic treadle pump designed to lift the water to an ingeniously designed 760-liter 'water basket') was tested and manufacturing established in Yangon. Careful attention to design details supported by continuous feedback from farmers' field-testing was key to bringing this low-cost drip system to market.

In addition to significant investment in drip irrigation engineering, the Myanmar drip team devoted considerable time and resources in establishing and training an extensive network of field-based promotion staff and private installation agents who could comfortably explain and demonstrate the technology in areas where it was previously unknown. Annual in-service trainings were provided for all field staff and drip team members who became problem solvers willing to experiment and adapt drip systems to a wide variety of crops and field layouts. Finally, the importance of product loans or other farm credit cannot be underestimated. Although time consuming and difficult to manage, product loans accelerated adoption and helped ensure more equitable access to drip irrigation after 2009.

iDE/Proximity Design's history of introducing drip in Myanmar was initially one of slow growth: custom systems were tested, improved, and demonstrated using imported pieces and parts while gradually substituting locally made components as they were designed and manufactured. Only in the final stages were drip sets or kits assembled and marketed with all locally made components. Private sector sales coupled with a strong on-farm demonstration program resulted in the eventual adoption of drip in almost all agro-climatic zones in the country for a variety of commercial vegetable, fruit, and flower crops. Over 22,000 small farm drip systems and drip set equivalents[12] were installed from 2008 to 2015. Now that drip has a foothold in many parts of the country, it appears that sales of component parts may soon outstrip the sales of pre-packaged kits (Rowell and Soe, 2015).

Conclusion

Low-cost drip irrigation has been widely tested in many environments. While in many cases the results have been disappointing, there are also examples of considerable success as shown in western India, Nepal and Myanmar. Key factors that explain the success of low-cost drip irrigation are: (1) smallholders with prior experience of furrow irrigating a large variety of crops including fruits, flowers, vegetables, cotton and sugar cane; (2) existing access to a limited supply of groundwater where switching from furrow to drip enabled the same water supply to greatly increase the area irrigated or save significant time and labour; (3) ready markets for high-value crops; (4) familiarity with the technology – sometimes promoted and supported by the government or an NGO through subsidy programs and/or well-trained field staff; (5) existence of a local manufacturing capacity and close proximity to large irrigation product manufacturing enterprises that small manufacturers can copy at lower cost, resulting in readily available drip irrigation products in local markets.

The examples from western India, Myanmar and Nepal clearly illustrate the need for integrating the design to local conditions. This requires keen observation and carefully listening to farmers' responses and then reworking designs to meet their needs. It also required widespread farmers' field demonstration, side-by-side with local irrigation practice, to confirm the benefit and cost-effectiveness in shifting to drip irrigation. Overall the lesson from these areas is clear: it takes a dedicated long-term research, development and extension effort to bring a new product to market that meets local needs at an affordable price.

Low-cost drip irrigation for many is synonymous with 'drip kits'. It is clear that low-cost drip has played an important role in assisting smallholders to profitably adopt drip irrigation in Western India, Nepal and Myanmar. However, drip kits primarily provided demonstration effect. The real impact in Western India was from the multitude of small manufacturers and robust distribution system already in place making drip components widely available for farmers to assemble their own systems. Considering the quality, ease of installation and expanded local access to drip lines with larger flow-path internal emitters, it may be time to reconsider the approach to low-cost drip. It may still be cost-effective to establish local production of off-take valves and fittings to connect locally available delivery pipe, but higher quality brand-name drip tape with internal emitters should be considered to replace microtube systems. For smaller high-value plantings that can accommodate 30–35 m maximum length drip lines, direct punched drip tape or even microtubes may continue to be a viable solution for smallholders.

Notes

1 Also frequently referred to as trickle or micro-irrigation.
2 Drip line also called 'drip tube', 'drip tape' or 'lateral line' refers to the pipe, usually thin walled and flexible, with emitters that deliver water to plants.
3 The 'emitter' is a device used to control the flow of water from the drip line to the soil. There are many types of emitters. External emitters are attached to the outside of the drip line and penetrate the wall to allow water to flow. Internal or built-in emitters are of many shapes and methods of attachment to the inside of the drip line. Drip lines are categorized by the flow rate of the emitter, some as low as 0.2 liters/hour and as high as 8 liters/hour.
4 This does not suggest that drip irrigation 'saves' water. Crops only use a finite amount of water; water applied to but not used by crops is, in most cases, recycled to ground and surface water and to the atmosphere and may be available for reuse.
5 'Crop water requirement' generally refers to transpiration from crop's plant leaves plus unavoidable evaporation from soil surface in the vicinity of the crop plants.
6 A drip irrigation zone refers to the group of drip lines that are operated as a unit. The zone size is adjusted to accommodate the flow characteristics of the pump, delivery pipe and drip lines so that the quantity of water delivered by each emitter is within the acceptable range of design uniformity.

7 Drip lines with less than 300 micron wall thickness are often call 'tape' and can be rolled into very compact rolls.
8 Microtube emitters provide excellent irrigation application uniformity, but because of their high flow rate (about 7 liters/hour with 1 m of water pressure for 125 mm long microtubes), the drip line length is much more limited (~25–35 m depending on the emitter spacing) than with lower flow internal emitters that normally have a flow rate in the range of 0.3 to 1.4 liter/hour at 1 m water pressure.
9 This thin-walled plastic tubing becomes round when filled with water at a slight pressure but flat when empty.
10 Later it was branded for sale in India as KB Drip and exported under the name IDEal Drip but will be referred to in this chapter as 'low-cost drip'.
11 KB brand Global Easy Water Products (GEWP) 2010 price list.
12 'Drip set equivalents' includes sales of drip line and replacement components for previous and new users and for those expanding previously installed systems.

References

Burt, C. (2011). *Drip Irrigation system cost sharing by irrigation districts for water conservation.* ITRC Paper No. P11–007. Available at www.itrc.org/papers/pdf/costsharing.pdf.

Keller, J., Adhikari, L., Petersen, M.R., and Suryawanshi, S. (2001). Engineering low-cost micro-irrigation for small plots. *Journal of International Development Enterprises,* iDE: Denver.

Keller, J., and Keller, A. (2005). Mini-irrigation technology for smallholders. *Proceedings of the World Water and Environmental Resources Congress,* May 15–19, Anchorage, Alaska.

Mikhail, M., and Yoder, R. (2008). *Multiple use water service implementation in Nepal and India – experience and lessons for scale-up.* International Development Enterprises (IDE), The CGIAR Challenge Program on Water and Food (CPWF) and the International Water Management Institute (IWMI). Available at www.ideorg.org/OurStory/IDE_multi_use_water_svcs_in_nepal_india_8mb.pdf.

Polak, P., Adhikari, L., Nanes, B., Salter, D., and Surywanshi, S. (2002). *Transforming rural water access into profitable business opportunities.* International Development Enterprises. Available at www.siminet.org/fs_start.htm.

Rowell, B., and Soe, M.L. (2015). Design, introduction, and extension of low-pressure drip irrigation in Myanmar. *HortTechnology,* 25(4): 422–436, August.

Simonne, E., Hochmuth, R., Breman, J., Lamont, W., Treadwell, D., and Gazula, A. (2015). *Drip-irrigation systems for small conventional vegetable farms and organic vegetable farms.* Horticultural Sciences Department, UF/IFAS Extension. Available at HS1144 http://edis.ifas.ufl.edu.

Thompson, E. (2009). *Hydraulics of IDEal drip irrigation systems,* MS Thesis, Paper 296. Available at http://digitalcommons.usu.edu/etd/296/

Verma, S., Tsephal, S., and Jose, T. (2004). Pepsee systems: Grassroots innovation under groundwater stress. *Water Policy* 6(4): 303–318.

12 Low-cost drip irrigation in Zambia

Gendered practices of promotion and use

Gert Jan Veldwisch, Vera Borsboom,
Famke Ingen-Housz, Margreet Zwarteveen,
Nynke Post Uiterweer and Paul Hebinck

Introduction

In the face of increasing global population, water scarcity and the need to tackle poverty in developing countries, various projects and organisations have promoted low-cost irrigation technologies for smallholder farmers. The non-governmental organisation (NGO) International Development Enterprises (iDE) is one well-known organisation that took up the challenge to re-engineer conventional irrigation technologies with the explicit objective to meet the needs of smallholder farmers. iDE perceives low-cost irrigation technologies for individual smallholder farmers as a potential solution to rural hunger and poverty. With these goals in mind, iDE started the Rural Prosperity Initiative (RPI), implemented in Myanmar, Nepal, Ethiopia and Zambia, financed by The Bill and Melinda Gates Foundation and the Dutch Ministry of Foreign Affairs. Re-design criteria for the irrigation technologies are low investment cost, rapid returns on investment, suitability for various small plots, and simple operation and maintenance (Postel et al., 2001; Keller, 2004; Polak et al., 2007). Technologies developed include low-cost plastic tanks for collecting and storing rainwater, pressure and suction treadle pumps for lifting water, low pressure sprinklers as well as drip emitters for efficient water application. Moving away from the traditional practice of simply 'handing-out' technological innovation to farmers, iDE operates with a market-based or business approach. The idea is to support farmers so that they can invest in and craft their own way out of poverty by increasing crop productivity and income from marketable surpluses. In this idea, the rural smallholder figures as a (potential) entrepreneur, producer and customer, rather than as a beneficiary and recipient of aid.

This market-based or business approach fits well with a wider neo-liberal trend in development cooperation, which proposes solving rural poverty by a combination of efforts to help small-scale farmers become more entrepreneurial on the one hand and to create the right conditions – in terms of transport, support services and finances – for markets to flourish on the other. This increased reliance on market mechanisms and private sector actors in bringing about development marks a definitive change from the more supply-driven food production focus of the past. Commercialisation and entrepreneurship are the magic key words of the new strategy. Commercialisation involves the introduction of the primacy of economic rationality into the agricultural sector. Its promotion is underpinned by the expectation that markets will effectively guide the distribution of incomes and assets to the rural poor. Informed by theorisations of human beings in terms of rational choice, another key assumption is that once smallholders are helped to become market players or entrepreneurs – something that among others depends on their full access to information – they will take decisions based on economic rationales, using their capacity to calculate all possible options to determine the most efficient, optimal and profit maximising one (Kabeer, 2000).

Low-cost irrigation technologies neatly fit such neo-liberal development cooperation strategies, both because they are privately owned and thus compatible with efforts to cut down public irrigation spending and because they are tools to help farmers become more productive. In this chapter we argue that one of the effects of making these technologies of such strategies is that – although specifically promoted and designed to reduce poverty in Southern Africa – they mostly end up on the farms of those who are already more entrepreneurial. In addition, we show how the individual focus of these strategies dangerously backgrounds the deeply relational and networked character of farming in Southern Africa. The studied organisation and its RPI programme recognize that men and women have different roles and responsibilities which should be accounted for as they strive for the programme to produce equal benefits for men and women. Gender equality, iDE believes, will make households, communities and society stronger (IDE, 2008, 2012). Yet, by approaching men and women as farmers in their own right, rather than as parties in sets of social relations of power that are governed by deeply gendered institutions, there is a risk that new production practices and marketing favour the intra-household bargaining position of men and make it more difficult for women to derive benefits from low-cost irrigation technologies. The uptake and use of the technology redistributes work, resources and incomes at farm household level and beyond. In the process, gender relations and identities are also sometimes renegotiated, providing opportunities for some women farmers.

We study agricultural production patterns in relation to the use of low-cost irrigation technologies and ways in which changes therein affect intra-household arrangements for sharing and distribution of resources, labour and incomes. The ways in which farmers talk about agricultural production and low-cost irrigation technologies are a starting point, that is: we look at farmers' narratives around investments in low-cost irrigation technologies and confront these with observations on practices of the mobilisation and sharing of resources, labour and incomes. We pay special attention to the gendered nature of these processes.

We understand gender as the social meaning given to the biological differences between men and women. Gender identities are contextual and hence worked out in society. They are under constant negotiation and review (Zwarteveen,

2006). As such, hegemonic forms of masculinity and femininity both constrain and influence the actual behavioural practices of women and men. However, they do not determine behaviour (Connell, 1987 in Zwarteveen, 2006). Gendered sociocultural and historically shaped norms and values play an important role in the practices and possibilities of male and female vegetable farmers in the studied villages of the Kabwe and Kafue districts in Zambia. Both gender and technology are social constructs, which derive meaning in a particular context, and are subject to change.

Fieldwork underlying this chapter was conducted through the International Development Research Centre (IDRC) sponsored project on 'Gender Differentiated Adoption of Low-cost Irrigation Technologies in Zambia' jointly implemented by IDE Zambia and Wageningen University. Field research was conducted by a number of jointly supervised MSc students conducting their thesis research as part of their MSc programmes at Wageningen University. This article particularly builds on the theses by Ingen-Housz (2009) and Borsboom (2012). They each conducted field research for about three months, the first one in 2008 and the latter two during 2011. Other student theses produced under this project are Magwenzi (2011), Mupfiga (2011), and Tuabu (2012). Research methodology consisted of a combination of study of project data, key informant interviews, value chain mapping, household-level case studies, and participant and non-participant observations in iDE pilot areas. Research questions focused primarily on the technological design process, the promotion strategies and appropriation practices, paying particular attention to gender aspects.

In the next section, the promotion strategies and promoted technological package are described. In the section 'gender as social process' we analyse the gendered nature of the programme and its appropriation by farmers. In the final section, we conclude with a reflection on what this implies for low-cost drip irrigation in Zambia and similar programmes in sub-Sahara Africa.

Promotion practices

In Zambia, iDE has been active since 1997 (IDE, 2012). At the moment of research, iDE Zambia was the main importer and developer of low-cost irrigation technologies in the country and had an office with around 20 staff members in the capital city of Lusaka, as well as several field offices throughout the country. The field officers were responsible for selling irrigation technologies to farmers in the field. In the head office in Lusaka, staff members were mostly concerned with market coordination, technology design and development, monitoring and evaluation, and financial administration.

In January 2007, iDE launched a project named the Rural Prosperity Initiative (RPI). The project ran for four years, and had three main goals: (1) the development of low-cost water control technologies, affordable to poor farmers and suitable for small plots of land; (2) the development of sustainable local supply chains, both for the manufacturing and distribution of the technologies; and (3) the development of value chains to increase the returns to the smallholder. The RPI was launched in four countries: Myanmar, Nepal, Ethiopia and Zambia. In all four countries, the objective was to increase the income of 40,000 households with at least US$200 annually. In Zambia, the target was set on 14,000 households. The idea is that the sales made to RPI participants will stimulate non-participants to also purchase micro-irrigation technologies. In 2009, under the Rural Prosperity Initiative, iDE Zambia developed the so-called Mosi-O-Tunya treadle pump. This is a low-cost version of the pressure pump, affordable for many more farmers.

During the RPI programme various system designs were promoted. The design that appealed to a majority of the farmers was the 200 m² version, which has the following system components (see also Figure 12.1): (1) a 200-litre poly tank in option; (2) a screen filter; (3) main lateral made from low density polyethylene; (4) laterals made from low density polyethylene and (5) micro-tubes.

The drip kits were sold to farmers with or without a tank, whereas the treadle pumps promoted along with the drip were sold separately. Offering loans for buying these technologies, through an associated micro-credit programme, was an important aspect of the project's strategy to promote sales.

'Contact farmers' played an important role in the implementation strategy. A contact farmer is a person in the village to whom iDE assigns the role of intermediary. When iDE, or iDE's field staff, are planning a training or workshop, the contact farmer is expected to assemble farmers interested in attending the meeting. Also, the contact farmer is supposed to be an exemplary farmer

Figure 12.1 Drip kit in Zambia as promoted under the Rural Prosperity Initiative.
Source: Obed Tuabu.

who can show the usefulness of new irrigation technologies to other villagers. Moreover, when a farmer needs assistance from iDE in any way, he or she can get in touch with the contact farmer, who can then inform iDE about the needs of the farmer.

The 2008 RPI annual progress report mentions that more than 798 treadle pumps were sold, and more than 1,442 drip kits were purchased by RPI farmers. In 2009, iDE found an expansion of small farmers' production area by 300 ha. Also, iDE field staff reported important changes in the livelihoods of smallholders. Farmers were, after adopting irrigation technologies, for example able to pay for school fees and medical expenses, to invest in new irrigation technologies, to build a new house and to purchase other goods like televisions, radios, solar panels or bicycles (iDE, n.d.).

The anchor of iDE's belief in the potential of low-cost irrigation technologies to alleviate rural hunger and poverty is the figure of the individual entrepreneur: he or she is the key unit and starting point of rural development. iDE calls its entrepreneur-based market-led approach to rural development PRiSM (Prosperity Realized through Irrigation and Smallholder Markets) (iDE, 2012). A key aspect of this approach entails the assumption that smallholder farmers are able to recuperate the money they invested in the technology, and multiply it to invest in more profitable irrigation technologies. For example, after investing in a treadle pump, farmers are expected to earn enough to be able to invest in a motorized pump. And after investing in a drip kit that covers 200 m^2, they are expected to raise enough money to invest in a drip kit covering 500 m^2. Hence, iDE views and approaches rural smallholders as business-oriented people, who increase their income by continuous investments in more profitable irrigation technologies.

iDE's approach recognizes that women play a central role in agriculture: it acknowledges that women generally conduct most of the work, but often do not have the same access to resources and control over outputs as men. iDE also admits that women tend to contribute more than men to children's food intake, health, and education, when their income increases. It is for these reasons that iDE strives for benefits from program interventions to be shared equally between men and women (IDE, 2012). Moreover, in its 'gender statement' (iDE, 2008), iDE explicitly states its belief that households, communities and society as a whole will become stronger as long as gender equality is ensured in income, education, health, asset ownership, and economic rights and influence. (iDE, 2008).

iDE presents the introduction of low-cost drip kits into Zambia as a practical response to the labour and productivity needs of smallholder farmers who use the laborious bucket irrigation method of irrigation, many of whom are women.

Use practices and farmers' narratives

In this section we describe two dominant narratives around irrigated agriculture, based on how irrigated agriculture is being practiced and talked about by both farmers and project implementers. To a large extent these narratives can be seen to have emerged in relation and response to iDE's approach of promoting agricultural development through the adoption of low-cost irrigation technologies. We show that these narratives are associated to investments and risk taking at *an individual level,* and thus are based on treatment of smallholder farmers as operating on the basis of individual preferences, resources and constraints. The effect is that it depicts gender questions mainly in terms of differences between men and women, mostly in terms of control over financial resources and priorities for expenditures. Women are thus portrayed as having less control over financial resources, while having greater responsibilities for family and household expenses. Because of this, so the analysis goes, they are less likely and less able than men to invest in technological or productive assets. We show how this assumption that farmers invest individually in their gardens is erroneous, as smallholder farming is a deeply relational and collective family affair centred on collaborations between men and women. The distribution of the burdens, costs, incomes and benefits of farming happens through intricate intra-household and wider kinship-based negotiations that are informed by culturally specific 'conjugal contracts' (Jackson, 2012; Crehan, 1997). Hence, decisions to invest or abilities to improve productivities and profits do not reside in individual persons, but come about in sometimes long-winded processes of bargaining and sometimes struggle between household members who have differential powers. Not recognizing this may inadvertently result in further strengthening the abilities and powers of men to bargain for intra-household shares in labour, resources and incomes.

Households and labour division

In Kafue and Kabwe, households can be characterised as semi-autonomous but overlapping production and consumption units. On the main household fields – which include gardens and rain-fed fields – in most rural households, cash crops and maize are grown collectively. Cash crops are sold locally or on an urban market, while most of the maize is retained for self-consumption. On these collective household fields, men, women and the older children work hand in hand and are almost equally involved in all farming tasks. Women and men in the visited rural communities indicated there is some division of tasks, however. Farming jobs which are not considered heavy, but are seen as needing patience and precision, such as weeding and planting, are generally considered female jobs. Tasks like ploughing, buying inputs on the market, the maintenance of irrigation equipment and the marketing of cash crops usually are considered as jobs for men. However, this division of tasks appeared to be flexible and dependent on the household labour relations, availability and organisation. The proceeds of the collective household fields are in principle used for the benefit of the entire household; utilized for example for irrigation inputs, equipment, school fees, clothes and luxury goods. Although both men and women indicate that decisions about expenditures are made in mutual deliberation, the head of the household, which is in Kabwe and Kafue usually a man, has the final control about what the money is spend on. Also decisions about what to grow in the garden or what inputs – such as what fertiliser or farming equipment should be bought – were mostly made by the household head.

In addition to the collective fields, about a third of the women in monogamous households also have their own individually controlled plots in the garden or rain-fed fields or both. On these plots, women usually grow crops which are used for home consumption, like groundnuts, maize, sweet potato, pumpkin and rape. Only the surplus of their plots is sold and therefore women do often not earn a lot of money with their own produce. The money they earn is however theirs; they can spend it without asking their husband's permission. Because women often have no money to buy fertiliser or chemicals themselves, they depend on their husband for buying farming inputs.

Our field research suggests that farming is very much a collective affair in which men and women collaborate and mutually depend on each other. They discuss and negotiate how to best allocate the available labour and resources over the different crops and fields, with the survival and reproduction of their own family as the main and shared objective. This shows that the treatment of men and women as farmers in their own right, as individuals who operate on the basis of personal preferences and constraints, is erroneous: farming is a collective business in which all household members are involved and in which the roles of men and women are interdependent (though not equal).

Narratives on irrigation

When examining the ways in which members of vegetable farming households practice and speak about vegetable production, we tried to understand irrigated agriculture. We found variations in how people talk about gardening, about their future expectations and plans for the garden and irrigation technologies, and about potential investments. We distinguish two typical and contrasting ways of speaking about vegetable production and call these *narratives* on irrigation. By relating them to the RPI implementation practices and iDE's ideas about low-cost irrigation they can also be seen as local responses to the project. We have seen that though some members of farming households have managed to run their garden as a business, continuously re-investing their profits in new irrigation technologies, most of these business-oriented farming households have relied on loans for the mobilisation of cash resources (narrative 'gardening is a business'). Such households relied on mobilisation of household labour and are characterized with a centralisation of decision-making over spending of the income in the head of the household. Farmers who do not engage in strongly commoditized forms of agricultural production lack the ability to mobilise the necessary (household) resources and are mostly uninterested to take loans (narrative 'I can't find the money'), hence do not use drip irrigation.

Narrative 1: 'Gardening is a business'

This quote is how Mr. Mutempa summarizes what iDE taught him. He used to be employed in Kafue Estates before he moved to Kabweza to start farming. When iDE came, he learned that with irrigation, you can grow crops the whole year round. Also, he realized that he could earn more money than when employed. It is characteristic for Mr. Mutempa and some other farmers to speak about their gardens in terms of investments, future plans, and the developments which their gardens and vegetable production have gone through in the past years. These changes are usually related to shifting to new irrigation technologies. Most of these farmers use more technologies such as a treadle pump, a motorized pump or drip irrigation. Changes which have occurred as a consequence of shifting to the use of a new technology are an increase in the portions of land under cultivation, growing different types of crops, as well as an increase in income and time saving.

Most farmers started irrigation with buckets, but under irrigation with a treadle pump or motorized pump, larger areas can be watered. They explain that each time they moved to the use of another irrigation technology, they were able to expand their garden. Some farmers now manage to cultivate up to 1 hectare of land. Most farmers also started growing a larger variety of crops. As Mr. Friday clarifies: 'Under bucket irrigation, I grew only tomatoes and [green leafy vegetables]. Now, I also grow cabbages and green pepper.' Overall, farmers report big changes in their income and consequently a variety of new expenses. School fees for children and the construction of a new house are often mentioned as expenses after the increased income. Mr. Kalenga explains: 'I now earn a lot more money. It depends on the crops I grow how much I earn and I do not know the exact numbers right now, but I even opened a bank account where I put money now.'

Apart from a higher income, new irrigation technologies also often result in time savings. This is particularly true in the case of the adoption of drip irrigation. As Mr. Friday explains:

> The drip kit benefitted me a lot. When I used bucket irrigation, I used a lot of water, had to walk a lot and carry heavy containers. Now that I have the drip kit, I can just pump water in the tank and open the pipes. I don't lose water and pumping takes me only 20 to 40 minutes. It means that I can do more works at the same time.

Mr. Mutempa tells a similar story:

> The main change brought by using the drip kit and the treadle pump is the time I spend on gardening. When I still used the bucket, it took me five hours to irrigate just a small portion of land. Now with the drip kit and the treadle pump, I can irrigate a big portion of land in less than an hour.

With respect to future plans and potential investments, these business-oriented farmers have a lot to say. Most of them intend to invest in sprinklers and a diesel pump to expand the area they can irrigate. Eventually, they want to extend their area under vegetable cultivation and diversify the crops they grow. During interviews or conversations with farmers, they often expressed their wish to be further assisted by iDE in finding loans or other possibilities to obtain credits. Indeed, most of the irrigation technologies of business-oriented farmers have been funded through a loan. These farmers have taken up loans already in the past, and for multiple reasons. Anita Mweemba, for example, repeatedly has loans to pay for her education, to construct a house for her parents, to invest in her fish business, and to purchase a treadle pump.

Hence, these farmers are used to reflecting on the chronology of developments related to their irrigated agriculture, and also to envisioning future expansion and improvement of their garden. They are used to speaking in terms of loans, and appear to be eager to find opportunities to access credits. Such a form of planned expansion depends on an ability to mobilise resource and to direct benefits into re-investments. Overall, the introduction of the treadle pump does not seem to enlarge the productive workload of women or men. The extra income from irrigation originates from a higher production of crops on the main household fields. This means the extra available money is spend on farming inputs, basic needs of the family, school fees, cloths and luxury goods and when there is disagreement about how to spend the household money, men have the final say in most households.

Most business-oriented farmers are men who have both a wife and children from which they can request labour. It should be noted that the relation between the size of the garden and the amount of labour needed to cultivate the land is not a direct relation and also depends on the crop choice and use of technologies.

Narrative 2: 'I can't find money'

The way of discussing loans and investments as described under the first narrative is observed only among a limited group of farmers. Other farmers seem to reflect on investments and loans in a very different way. Often, loans are perceived with negativity or fear. Mr. Mwaanza recalls his father's experiences:

> I don't like loans, because the interests are too high. Once, my father almost lost his farm because of a loan. He got a loan to buy fertiliser, but he didn't manage to pay back the money including interest on time. So, he had to sell cattle and goats. Now, to buy a petrol pump, I don't want to have a loan. I just want to raise the money myself, by selling vegetables.

Mr. Hamweene, contact farmer for Mungu, remarked:

> They [the farmers in Mungu] are afraid of loans. [They have had a negative experience with a microcredit organization in the past.] The interest rate

was high, and [the staff of the organization] threatened to take personal properties from people that were unable to pay. Now, many people fear loans because they might not be able to return the money and they are afraid even to end up in prison.

Mr. Bimu, in Kabweza, complains that he missed out on his opportunity to purchase a treadle pump when he was not present at iDE's meeting where discount vouchers were distributed to those participating in the meeting. He cannot find money now to buy a treadle pump.

Some farmers are in a difficult situation. Exemplary is the situation of Esther Cheelo, who explained that her husband is employed in a commercial farm. They use some of his salary to pay for food and school fees. However, she does not know what the rest of the money is spent on. When she asks her husband, he will reply that he is the one that earned the money and she has no right to ask where it goes. To earn money for herself, she started gardening about eight years ago. She started with a relatively small garden, but now that she has four children to raise, she needs more money, hence she has expanded her garden throughout the years. She still irrigates with buckets, and would like to have a treadle pump and pipes to make irrigation easier. However, each time she managed to save some money, a problem arises for the family on which all her money is spent. The field is considered her individual plot and thus she cannot direct household investment toward it. Also she fails to mobilise sufficient labour.

Hence, these farmers might have the intention to develop their garden, expand their area under cultivation and grow more crops with a technology that makes the work lighter. However, they seem to be unwilling to take the risk of a loan. Some female farmers, however, express their wish to obtain a loan and invest in a new irrigation technology. Mrs. Chipoma, who manages a kitchen garden, for example, would like to have a pump so that she can extend her garden. But, she says, she does not have the money and is unable to find a loan. 'Tradition says that my husband is the boss, and I can only have a loan when my husband agrees with it.' Also Winfrida appears to have been held back by her husband in developing her vegetable garden. In the past years, her husband used to drink and therefore spend a lot of money on alcohol. Winfrida used to manage the garden by herself. Since there was not much money left to invest in the garden, she was bound to use buckets for irrigation and cultivate only a small portion of land. A year earlier, Winfrida's husband had stopped drinking and started working in the garden. Since then, they have invested in a treadle pump and later in a motorized pump, and they have expanded the garden.

Hence, the farmers who do not manage to develop their garden like business-oriented farmers, lack the resources which a loan can provide. For farmers with limited ability to mobilise resources this is risky, as they may not succeed in paying it back including interest. Especially female farmers have difficulties mobilising the necessary resources, as they cannot claim labour contributions by their husbands and often have limited cash incomes from their individual fields. Farmers who maintain their garden without making substantial investments, generally have a limited labour force to assist them; female farmers who have children who can assist them and male farmers who mobilize the labour of their wives, but generally have fewer children to request labour from.

Does irrigation make a difference?

iDE's understanding of low-cost irrigation technologies assumes that affordable, individual irrigation technologies can create possibilities for rural dwellers to escape poverty, by increasing their vegetable yields. We observed that low-cost irrigation technologies indeed make a difference. The farmers who have adopted irrigation technologies are those households who produce more crops, cultivate a larger piece of land, and continue to invest in more advanced technologies. These farmers report that since they started investing in irrigation technologies for their garden, their income has increased and they can now for example afford school fees for their children and the construction of a new house. Hence, irrigation technologies can make a difference in increasing the income of farmers and can serve as a catalyst to develop the garden into and as a business. This observation supports Polak's belief in irrigation technologies as a potential 'path out of poverty' (Polak, 2005: 134).

However, caution is needed when stating that affordable irrigation technologies are indeed the catalyst of poverty reduction. Firstly, irrigation technologies, even those that have been transformed into low-cost affordable versions, are not available to everybody. As the programme manager of iDE puts it: 'The people that buy irrigation technologies are already better off than other people without money.' iDE assisted poor farmers by distributing vouchers which allowed them to purchase drip kits for a top-up of 30,000 Zambian Kwatcha (approximately 4.50 euro) but most people had to find other ways to collect money.

Gender as social process

Many of the differences among vegetable farmers appear to be related to gender struggles. Whereas both men and women in charge of their gardens use a treadle pump for irrigation, motorized pumps and drip irrigation are almost exclusively invested in for the main household field, and in male-headed households that have access to labour. Most of the irrigation done by buckets is carried out by women. Women produce vegetables on their own fields, but these are often primarily for home consumption. Business-oriented vegetable production is almost exclusively controlled by men.

This can be explained by the gendered patterns in income expenditure and labour mobilization, as explained above. Male-controlled income from the main plots can be used for investment in productive assets while they also have the potential to mobilize labour for jobs in the garden. In our study men, in their role as head of the household, are in a better position and therefore more likely to take the risk of obtaining a loan, return their debts and interests, and re-invest in new irrigation technologies. Where in practice these men do this representing a household unit, iDE's market-based approach depicts them as individual entrepreneurs. Women, who generally cannot claim labour from other household members and who have relatively small individual plots from which they do control benefit flows are less likely to be able to accumulate the money required to pay back a loan and interest. Some female farmers, who have the desire to take loans and envision higher yields and productivity, are prevented by their husband from doing so.

The iDE contact farmer for the Kabweza community is frequently put forward by iDE as an example to show how low-cost drip benefits women. Although she is heralded as a model and makes use of this status herself, she does not seem to be representative of other women in the community. Yet, it does show that gender relations and identities are also renegotiated in the light of new agricultural practices and irrigation technologies, providing opportunities for some women farmers.

Conclusion

Irrigation technologies are used differently by different smallholder farmers. We identified two narratives, two different ways in which farmers speak about and reflect on irrigated vegetable production. Some farmers are seen as business-oriented. They regularly borrow money to invest in irrigation technologies and other farming inputs to increase their income. They themselves also reflect on their vegetable production in terms of past and future changes and developments, and narrate clear plans on how to develop their garden in the future. This is not only a story about an orientation of development but also a practice rooted in social relations of power and ability to mobilize resources and labour. Other farmers do not speak in terms of loans and investments. They are not willing to take the risk of a loan, or are unable to find a way to obtain one. They have ideas on how to make the work in the garden less intensive, and to expand the area under cultivation, but want to save up money to invest in irrigation technologies which can achieve these changes rather than take a loan to finance these. This often does not happen. Where 'business-oriented farmers' have the power and ability to mobilise resources and labour from household and/or community members, people in this second group lack such abilities.

iDE is right to assume that irrigation technologies potentially make a large difference in the income of smallholder farmers. However, analysing farmer's narratives on vegetable production and irrigation technologies reveals that not everybody is inclined to take the risk of a loan to invest in new technologies. This is linked to the ability of farmers to put demands on larger social structures such as the household, which in practice remains an important unit of agricultural production, but also beyond the household. Men rather than women are more likely to be able to mobilise such resources, meanwhile diverting benefits in the same direction. As Namara et al. (n.d.) argue, the diversity in households, needs, capacities and production strategies is large and results in an unequal benefit of project implementation to local people.

iDE and its market-based approach are aimed at individual farmers who are perceived as entrepreneurs. The assumption is that rural hunger and poverty can be solved when rural smallholders take a loan and invest in irrigation technologies. The yields of vegetable farming with the help of these technologies will enable these farmers to recoup the money of the loan and increase their income. However, the narratives on irrigation which we examined suggest that the understandings of vegetable production by local farmers do not entirely fit iDE's approach. Our analysis shows that seeing individual farmers as potential entrepreneurs has two important implications. First, low-cost irrigation technologies are mainly adopted by farmers already producing for a market. Second, aiming at individuals' 'backgrounds' the deeply relational and networked character of farming in Southern Africa whereby mobilizing resources is a household gendered affair rather than an individual one. The poverty impacts of low-cost water technologies are thus potentially skewed, favouring the intra-household bargaining positions of men and make it more difficult for women to derive benefits from low-cost irrigation technologies.

References

Borsboom, V. (2012). *"You have to dig for the money in the soil": A discourse analysis of low-cost irrigation technologies in rural Zambia*. MSc Thesis, Wageningen University, Wageningen.

Crehan, K. (1997). *The fractured community: Landscapes of power and gender in rural Zambia.* Berkeley, CA: University of California Press.

Ingen-Housz, F. (2009). *A gender perspective on treadle pump profit making a case study of treadle pump irrigation in IDE's project areas Kafue and Kabwe, Zambia*. MSc Thesis, Wageningen University, Wageningen.

International Development Enterprises (IDE). (no date). *2008 Annual progress report*. Lusaka: IDE Zambia.

International Development Enterprises (IDE). (2008). Gender statement. Available at www.iDE-canada.org/OurMethod/Gender (Accessed February 17, 2012).

International Development Enterprises (IDE). (2012). *IDE, cultivating potential*. Available at www.iDEorg.org (Accessed January 25, 2012).

Jackson, C. (2012). Conjugality as social change: A Zimbabwean case. *The Journal of Development Studies* 48(1): 41–54. doi:10.1080/00220388.2011.629649

Kabeer, N. (2000). *The power to choose*. London and New York: Verso.

Keller, J. (2004). *Irrigation technologies for small holders. Limits to increasing the productivity of water in crop production*. In J. Keller, K. Andrew, and D. Seckler (Eds.), Discussion Paper. Available at www.winrockwater.org/docs/Irrigation%20Technologies%20for%20Small%20Holders_rev.doc.

Magwenzi, K. (2011). *An analysis of adoption of low cost drip irrigation kits in Zambia*. MSc Thesis, Wageningen University, Wageningen.

Mupfiga, M.T. (2011). *The farmer and the drip planner chart: A user based evaluation of the applicability of a smallholder drip irrigation scheduling tool*. MSc Thesis, Wageningen University, Wageningen.

Namara, R., Awulachew, S.B., and Merrey, D.J. (no date). *Review of agricultural water management technologies and practices.* Paper for MoWR/MoARD/USAID/IWMI Workshop.

Polak, P. (2005). Water and the three other revolutions needed to end rural poverty. *Water Science and Technology* 51(8): 133–143.

Polak, P., Nanes, B., and Adhikari, D. (2007). A low cost drip irrigation systems for smallholder farmers in developing countries. *Journal of the American Water Resources Association* 33(1): 119–124.

Postel, S., Polak, P., Gonzales, F., and Keller, J. (2001). Drip irrigation for small farmers. *Water International* 26(1): 3–13.

Tuabu, O.K. (2012). *The innovation of low-cost drip irrigation technology in Zambia. A study of the development of KB low-cost drip by International Development Enterprises and smallholder farmers.* MSc Thesis, Wageningen University, Wageningen.

Zwarteveen, M.Z. (2006). *Wedlock or deadlock? Feminists' attempts to engage irrigation engineers.* PhD thesis, Wageningen University, Wageningen.

13 The conundrum of low-cost drip irrigation in Burkina Faso

Why development interventions that have little to show continue

Jonas Wanvoeke, Jean-Philippe Venot, Margreet Zwarteveen and Charlotte de Fraiture

Introduction

Research and development efforts concerning drip irrigation have traditionally been oriented toward intensive commercial farming in developed economies, focusing on ways to improve efficiencies and productivities. From the mid-1990s onward, an increasing number of research institutes and non-governmental organizations (NGOs) have engaged in efforts to make drip irrigation also accessible to smallholder farmers in developing countries. This entailed attempts to design much smaller drip systems, which would be cheaper and easier to operate than the high-tech systems used by farmers in developed economies. These systems would quickly become known under the generic term of 'drip kits'; they operate under low pressure to provide localized irrigation to small plots, with sizes extending from a few square meters to a few hundred square meters. Apart from the water reservoir (tank), 'kits' come in pre-packaged plastic bags (smaller kits) or carton boxes (larger kits) (see Chapter 11, this volume and Box 13.1). In the early 2000s, the idea of making a modern technology suitable for use by poor smallholder farmers acquired wide resonance among a diverse group of actors working in the fields of agriculture, water governance, irrigation, and more generally the environment and development (Cornish, 1998; Kay, 2001; Polak et al., 1997; Postel et al., 2001). Smallholder drip irrigation has now become one among a number of popular 'development technologies', promoted by many organisations that aim to improve smallholder farming in the South. Yet, the use of drip kits outside of development projects arena remains at best sketchy (Chapter 14, this volume).

What we find remarkable, and will attempt to explain in this chapter, is that the absence of sustained drip kits use by farmers did little to temper the enthusiasm of their promoters, and has not dented the reputation of the technology as a promising development device. It seems that when one project ends, a new one is implemented to promote and disseminate drip kits, even when evidence of use by farmers is lacking. We use the rest of the paper to provide an explanation of why development actors continue promoting drip kits even when there is little empirical evidence to support the belief that, indeed, they deliver on the promises of poverty alleviation at household, national and continental scale.

Our objective here is not to dismiss a specific technological package, the 'drip kit', nor is it to analyse why smallholder farmers are not interested in the kits (Dittoh et al., 2013; Friedlander et al., 2013 as well as Kulecho and Weatherhead, 2015, 2016 provide insightful analysis on the constraints smallholder face to use such kits), nor do we aim to compare how different dissemination approaches (e.g. handing them out for free versus having farmers pay for them) influence rates of adoption and sustained use. Instead, we use our analysis of projects promoting drip kits to shed light on the workings of development cooperation, considering our case as symptomatic for the sector whereby 'development projects' have many flaws including short time line and a lack of quality supervision and follow-up.

In the section that follows, we start by describing the enthusiasm for drip kits in Burkina Faso and our elusive search to find these in farmers' fields. We then describe the analytical framework that has guided our analysis and the methodology used. The third section identifies the five pillars on which development actors rely to bring into being and indeed perform 'drip kits' as a success. A short conclusion summarizes our main finding, which is that drip irrigation in Burkina Faso is an exemplary case of how development cooperation works: rather than experiences of use by farmers, its success is the result of the active storytelling, designing and staging interventions, analysing and reporting by the parties promoting drip irrigation.

Box 13.1   An insider quarrel: which drip kits are best?

Broadly speaking, two types of smallholder drip irrigation systems (also called 'drip-kits') can be identified. The first type of system, designed and marketed by large drip irrigation equipment companies such as NETAFIM and NaanDan-Jain, are 'classic' drip irrigation systems that have been 'down-sized' to fit smallholder plots (the Family Drip System – FDS- of NETAFIM was, for instance, originally designed to cover a square plot of 500 m^2 but can be adapted to variable plot sizes). These systems are made of relatively advanced components such as in-line drippers that require good water filtration and a minimum water pressure of 0.3 bars (Kay, 2001). Though they are widely seen as being the best quality low-cost drip irrigation systems on the market, they are also criticized for being difficult to operate for smallholders and rather expensive. The second type of systems is the drip kits promoted by NGOs and social enterprises. These kits are often manufactured by small companies in India and China and are the results of a quest to decrease manufacturing costs and improve the ease of use (Polak, 2008). They are made of cheaper (and often more fragile) plastic pipes, use micro-tubes as emitters (so as to avoid clogging) and can be adjusted to plots as small as 20 to 100m^2 (though they are most often promoted in a range of size going from 250 to 1,000 m^2). Regardless of the system, debates among insiders reveal that each of the two options has specific requirements for use, their advantages and drawbacks.[1]

Early enthusiasm, a wealth of projects and illusive drip kits

In the late 1990s and early 2000s, several books intended for a wide audience (Polak, 2008; Postel, 1999) and specialized publications (Polak et al., 1997, Postel et al., 2001) voiced enthusiasm for smallholder drip irrigation systems, often on the basis of a limited number of experiments conducted in South Asia. These publications highlighted that smallholder drip irrigation offered the prospect to save labour and water, while improving yields and food security, as well as boosting incomes at household level. This was followed by a series of projects implemented by different consortiums (funded by different organizations) in different regions of the world during the 2000s.[2] In the early 2010s, a new series of publications reporting on some of the major development projects that had been conducted in sub-Saharan Africa voiced the same enthusiasm regarding the prospects offered by low-cost drip irrigation for smallholders (Burney et al., 2010; Burney and Naylor, 2012; Woltering et al., 2011a, 2011b).

These publications reflected a shared eagerness to test and promote low-cost drip irrigation system in developing countries among research institutes, development agencies, NGOs, and national governments alike, who widely advertised the successful dissemination of thousands of drip kits to smallholders in sub-Saharan Africa (for instance, Abric et al., 2011; ICRISAT, 2005; Kabutha et al., 2000). Burkina Faso is no exception to this trend and from 2004 to 2015, there were no less than eight projects aimed at promoting smallholder drip irrigation in the country (see Table 13.1). At the time of conducting fieldwork

Table 13.1 List of drip irrigation projects initiated since 2004 in Burkina Faso

Dates	Project Name	Funding agencies	Main implementing agencies
1. 2004–2007	African Market Garden (AMG)	USAID/ Africare/ Swiss Development Cooperation (SDC)	ICRISAT, INERA
2. 2007–2010	Approche Intégrée pour le Développement de la Maraîcherculture (AIDEM)	Swiss Development Cooperation (SDC), Burkina Faso Office (BuCo)	Optima Conseils Services (OCS), GEDES
3. 2008–2012	Drip irrigation promotion	IFAD	IFDC

Dates	Project Name	Funding agencies	Main implementing agencies
4. 2008–2014	Projet d'Irrigation et de Gestion de l'Eau à Petite Echelle (PIGEPE)	IFAD & Government of Burkina Faso	Ministry of Agriculture (MAHRH)
5. 2009–2012	Enhanced Homestead Food Production	USAID	Helen Keller International (HKI)
6. 2010–2014	Programme de Développement du Maraichage par l'Irrigation Goutte à goutte (PDMIG)	BuCo/SDC	GEDES, OCS, CSRS, Kali Service
7. 2012–2013	Water use and sustainability in market gardening in Burkina Faso	Self Help Africa (SHA)	SHA, ADECCOL NGO, and iDE
8. 2011–2015	Scaling Up Micro Irrigation (SUMIT)	SDC	iDE

Source: The authors, 2015.

for this research (2011–2014), enthusiasm for smallholder drip irrigation in Burkina Faso was high, with several ongoing projects and others starting. We interpret this multiplication of projects as testimony of a continued belief in and enthusiasm for drip irrigation as a promising development technology, and also take it to mean that previous dissemination efforts were held as successful (or at least promising).

Our own observations and a review of the few studies documenting efforts to disseminate drip irrigation to smallholders, however, provided little support for such enthusiasm. When we first ventured into the field – nearly 10 years after the first projects promoting drip irrigation in Burkina Faso had started – we had difficulties to locate sites in which drip irrigation had been promoted in the past. There seemed to be little institutional history of drip irrigation projects. As far as ongoing projects were concerned, the sites we were told to visit appeared to be demonstration plots, in which farmers used drip kits with significant external support. With the help of individuals that had been involved in projects promoting drip irrigation in Burkina Faso over the last decade, the first author of this article visited all 87 project sites he could identify and in which drip kits had been promoted. Out of these 87 sites, drip kits were in use in 28; these were sites of on-going projects. All other sites had once seen drip kits (maybe) but farmers had stopped using them as soon as projects ended and no sign of it could be found (Wanvoeke et al., 2016). For instance, of the several hundreds of farmers that were said to have benefitted from the African Market Garden Project (AMG), we could only identify one who had continued using drip irrigation after the project ended in 2007 (see Wanvoeke et al., 2015b for a detailed description of the history and results of the AMG project). Likewise, in February 2015, just after the Programme de Développement du Maraichage par l'Irrigation Goutte à Goutte (PDMIG) ended in December 2014, none of the 15 demonstration sites that the NGO implementing the project had said to have set up seemed to be in use. At the peak of the vegetable gardening season, fields were idle, with drip lines lying in messy heaps in the corners of the fields. As for the PIGEPE project, the main government initiative promoting drip irrigation at the time of the study, field visits indeed revealed drip lines dotting the landscape (in 2012, 488 kits out of the 15,000 originally envisioned in 2006 had been disseminated to farmers).[3] Yet, although farmers had left them in the fields, they preferred using watering cans and hoses to irrigate their plots (see Plate 13 in the colour plates). These observations confirm what other scholars have said about the difficulties faced by smallholders to use drip kits and their rapid abandonment once the projects end, for instance in Kenya and Zimbabwe (Belder et al., 2007; Kulecho and Weatherhead, 2005; Chapter 14, this volume).

Theories and methodology

Within the broad discipline of the anthropology of development, there is a tradition of work that investigates the narratives and practices of development projects and aims at explaining how projects come to be seen as failures or successes (Mowles, 2010; Olivier de Sardan, 2005; Rottenburg, 2009; van Assche et al., 2012). For our analysis, we notably draw on Mosse (2005). His argument is that writings about development project outcomes do not merely report a reality that is 'out-there', but instead actively construct and even help perform it. Projects realities thus come into being 'through the interpretive work of experts who discern meaning from events by connecting them to policy ideas, texts, log-frames, and project documents' (Mosse, 2005: 157). According to Mosse, 'development success is not merely a question of measures of performance, it is also about how particular interpretations are made and sustained socially. It is not just about what a project does, but also how and to whom it speaks, [and] who can be made to believe in it' (Mosse, 2005: 158). In the same vein, and adopting a science and technology study perspective, development projects can be seen as arenas, or knowledge infrastructures in which realities are generated through the production of specific data, results, outcomes, and are interpreted through specific indicators by actors coming from specific networks.

We use these ideas for analyzing the travels and travails of smallholder drip irrigation. They turn the analytical attention to how a technology such as drip irrigation is made 'successful' through the efforts (reports, information, data, interpretations, stories) of the coalition of development practitioners, funding agencies, scientists and policymakers who are involved in its design, testing, promotion and dissemination. Together these efforts help bring a specific smallholder drip irrigation reality into being, one in which the technology figures as something promising, even though it is little used by farmers themselves. We have identified five specific ways through which this happens: (1) fitting smallholders to the technology; (2) demonstrating success through the use of experimental plots and pilot farmers; (3) highlighting results 'at scale' through specific metrics; (4) aligning the technology to broader development narratives; and (5) creating a network of actors supporting the dissemination of the technology. We further describe these in the following sections.

We base our analysis on a broad literature review of past experiences with and current trends in the promotion and dissemination of drip kits in developing countries, and on an in-depth analysis of the grey literature (project documents, reports, Web-based information) of projects implemented in Burkina Faso. This yielded a comprehensive list of all actors involved in the sector in Burkina Faso, most of whom we contacted for interviews and discussions. In addition, we conducted fieldwork in Burkina Faso, in four phases. First, from June 2011 to November 2012 we interviewed 44 agents from international and national development agencies, government officials, non-governmental organizations, and private companies involved in the promotion of smallholder drip irrigation in Burkina Faso. Second (November 2012 to November 2013), we visited 28 of the 87 drip irrigation project sites we had identified through these interviews. Through these visits, we aimed at gaining a better understanding of the implementation modalities of drip irrigation projects and of drip irrigation-in-use. Third and during the same period (November 2012 to November 2013), we selected three projects (PIGEPE, SUMIT and PDMIG) for more detailed cases studies. We also used this period for additional interviews with staff and partners, visits to projects sites and for participatory observations during project events (promotional campaigns). Fourth, in February 2015, and after some of our initial critical conclusions had raised questions among the organisations promoting drip irrigation in Burkina Faso, we conducted a short visit to assess how the situation had evolved in some of the sites we had previously visited and documented.[4]

Beyond events: staging interpretations, performing realities

Fitting smallholders to the technology

The engineers working to re-design and re-adjust drip irrigation to smallholder farmers' needs were genuinely committed to helping solve problems of poverty and hunger. They were, however, guided by (largely untested) assumptions about smallholder farmers. These were thought to be poor, uneducated, technologically unskilled, and to have (very) small (square) plots in which they wanted to grow vegetables. Hence, the systems were to be low-cost, requiring a low initial capital investment,[5] function at low pressure (to decrease energy cost), whereas they were also designed and tested in the form of ready-made kits for ease of operation and maintenance as smallholders, it was considered, lack the necessary know-how to operate drip systems (Cornish, 1998; Keller and Keller, 2005). An additional requirement that guided the design of such a system was that it had to be readily reproducible in a variety of regions

(Polak, 2008), preferably with locally available material and by local artisans or manufacturers.

The design principles may appear to be sound, but perhaps more than referring to any living rural person in Africa, the poor farmer for whom drip kits were designed exists above all as the iconic rural development client that justifies the existence of aid organizations and donors. We do not want to make any claim about whether or not this client really exists (poor farmers are indeed in the millions); our point is that the story of a poor farmer as the grateful recipient of a simple and appropriate irrigation technology was, and is, a very attractive one to market to a donor audience. In this respect, it is telling that almost irrespective of where drip kits are disseminated, similar documentaries and short press articles with photos appear that tell strikingly comparable stories of how drip kits were the steppingstone to help a particular poor farmer family escape poverty (for a particularly striking video see the 'personal story' of the first CEO of iDE-India in the *Uncommon Heroes* film series of the Skoll foundation; www.skollfoundation.org/approach/uncommon-heroes/). Against the near absence of any farmer actually using smallholder drip systems in their fields, the continued persistence and production of such stories is ironic.

In actor-network terms, one could say that in the process of designing the 'appropriate' technology, also an appropriate user was invented: the small, poor farmer nevertheless capable of making good use of this technology with proper training. An even more appropriate user is the woman farmer responsible for household management, hard working and more entrepreneurial than her male counterpart, and whose workload needs to be diminished so that she can take care of her other chores. What is problematic here is not to say that poor farmers exist, it is the tendency to standardize smallholders the same way drip kits have been standardized, giving the illusion that a unique technology can meet the needs of all smallholders alike, independently of their specific situation. The configuration of the user, though, is key to sustain the appeal of drip kits: the existence of this much advertised 'iconic farmer' who realizes the benefits of drip kits also provides a way to explain why certain projects 'fail' and to call for further funding: this is because smallholders fall short of the icon and need to be further helped. This iconic farmer often operates in specific places: the experimental plots, which also play a key role in sustaining the positive imagery of drip kits as described below.

Performing success through experimental plots and pilot farmers

One oft-used strategy to demonstrate the success of drip kits is to literally 'create the context/conditions' for these to be effective and successful through the establishment of experimental plots involving 'pilot' farmers. Dedicated field extension agents actively support these farmers during the project implementation phase, helping them to become the entrepreneurs of development narratives and project documents, in other words the 'public face' of drip irrigation (see Wanvoeke et al., 2016 for a description of one of these public face farmers).

There, researchers and development agents use the realities that they themselves actively helped construct to stage the promises of drip kits, which are then reported in scientific publications or dramatized and staged in promotional campaigns. Our argument here is not that it is wrong to run pilot sites and demonstration plots. We want to show how these crucially figure in the dramatic staging of particular approaches or technologies – in our case drip irrigation kits – as successful. More importantly, thanks to the heavy financial and human investment, experimental sites look attractive and well-organized: because they literally perform the reality that projects strive after, they function as billboards to advertise the success of the projects as well as of the technology. Just as many other development projects do in sub-Saharan Africa, projects promoting low-cost drip irrigation like bringing outside visitors and potential donors to these sites, using them as successful examples of what drip irrigation can help achieve (see the foreword by Jean Marc Faurès). In the near absence of the use of drip kits by farmers, outside of any development arenas, this vision often becomes the only drip irrigation reality that is being portrayed.

This specific drip reality performed on experimental sites by pilot farmers is also the one against which what is happening in farmers' field is assessed. Hence, if results obtained in real-world (uncontrolled) conditions are disappointing, this tends to be interpreted as a failure of farmers or, in some instances, as a failure of projects to provide an adequate follow-up, never a failure of the technology. But even those results would not be enough to explain the attention that drip irrigation kits has received without a strategy highlighting that drip kits could achieve change 'at scale'; this is when specific metrics come into the picture.

Highlighting 'results at scale' through specific metrics

A common way to establish and perform low-cost drip irrigation as a positive and effective development technology is to choose specific indicators for measuring success. All drip irrigation projects in Burkina Faso used the number of kits disseminated and/or the number of farmers trained as their primary measure of success (see Table 13.2). In most cases, as for many development projects,

Table 13.2 The metrics of success of different drip irrigation projects in Burkina Faso

Projects names	Metrics of success
1. African Market Garden (AMG)	2000 FDS type kits distributed among farmers (500 in Burkina)
2. Approche Intégrée pour le Développement de la Maraîchiculture (AIDEM)	– 14 drip sites installed – 28 drip kits of 500m^2 distributed
3. Drip irrigation promotion (IFAD Grant)	– 60 kits of 10 m^2
4. Projet d'Irrigation et de Gestion de l'Eau à Petite Echelle (PIGEPE)	– 4200 kits are acquired – 1200 kits are disseminated

(Continued)

Table 13.2 (Continued)

Projects names	Metrics of success
5. Enhanced Homestead Food Production	– 300 drip kit of 20 m² installed
6. Programme de Développement du Maraichage par l'Irrigation Goutte à goutte (PDMIG)	– 15 drip sites installed – 30 kits of 500m² distributed – 450 farmers trained – 60 000 farmers have visited the pilot sites – 3450 farmers have adopted drip kits
7. Water use and sustainability in market gardening in Burkina Faso	– 7 gardens for drip are installed – 21 kits of 500 m² and 7 kits of 100 m² distributed
8. Scaling Up Micro Irrigation(SUMIT)	– 4000 kits of different sizes are sold

Source: Authors, 2015 (Data collected form project reports).

these numbers are self-generated and not checked by any independent organization, or backed up by feedback from users. If kits are said to have been sold (as opposed to given away), this is meant to be further proof of their potential: which poor farmer would indeed purchase a drip kit if she/he had not identified its value and was not planning on using it. Yet, and irrespective of formal philosophies, the mode of operation of most projects is that the vast majority of drip kits are sold to other development agents (NGOs, government agencies) as opposed to being sold directly to individual farmers (Venot, 2016). These agents then disseminate the kits to farmers, often heavily subsidized or for free, as part of their development projects. Indeed, direct purchase by farmers is rare. For example, iDE reported that since 2012 they sold 4000 drip kits in Burkina Faso. Of these, 85% was sold to 'institutional clients', that is, development organisations who conduct their own drip irrigation promotion.

Using the number of drip kits sold or disseminated as the prime indicator of their effectiveness is, in other words, yet another way to help bring about or perform the reality of development cooperation actors, a reality in which low-cost drip irrigation is what farmers need and want (as they accept or buy them), which is why it is worthwhile to continue financing its promotion.

Aligning the technology with development buzzwords

Beyond pilot farmers and metrics such as the number of kits sold or distributed, it is the ability of drip kits to lend themselves to multiple development buzzwords and development narratives that also explain why and how smallholder drip irrigation has come to be seen as a successful development device. In other words, drip kits are 'successful' because they hold the promises of addressing some of the most pressing challenges facing the African farmer.

Smallholder drip irrigation has, right from the start, been discussed in relation to its potential to alleviate hunger and poverty (Bala, 2003; Postel et al., 2001) and to enhance food security (Burney et al., 2010; Lewis, 2010). This is because drip irrigation has been promoted as a way to boost vegetables production during the dry season, leading to improving nutrition though self-consumption and improving revenues through sales. Results from experimental plots obtained as part of cross-country international research projects have been central to legitimize these promises (see for instance Pasternak et al., 2006 and Woltering et al., 2011a).

In addition to poverty alleviation and food security, drip irrigation has also been discussed in relation to gender equity and women empowerment that are high on national and international development agendas. The drip kit discourse indeed highlights the small size of the kits and their ease of operation, which make them 'particularly suitable' to women who are said to face particularly acute constraints in terms of means of investments and access to land (on the constraints women face to access land, see notably van Koppen, 2000). Given their size and mobility, it is said that these kits can be installed virtually anywhere and do not require secure land access per se. The 'investment barrier' is, on the other hand, often lifted through micro-credit schemes that are also a favourite modus operandi of development actors to help women to be independent. This suitability of the kits for women is reinforced by the fact that they are mostly used for vegetable production and home gardening, seen as a 'feminine' component of African farming systems (Dittoh et al., 2013).

A third discursive dimension which plays out in the background relates to the preservation of natural resources. Drip kits share enough commonalities with their 'hi-tech brother' promoted in developed countries to be associated with relatively high water use efficiency. This lends them an aura of 'greenness' and 'efficiency' that makes them attractive to many development actors. These indeed have to juggle with two overwhelming narratives, the imperative needs for speeding up irrigation development (which is said to lag behind that of all other continents) in a context of little availability and high unreliability of water, and for curbing sub-Saharan African environmental degradation (for a critique of the latter, see Keeley and Scoones, 2003).

Finally, what has made drip kits attractive over recent years is the close association with a notion that is gaining importance in the international development sector, that of 'entrepreneurship'. The entrepreneurship narrative plays at two distinct levels as far as drip kits are concerned, the level of the organisation promoting the drip kits and the level of the farmer. At the level of the organisation, widespread disappointment with development *aid* has led many actors to call for more private sector involvement and for mainstreaming a 'business approach' to development and public policy (see, for instance Chapter 18 for a study of a privately led mechanism of managing public policy in the state of Gujarat, India). This notably translated in the multiplications of 'social enterprises', that is, organisations aiming at making a profit while pursuing a social or environmental goal. These have come to be widely seen as alternatives to 'traditional' development agencies and big businesses (see Venot, 2016 for a discussion on social enterprises and drip irrigation). In the field of drip irrigation, international Development Enterprises (iDE), maybe the most well-known advocate of drip kits, falls in this category and strongly communicates about its 'business approach to poverty' (for instance Polak, 2008; see also Box 13.2). iDE for instance insists that it sells drip kits, rather than giving them away; the story then goes that if sold, it means the drip kits are bought by farmers hence that they are needed, which lends them legitimacy (for a critique of this narrative, see Venot, 2016). For social enterprises, and the powerful private foundations who support them (including financially), African farmers are not anymore the mere beneficiaries of 'hand-outs' but rather 'clients' free to choose whether or not they want to buy a particular product. At the level of the farmer, drip kits – because they need to be purchased and allow for market-oriented vegetable gardening – are the means through which poor smallholder farmers, with adequate support and tools, will evolve into and successful profit-making rural entrepreneurs.

Box 13.2   The start of a business approach: low–cost drip irrigation promotion in Nepal

Nepal (together with India) is one of the first countries in which drip kits have been tested; the approach used there by iDE in the mid-1990s has strongly inspired their later operations in other countries. In 1994, iDE signed an agreement with the Agricultural Development Bank of Nepal to promote small–scale drip systems (Upadhyay, 2004) and started developing its first kits in 1995 to irrigate home gardens. The particularity of these kits is that they are made of PVC, rather than polyethylene like in other countries; this is because PVC was readily available in Nepal and iDE wanted to source equipment locally as much as possible.

Right from the start, iDE did not solely envision the promotion of drip kits; they aimed at establishing a supply chain for (1) the drip irrigation equipment but also for (2) the agricultural (vegetable) products. They notably supported the establishment of a drip irrigation equipment factory in Kathmandu and set up small market centres, This happened through collaboration with Winrock International and the USAID-funded Smallholder Irrigation Market Initiative and its Commercial Pocket Approach. In practice, this meant that drip kits were mostly promoted and installed in areas that were deemed favourable for commercial farming activities, including having a reliable source of water and being close to roads and market centres. Starting in 2003, Vegetable Collection Centres, which are meant to function as entrepreneurial cooperatives and trading centres for up to 300 smallholders, have been set up and Community Business Facilitators, partly paid on the basis of commission, (now dubbed Farm Business Advisors in some iDE operations in sub-Saharan Africa) identified to facilitate marketing of both agricultural inputs and outputs.

Another key element of iDE operations in Nepal, which partly explains the large number of kits that have been disseminated there, is the early association of drip kits with a new type of rural water supply known as Multiple Use Systems (MUS), which was and still is strongly supported by Winrock and the International Water Management Institute. The idea is to design rural water supply systems so that both domestic and productive water needs can be met, but through separate pipe systems. Such a design, is the idea, enables households to meet water needs for domestic purposes first (using household outlets), and then to make more risky, market decisions per household on the use of water for commercial purposes (using productive outlets) (see Plate 12 in the color plate). Recently, the commitment to purchase drip kits has often been a prerequisite for communities to benefit from a MUS system that also provides water for domestic purposes.

In spite of iDE's enduring commitment to promote low-cost drip irrigation in Nepal, the rate of adoption among smallholders has been very low. By 2015, approximately 250 MUS systems had been installed by iDE at the village level, but not all smallholders in these systems opt for using drip irrigation, and many abandon the kits after two or three years of use (Clement et al., 2015). Reasons are that low-cost drip systems are still too expensive for many smallholders, especially because the gains of vegetable gardening are not that high, and that alternatives for home garden irrigation are cheap and readily available, for instance, through hand watering, or by using a bucket or the hose of the MUS system for irrigation (von Westarp et al., 2004; Liebrand, 2015). In this regard, the promotion of low-cost drip kits in Nepal by iDE is emblematic of a technical intervention that does not travel outside the boundaries of the project development arenas.

Janwillem Liebrand and Jean-Philippe Venot

Creating a supportive coalition

The above stories in which smallholder drip irrigation figures as a successful device to help meet a large range of development objectives are not free-floating tales. They are actively produced, articulated and circulated by the individuals and organisations involved in promoting and disseminating drip kits. Their existence, in turn, influences what happens in projects: the narratives and the ideas that they contain are connected, circulated and indeed made true through networks of people, flows of money and equipment (Mosse, 2005).

Our research shows that since 2004, when the first drip kits were introduced in Burkina Faso by two research centres (ICRISAT and its national partner INERA) in the framework of the AMG research for development project (see Wanvoeke et al., 2015b), the network of actors promoting this technological package has significantly extended, further framing the technology as a success in what may appear as a self-predicating prophecy. After all, if renowned organizations, acting for the right reasons, promote a modern technology, it can only be positive.

Figure 13.1 highlights the far-reaching nature of the actor-network at play around drip irrigation promotion in sub-Saharan Africa, and Burkina Faso more specifically. It now includes several funding agencies, international and national NGOs, research organizations, private companies and small-scale entrepreneurs as well as the national government (see Figure 13.1). These actors are closely connected to each other through the different smallholder drip projects that have been implemented in the country over the last 10 years. The coalition started 'small' but with 'big international players' brought together by an Israeli scientist. The World Bank, Netafim, ICRISAT (and later SDC and USAID) got together to promote drip kits and vegetable gardening in the framework of the AMG project, whose results were published in scientific journals and widely advertised (see above). INERA, the Burkinabè national research institute, was in charge of implementing the project in Burkina Faso (see Wanvoeke et al., 2015b for a detailed description of the genesis and history of the AMG project). A bit later and surfing on the massive support they had received from the Bill and Melinda Gates Foundation in the late 2000s, iDE was looking to extend its operation in sub-Saharan Africa. It opened an office in Burkina Faso in 2011 with the support of one of its historical partners, the Swiss Development Cooperation (SDC), and quickly extended its collaborations with other actors

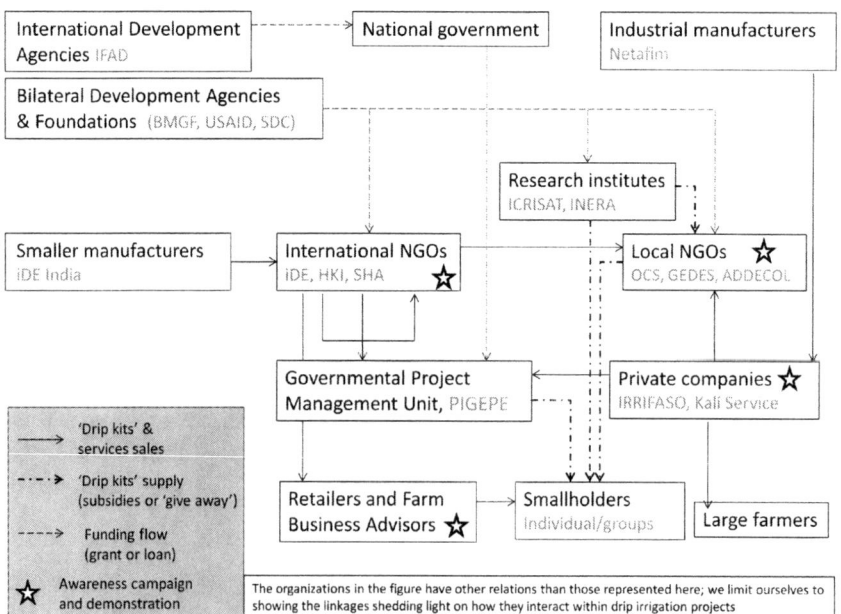

Figure 13.1 The actor-network supporting the promotion of drip irrigation in Burkina Faso.
Source: the authors.

that were also looking at promoting drip irrigation (notably IFAD). This largely happened on the basis of prior professional and personal relationships between individuals working for different organizations.

At present, the drip irrigation coalition counts 'international players' that give a global resonance to drip irrigation projects implemented in Burkina Faso. These are international aid agencies, private foundations or large private businesses involved in manufacturing and selling drip irrigation equipment worldwide. International aid agencies and private foundations are all too happy to jump on the bandwagon of funding a new technology often presented as a 'silver bullet' and that has acquired a scientific and technical legitimacy through the involvement of renowned engineers, development practitioners and researchers alike. By funding programs that promote low-cost drip irrigation, these international institutions largely contribute to shaping the positive connotation that surrounds low-cost drip irrigation. For large private companies (such as NETAFIM and Jain Irrigation Inc.), drip kits are a means to make tangible their corporate social responsibility rhetoric and policy (alleviating poverty while preserving natural resources). It is also a way to enter and tap new markets, whose actual dimension still remains to be assessed but has come to be seen as a potential El Dorado (see for instance, Prahalad and Hart, 2002).

At a more local level, the 'reality' of drip kits is also defined by the multiple actors involved in project implementation. As far as these are concerned, several main categories of actors can be identified (see Figure 13.1). First, research institutes that are focused on demonstrating the suitability and performance of drip irrigation through pilot studies; they play a key role by building the (scientific and technical) legitimacy of drip kits as a 'development technology' and were among the first to promote drip kits. Second, national and international NGOs as well as government-managed projects who, by using specific metrics and staging the potential of low-cost drip irrigation (see above), 'demonstrate' that results can be achieved at scale hence attract the attention of potential funders. In some instances, these also 'refine' the technological package providing another alternative reality: that of a technology that is standard enough to have widespread impact but adjustable to context, making it 'locally relevant'. Third, private companies and consulting firms who by retailing equipments and/or providing services on the design and use of drip irrigation systems contribute to giving an image of a vibrant sector that takes the form of farm business retailers and small rural business (still largely dependent on funds from development organizations) at the most local scale (i.e. in the vicinity of the pilot sites established by NGOs and other development actors). Finally, as stated above, farmers (male/female) using drip irrigation (be they small- or large-scale farmers) provide the 'iconic model' of the entrepreneur that is sought after by many in the development cooperation sector.

The precise terms about which these actors engage with smallholder drip irrigation may differ, but their stories and projects are nevertheless compatible and complementary (see table 2). Because all actors largely depend on finance provided by international development agencies, and are closely connected through personal relationships, flows of funds, equipment and staff, they are accountable to each other and united in their desire to bring smallholder drip irrigation into being as a successful and effective development tool, in the process also actively creating the context through and in which this becomes true.

Conclusion

Development agencies, NGOs and government actors have been instrumental in presenting smallholder drip irrigation as a promising technology that has the potential to boost incomes and improve food security. The technology figures in multiple development efforts as a tool to alleviate poverty. In Burkina Faso, and more generally in sub-Saharan Africa, these development efforts that started almost two decades ago have not led to the widespread use of drip irrigation by smallholders that was prophesized and hoped for. Indeed, the use of smallholder drip irrigation is not spreading beyond the experimental settings in which development interventions take place. Remarkably, this has done little to temper enthusiasm regarding the technology among development agents and funding agencies, which still (want to) believe in its promises.

This paper is an attempt to provide an explanation to this conundrum: why, after almost two decades of efforts that have yielded little, do many actors in the sector continue to speak of smallholder drip irrigation and drip kits in particular in such an enthusiastic way? We do so, by turning our attention away from the technical characteristics (and potentialities) of the technology per se toward the actor-network in and through which it is being enacted.

We show that the positive imagery associated with smallholder drip irrigation is the result of conscious and carefully crafted strategies by development actors, united in a supportive coalition (the first pillar of the discursive success of drip kits). This positive imagery also rests on four other key features. First is the 'engineering soundness' of drip irrigation that several contributions in this volume question (i.e. the largely advertised fact that drip irrigation leads to productive and efficient use of water). Second is the careful selection of indicators against which the success of drip irrigation projects is measured: experimental results on yields and revenues and number of kits disseminated, which say little on how farmers perceive and may use drip irrigation kits. Third, the discursive success of smallholder drip irrigation is also linked to a specific vision of African smallholders (small, poor, nevertheless capable of making good use of a technology to grow crops for the market given appropriate training). Finally, the positive connotations of drip kits hinges on the ability that actors in a wide-reaching supportive coalition have to strategically present them in relation to dominant environmental and development discourses. By doing so, smallholder drip irrigation becomes an option to solve grand challenges, which makes it attractive to international funding agencies, but also to the wider public: who indeed would be against alleviating poverty in an environmentally responsible way? As previously argued by Mosse (2005) in another field and context, it is less the events (in our case, the – limited – extent to which drip kits are used) than the interpretation of the events (the promises of drip irrigation) that matters: shortcomings and failures, instead of discrediting past interventions, call for more as 'next time, we will do better'.

As such, smallholder drip irrigation, and in particular drip kits, shares many similarities with other development 'silver bullets' that attract a lot of attention, in relation to the hope they give rise to. It is also exemplary of the workings of an international development cooperation that is little accountable to the people it is meant to serve, and largely functions on the basis of short-term projects even though most of the changes it aspires at supporting or triggering are long term trends. What is remarkable – and maybe unique – to drip kits is the in-commensurate attention they have received among development practitioners and researchers (especially at the end of the 2000s when iDE convinced the Bill and Melinda Gates Foundation to support their activities) even though, in relative terms, (investments in) drip irrigation remains marginal when put in the broader perspective of initiatives that support the development of the irrigation sector in sub-Saharan Africa. These indeed remain dominated by rehabilitating and developing large-scale irrigation systems. This is partly because drip kits had skilful spokespersons who, to paraphrase Akrich (1998a, 1988b), created interessement but also because they were both 'mainstream' and 'alternative' hence could receive the support of a wide-reaching coalition. Indeed, drip kits could be framed as contributing to a major policy priority of sub-Saharan African countries, that of irrigation development, but with a 'twist' as this was predicted to happen on the basis of widespread individual decisions from business-minded farmers rather than through public investment, which have long been maligned for its inefficiency.

Acknowledgments

We thank the anonymous reviewers whose thoughtful, though sometimes challenging, comments help us improve the quality and clarity of our argument.

Notes

1 Interestingly, due to a significant decrease in the cost of the equipment proposed by major manufacturers (such as NETAFIM) and because of difficulties to ensure a steady supply of standard quality material from small-scale producers, iDE is currently considering partnering with large manufacturers to use their material in their programs.
2 Large-scale projects include the African Market Garden project (AMG, 2004–2008; see Wanvoeke et al., 2015a for a detailed description) and the Bill and Melinda Gates Foundation's support to iDE (2007–2010). There were also smaller scale initiatives implemented by NGOs such as Catholic Relief Services, CARE, AVSF, etc. (see Chapter 14, this volume).
3 The problems faced to reach the initial targets are largely linked to difficulties (1) regarding procurement and (2) ensuring a steady supply of good quality drip kits. This would lead IFAD, the funding agency, to lower its targets and redefine the project institutional set-up (see Wanvoeke et al., 2016 for further details). By 2015, more kits had been 'procured' but the end of the project and absence of follow-up makes it unclear if these kits have reached farmers, let alone are used.

4   The main critic our analysis raised (and still raises) among smallholder drip irrigation proponents in Burkina Faso was that we had mostly documented development initiatives that had finished and were well known among insiders for has having been 'ill-designed'. On this basis, we hastily dismissed drip irrigation as ill-suited to smallholders in Burkina Faso while we should rather be critical about 'the projects', not about the technology per se. Our point of view is that it is impossible, in the sub-Saharan context, to separate 'drip kits' (that would have an inherent 'potential') and 'development projects' as the former only come into being through the latter. For us, the potential of drip kits is also defined 'through' the development projects that promote them and does not exist per se. We were also asked to pay more attention to on-going projects, in which it was said drip irrigation was successfully adopted on a large scale. This new visit actually confirmed our first assessment, that is, that projects promoting drip kits are mostly discussed and assessed on the basis of what happens in pilot sites, with the prime indicator being the number of kits disseminated even if this does not give any information on whether these are used or not and with which results in farmers' fields (see Venot, 2016 and Wanvoeke et al., 2016).

5   At most a few hundred dollars, with a basic unit cost at or under 1USD/m^2, and a plot size under 1000 m^2.

References

Abric, S., Sonou, M., Augegard, B., Onimus, F., Durlin, D., Soumaila, A., and Gadelle, F. (2011). *Lessons learned in the development of smallholder private irrigation for high-value crops in West Africa.* Discussion Paper No. 4. Washington, DC: The Worldbank.

Akrich, M., Callon, M., and Latour, B. (1988a). A quoi tient le succès des innovations? 1: L'art de l'intéressement. *Gérer et comprendre, Annales des Mines* 11: 4–17.

Akrich, M., Callon, M., and Latour, B. (1988b). A quoi tient le succès des innovations? 2: le choix des porte-parole. *Gérer et comprendre, Annales des Mines* 12: 14–29.

Assche, K. van, Beunen, R., and Duineveld, M. (2012). Performing success and failure in governance: Dutch planning experiences. *Public Administration* 90(3): 567–581.

Bala, S.K. (2003). Low-cost drip irrigation technology for poverty alleviation. *Journal of International Affairs* 7(2): 30.

Belder, P., Rohrbach, D., Twomlow, S., and Senzanje, A. (2007). *Can drip irrigation improve the livelihoods of smallholders? Lessons learned from Zimbabwe.* Global Theme on Agroecosystems Report No. 33: International Crops Research Institute for the Semi-Arid Tropics (ICRISAT).

Burney, J., and Naylor, R. (2012). Smallholder irrigation as a poverty alleviation tool in sub-Saharan Africa. *World Development* 40(1): 110–123.

Burney, J., Woltering, L., Burke, M., Naylor, R., and Pasternak, D. (2010). Solar-powered drip irrigation enhances food security in the Sudano – Sahel. *Proceedings of the National Academy of Sciences* 107(5): 1848–1853.

Clement, F., Paras, P., and Sherpa, T.Y.C. (2015). *Sustainability and replicability of multiple-use water systems (MUS).* Kathmandu: International Water Management Institute (IWMI).

Cornish, G.A. (1998). Pressurised irrigation technologies for smallholders in developing countries – a review. *Irrigation and Drainage Systems* 12(3): 185–201.

Dittoh, S., Bhattarai, M., and Akuriba, M.A. (2013). Micro irrigation-based vegetable farming for income, employment and food security in West Africa. In M.A. Hanjra (Ed.), *Global food security*. Hauppage, NY: Nova Science Publishers, Inc., pp. 178–199.

Friedlander, L., Tal, A., and Lazarovitch, N. (2013). Technical considerations affecting adoption of drip irrigation in sub-Saharan Africa. *Agricultural Water Management* 126: 125–132.

ICRISAT. (2005). *Market Garden in the Sahel. Fruit-trees and vegetables for land regeneration and management*. Final technical Report. Niamey, Niger: International Crops Research Institute for the Semi-Arid Tropics (ICRISAT).

Kabutha, C., Blank, H., and van Koppen, B. (2000). *Drip irrigation kits for smallholder farmers in Kenya: Experience and a way forward*. Paper presented at the 6th International Micro-Irrigation Congress (Micro 2000), Cape Town, South Africa.

Kay, M. (2001). *Smallholder irrigation technology: Prospects for sub-Saharan Africa*. Knowledge Synthesis Report No. 3. Rome: FAO/IPTRID.

Keeley, J., and Scoones, I. (2003). *Understanding environmental policy processes: Cases from Africa*. London: Earthscan.

Keller, J., and Keller, A. (2003). *Affordable drip irrigation systems for small farms in developing countries*. Paper presented at the Proceedings of the Irrigation Association Annual Meeting in San Diego, CA.

Koppen, B. van. (2000). *Gender and poverty dimensions of irrigation techniques for technical people*. Paper presented at the 6th Micro Irrigation Congress, Micro-irrigation Technology for Developing Agriculture, Cape Town, South Africa.

Kulecho, I.K.; Weatherhead, E.K. (2005). Adoption and experience of low-cost drip irrigation in Kenya. *Irrigation and Drainage* 55(4):435-444.

Kulecho, I.K., and Weatherhead, E.K. (2005). Reasons for smallholder farmers discontinuing with low-cost micro-irrigation: A case study from Kenya. *Irrigation and Drainage Systems* 19(2): 179–188.

Lewis, D. (2010). Can drip irrigation break Africa's hunger cycles? *Features Reuters*, May 6.

Liebrand, J. (2015). *Research report on drip irrigation in Nepal*. Drip irrigation Realities in Perspective (DRiP), Wageningen University. Unpublished research report.

Mosse, D. (2005). *Cultivating development: An ethnography of aid policy and practice*. London: Pluto Press.

Mowles, C. (2010). Successful or not? Evidence, emergence, and development management. *Development in Practice* 20(7): 757–770.

Olivier de Sardan, J.P. (2005). *Anthropology and development: Understanding contemporary social change*. London: Zed Books.

Pasternak, D., Nikiema, A., Senbeto, D., Dougbedji, F., and Woltering, L. (2006). Intensification and improvement of market gardening in the Sudano-Sahel region of Africa. *Chronica Horticulturae* 46(4): 24–28.

Polak, P. (2008). *Out of poverty: What works when traditional approaches fail*. San Fransisco, CA: Berret-Koehler.

Polak, P., Nanes, B., and Adhikari, D. (1997). A low cost drip irrigation system for small farmers in developing countries. *Journal of the American Water Resources Association* 33(1): 119–124.

Postel, S. (1999). *Pillar of sand: Can the irrigation miracle last?* New York: WW Norton & Company.

Postel, S., Polak, P., Gonzales, F., and Keller, J. (2001). Drip irrigation for small farmers: A new initiative to alleviate hunger and poverty. *Water International* 26(1): 3–13.

Prahalad, C.K., and Hart, S.L. (2002). The fortune at the bottom of the pyramid. *Strategy + Business* 26: 1–15.

Rottenburg, R. (2009). *Far-fetched facts: A parable of development aid-inside technology*. London: The MIT Press.

Upadhyay, B. (2004). Gender aspects of smallholder irrigation technology: Insights from Nepal. *Journal of Applied Irrigation Science* 39(2): 315–327.

Venot, J.P. (2016). A success of some sort: Social enterprises and drip irrigation in developing countries. *World Development* 79: 69–81.

Wanvoeke, J., Venot, J.-P., Fraiture, C. de, and Zwarteveen, M. (2015a). Smallholder drip irrigation in Burkina Faso: The roles of development brokers. *Journal of Development Studies* 52(7): 1019–1033.

Wanvoeke, J., Venot, J.P., Zwarteveen, M., and Fraiture, C. de. (2015b). Performing the success of an innovation: The case of smallholder drip irrigation in Burkina Faso. *Water International* 40(3): 432–445.

Wanvoeke, J., Venot, J.P., Zwarteveen, M., and Fraiture, C. de. (2016). Farmers' logics in engaging with projects promoting drip irrigation kits in Burkina Faso. *Society & Natural Resources* 29(9): 1095–1109.

Westarp, S. von, Sietan, C., and Schreier, H. (2004). A comparison between low-cost drip irrigation, conventional drip irrigation, and hand watering in Nepal. *Agricultural Water Management* 64: 143–160.

Woltering, L., Ibrahim, A., Pasternak, D., and Ndjeunga, J. (2011a). The economics of low pressure drip irrigation and hand watering for vegetable production in the Sahel. *Agricultural Water Management* 99: 67–73.

Woltering, L., Pasternak, D., and Ndjeunga, J. (2011b). The African market garden: The development of a low-pressure drip irrigation system for smallholders in the Sudano Sahel. *Irrigation and Drainage* 60(5): 613–621.

14 The mysterious case of the persistence of donor-and NGO-driven irrigation kit investments for African smallholder farmers

Douglas J. Merrey

Introduction

Over the past twenty years, numerous organisations have been developing and promoting low-cost, low-pressure drip irrigation kits for smallholder farmers in developing countries. These organisations include donors, governments, non-governmental organisations (NGOs), private firms and religious organisations. Large investments have gone into trying to make the kits low in cost, easy to use and reasonably efficient in terms of water use. An oft-cited early paper touted drip irrigation as a technology that 'can form the backbone of a second green revolution, this one aimed specifically at poor farmers in sub-Saharan Africa, Asia, and Latin America' (Postel et al., 2001: 3). Many companies and NGOs promote these kits throughout sub-Saharan Africa, Asia and parts of Latin America.[1] Some proponents are highly committed to promoting drip irrigation kits as a key strategy to reduce poverty and malnutrition and increase the incomes of smallholders. The kits are often touted as being especially appro-priate for women, and as relief for poor people affected by droughts.

This chapter reviews the available evidence on the promotion, performance and uptake of low-cost, low-pressure drip irrigation kits focused on sub-Saharan Africa. These kits are usually combined with either a raised bucket or a drum to hold the water supply. The paper does not examine the use of better-quality, relatively costly high-pressure drip irrigation systems used by small and medium commercial farmers producing for well-developed local and export markets. The evidence presented shows that the uptake and sustainability of low-pressure drip irrigation kits have been disappointing. The chapter then tries to answer the following questions: Why has the uptake and sustainability of use been so disappointing in sub-Saharan Africa? Why have the proponents not commissioned proper evaluations of the results of their investments? Why, in the face of this apparent lack of take-off and sustainable use, do donors, govern-ments and NGOs persist in repeating the same types of projects? And given this lack of uptake, what should donors, governments, and NGOs do differently?

The next section briefly explains the methodology and approach to writing this paper. An analytical review of the available evidence from the literature on the performance and uptake of drip irrigation kits follows. The next section discusses possible alternative explanations of the lack of sustained substantial uptake. The paper then examines the case of one major donor's program for further insights. The final section offers conclusions and specific recommendations aimed at donor agencies, NGOs and governments.

Methodology

This essay draws on what the author has learned over the past ten years reviewing experiences with a large variety of small-scale agricultural water management technologies, mainly in sub-Saharan Africa. These studies focused on a range of small-scale individualized water management technologies, and reviewed the strategies for promoting them, their performance, uptake by farmers, and the sustainability of their use. A few of these studies involved surveys and fieldwork (e.g. Rohrbach et al., 2006; Merrey et al., 2008); but most involved reviews of the available literature (e.g. Merrey, 2013; Merrey and Langan, 2014). These studies were financed by the Food and Agriculture Organisation of the United Nations (FAO), the International Fund for Agricultural Development (IFAD) and United States Agency for International Development (USAID), through international agricultural research institutions (International Crops Research Institute for the Semi-Arid Tropics [ICRISAT] and the International Water Management Institute [IWMI], respectively).

These reviews form the basis for much of this chapter, complemented by reviews of other literature. The paper also draws on recent reports from USAID's 'Feed the Future' program to substantiate and reinforce conclusions emerging from the literature. The next section reviews and synthesises the findings from this decade of research to identify the main results and patterns as they relate to drip irrigation kits.

Review of the available evidence

The literature on drip irrigation for smallholder farmers in Asia and Africa reflects the strong interest over the past decade or so by manufacturers, NGOs, governments, donors and farmers. For commercial farmers, there is little doubt about the relatively high performance of drip irrigation: in general, it can save water and labour, increase yields, and improve the quality of the produce. Because of its demonstrated high performance for commercial farmers, there has been a long-standing interest in developing and promoting low-cost, low-pressure drip irrigation kits appropriate for smallholders. Indeed, some prominent scholars have argued strongly for a major international initiative to promote low-cost drip irrigation kits as an avenue to achieve significant gains in food security rapidly and cost effectively (e.g. Postel et al., 2001).

By about the year 2000, several NGOs had developed and were actively promoting 'affordable' drip systems. Chapin Living Waters, a religious NGO, was a pioneer; its kits were tested by the Kenya Agricultural Research Institute (KARI), which still promotes kits of various sizes. The largest and most active

NGO is iDE, an international NGO founded and initially led by a charismatic proponent of low-cost water management technologies for smallholders. In India, for example, a bucket and drip kit to irrigate 100 plants over 25 m² was being marketed for about US$5.00. Shah and Keller (2002), based on experiences in South Asia, argued that it is an effective way both to help women improve their food security and incomes (Nepal) and to help small commercial farmers in very dry areas (India).[2]

The proponents of drip irrigation systems have raised high expectations with their claims that this technology could be a game-changer for smallholders in Asia and Africa, (e.g. Postel et al., 2001; see Venot, 2016a for other references). Many claims have been made regarding the capacity of drip irrigation to save water and labour and increase yields – i.e. to increase water productivity and, therefore, also improve incomes and nutrition (see multiple references cited in van der Kooij et al., 2013). However, fifteen years after the article by Postel et al. (2001) and millions of dollars of expenditure, the actual experience with low-pressure drip irrigation kits for poor rural farm households remains mixed, with a few modest successes and many disappointments.

This author's views have evolved over time, as the evidence from various studies has accumulated. A broad study of experiences with a range of agricultural water management interventions, including drip irrigation kits, in nine Southern African countries concluded that many of these technologies were promising and should be made more widely available (IWMI-Southern Africa Regional Office, 2006; Merrey and Sally, 2008). This included drip irrigation kits: in some countries, some farmers were using them successfully, though they faced a number of technical and other problems. That study was supported by USAID and implemented by IWMI. It recommended that projects avoid pushing a single water management technology and advocated the promotion of a competitive regional private-sector driven support system to make a wide range of small-scale individualized water management technologies available to farmers.

A subsequent in-depth survey led by the ICRISAT evaluated the experiences with drip irrigation kits in Zimbabwe (Rohrbach et al., 2006; Belder et al., 2007).[3] From 2000 to 2006, about 70,000 drip irrigation kits had been distributed in Zimbabwe by aid agencies through NGOs as a strategy for drought relief. This study, supported by FAO and based on a sample survey of 232 beneficiaries and 135 neighbours who continued using watering cans, found that few drip kits were used after the first season. This finding was confirmed by other studies (e.g. Merrey et al., 2008; Moyo et al., 2006).

It is important to note that these kits were distributed free of charge as a way of providing drought relief. The implementing NGOs provided training in their use, but there was no continuing institutionalised support system to provide training and advice, and spare parts were not available. All the cited Zimbabwe studies agreed that while drip irrigation may well be an appropriate technology, where there is sufficient institutional support, it is not suitable as part of drought relief programs. However, these studies all suggested drip irrigation kits have good potential, and claimed the main problems to be related to such issues as having a suitable water supply available, having access to continuous training and technical support, and being able to sell produce in functioning output markets. They also found NGOs are not good at targeting the kinds of households the donors were most interested in helping – precisely because these households had little capacity to use the kits. The only study from that period (i.e. 2000–2008) that questioned their relevance to Zimbabwe was an earlier comparison of Indian and Zimbabwe experiences with drip kits (ITC et al., 2003).

This author led a more recent study implemented through IWMI and funded by IFAD that was intended to provide advice to IFAD on how it could maximise returns on its substantial agricultural water management investment portfolio (Merrey, 2013). The review made numerous recommendations, but for the purpose of this chapter it suffices to just reiterate that the major recommendation was as follows: 'the greatest potential to achieve substantial and sustainable benefits is by investing in the development of competitive private sector markets for the sale and service of small scale technologies, especially low-cost pumps' (Merrey, 2013: 30). The report emphasized that donors should avoid promoting specific water management technologies; rather they should focus on the entire value chain, testing innovative and comprehensive approaches, and scaling up those that show promise.

In 2014, IWMI commissioned the author to carry out a comprehensive literature review of home garden experiences, with an emphasis on water management practices, especially drip irrigation kits (Merrey and Langan, 2014). That paper reviewed the available literature on the performance of low-pressure drip irrigation kits. It focused on several dimensions: technical performance, (e.g. water use efficiency, uniformity and adequacy of water to meet crop water requirements); economic and social performance, (e.g. costs and benefits, profitability and gender equity); customer satisfaction and sustainability; and overall impacts on livelihoods and well-being. Here, the main conclusions of this review are summarized.

Most studies of the technical performance of drip irrigation kits have been done in laboratory settings; few studies have been done on farmers' fields. Overall, results are mixed. The evidence for the water-saving potential of drip irrigation is not conclusive; and more important, studies of farmers' use of irrigation show clearly that saving water is rarely an important goal for smallholders.[4] There are only a few studies of the economic and social benefits of small drip irrigation kits, but those reviewed from Malawi, Eritrea and Zimbabwe all found that drip irrigation kits are only rarely profitable for smallholders, though they may be profitable for larger farmers (Merrey and Langan, 2014).

A similar finding comes from the African Market Gardens (AMG) program in West Africa. ICRISAT and its partners have been testing a horticultural production system for small producers originally developed in Israel. It combines a crop management package with high-quality low-pressure drip irrigation kits (Woltering et al., 2011a, 2011b).[5] These kits come in four sizes, 80 m² ('Thrifty'), 500 m² ('Commercial'), 'Cluster' (modules of 500 m²) and 'Community'. The latter two models are for groups of farmers. The Thrifty System is aimed at women who cultivate gardens in the cool dry season. The capital cost of the drip irrigation equipment is 20–25% of the total package. Despite finding that the returns to the Thrifty System are in principle quite good, all 600 of the Thrifty Systems distributed in Niger were abandoned within a year; on the other hand, the larger packages adopted by professional gardeners continued to be used profitably (Woltering et al., 2011a). Citing Dittoh et al. (2010), Wanvoeke et al. (2015) report a similar phenomenon in Burkina Faso: of 500 AMG kits distributed, only 1 was said to be in use in 2012.

Overall, Merrey and Langan (2014) concluded that more studies are needed to understand the conditions under which these small low-cost drip irrigation kits are economical for farmers and likely to be sustained; and studies are needed to understand the gender and equity impacts. Because of their low cost, potential for labour and water savings, proponents argue that low-cost drip irrigation kits could be an entry point for enhancing women's well-being and status within the household. At present there is no evidence on this from Africa[6] – it may well be wrong. Furthermore, there is strong evidence from multiple countries that smallholder households quickly stop using these kits, a finding supported by other recent reviews (e.g. Friedlander, Tal and Lazorovich, 2013; Garb and Friedlander, 2014; Wanvoeke et al., 2015).

The evidence is clear that drip irrigation kits are not an appropriate form of drought relief. Attempts to introduce drip irrigation kits as part of a package to enhance the profitability of smallholder agriculture have also been disappointing, at best. However, this technology continues to attract support: Wanvoeke et al. (2015) provide a long list of donors and NGOs promoting these kits in the Sahel. They include the World Bank, USAID, Swiss Agency for Development Cooperation (SDC), MASHAV (Israel's Agency for International Development Cooperation), United Nations Environmental Program (UNEP) and Africa Care. The next section turns to explanations that have been proposed for this continued support for promoting drip irrigation kits.

Proposed explanations and solutions: promoting uptake of drip irrigation kits

Several hypotheses have been proposed to explain: 1) why there has been limited uptake of low-cost, low-pressure drip irrigation kits to date; and 2) what needs to be done to ensure successful large-scale adoption. A third, more critical question is, why do donors, governments and NGOs continue to promote them following similar strategies in successive projects? This question is addressed in the next section. Proposed answers to the first two questions can be classified as follows:

Two explanations with some similarities but different priorities focus on the contextual support for *transfer* of the technology; these are the most common:

1. The *technology cum education* hypothesis: investing in improving the technology (including cost reduction), combined with better education of farmers will lead to successful adoption on a large scale (e.g. Friedlander et al., 2013; Moyo et al., 2006; Sijali, 2001).
2. The *policy-institutional-market support* hypothesis: getting the policies, markets and institutional support systems right will lead to successful adoption on a large scale (e.g. Merrey and Sally, 2008; Merrey, 2013; Dittoh et al., 2010; Belder et al., 2007; Magistro et al., 2007; Woltering et al., 2011a, 2011b; Abric et al., 2011; Giordano and de Fraiture, 2014).

There is a third type of explanation:

3. The '*technology translation*' or '*innovation systems*' hypothesis. This focuses on strengthening dialogue and learning processes in an innovation systems framework, expected to lead to successful adaptation and adoption on a large scale (e.g. Garb and Friedlander, 2014; Venot et al., 2014).

This is a more nuanced hypothesis. While the hypotheses focusing on 'getting the context right' are linear – education and markets will *ipso facto* lead to uptake of drip irrigation kits – the innovation systems hypothesis emphasizes the role of creative interactions among stakeholders (e.g. manufacturers, retailers, extension staff, farmers) as they learn from each other and 'co-create' a context in which uptake occurs.

These three hypotheses seek to solve a problem defined as 'how to be more effective in promoting drip irrigation kits'. They share a fundamental assumption that drip irrigation kits are 'good' and the problem is to find ways to overcome impediments that are preventing large-scale use.

The *technology cum education hypothesis* assumes that some combination of further technical improvements and training of farmers will lead to more rapid and sustained uptake or cost reduction.[7] Friedlander et al. (2013) present this rationale clearly. Their study is based on interviews of sixty-one drip kit users in four African countries. All users had faced serious technical problems. Overall, their findings support those of other studies on the problems with drip kits that often lead to high rates of dis-adoption. They recommend the following: redesign the kits to solve technical issues; educate users, especially in the use and repair of the kits; and encourage the adoption of complementary technologies to support their functioning such as water storage, water purification and defense of the crop against animal predation. While admitting 'drip irrigation is not a panacea', Friedlander et al. (2013: 132) conclude that 'it can be highly effective at improving health and livelihoods if successfully adopted and correctly implemented'.

The *policy-institutional-market support hypothesis* also accepts that the technology is appropriate and potentially useful, but encouraging widespread and sustained use requires more than just training and adjustments to the technology. Referring to private small-scale irrigation technologies in general (i.e. pumps, drip kits, etc.), and based on evidence from six countries, Giordano and de Fraiture (2014) identify four elements of a successful program: 1) enhancing access to technology; 2) strengthening smallholder value chains; 3) fostering supportive policies; and 4) strengthening institutional capacities to manage trade-offs among competing water users water at the watershed level. Both Magistro et al. (2007) and Woltering et al. (2011a, 2011b)[8] describe specific intervention programs aimed at linking smallholders using drip irrigation kits to high-value produce markets. Both claim that smallholders can earn substantially higher profits from adopting the respective market-oriented models ('Poverty Reduction through Irrigation and Smallholder Markets' [PRISM] and 'African Market Garden' [AMG]). Dittoh et al. (2010) argue that the AMG model is indeed a potential game-changer but that the only way to scale it up will be through 'modified public-private partnerships' to address the initial high costs involved. An explicit assumption is that large numbers of smallholders are entrepreneurs.

Based on the Southern Africa study cited above, Merrey and Sally (2008) focus on policy reforms needed to encourage the development and dissemination of these technologies, including drip irrigation. They argue that the way forward could be to encourage the private sector to offer a wide range of irrigation technologies; to create economies of scale, they suggest a large regional market, as national markets may be too small to encourage adequate investment (Box 14.1 also stresses the importance of 'starting big' to initiate a virtuous circle). They make seven recommendations aimed at policymakers. Their major overall message is that 'approaches based on experimentation, innovation, testing, adapting, and shared learning will be essential for successful application of these technologies' (Merrey and Sally, 2008: 524). Although not referred to as such, this message is the essence of an 'innovation systems' hypothesis.

Box 14.1   The SCAMPIS project: starting big to generate an adoption momentum

Inspired by the iDE experience in India, the working hypothesis underlying the design of the SCAMPIS project (SCAling up Micro-irrigation for the Poor – operated between 2009 and 2012) – was the necessity to reach a 'critical mass' of up-takers in order to sustain a viable supply chain of low-cost irrigation systems. This would, in an assumed virtuous cycle, allow sustaining drip irrigation use by farmers and further dissemination of the technology. In other words this initial 'critical mass' was expected to provide a demonstrative effect strong enough to fuel a demand flow for acquisition and repair/maintenance of equipment that would make the supply chain viable. The SCAMPIS promoters were also conscious that building the capacity of smallholders in irrigated production and marketing of high-value crops (mostly vegetables) was an indispensable ingredient to make the adoption of drip irrigation a profitable decision for farmers – a key factor for the generation of demand. The referred-to 'critical mass' was thus a notion applied not only to the number of farmers enrolled but also to the quantity and quality of the required complementary interventions. SCAMPIS ambitioned to demonstrate the feasibility of such 'critical mass' approach to low-cost drip irrigation dissemination in different contexts. Three countries in which smallholders represent the majority of farmers were selected for this purpose: Madagascar in Africa, the state of Orissa in India, and Guatemala in Central America. It was planned that the project would interlink with other ongoing rural development initiatives funded by the same organisation (IFAD), thereby benefiting from their institutional set-up and additional support (in terms of inputs supply, capacity development etc.).

The assumed 'critical mass' of up-takers was 'guestimated' at a bare minimum of 10,000 families to start with – a target that was thought to be attainable within the three-year timeframe and the resources imposed by the financier.[9] Although it was acknowledged that this timeframe was insufficient to anchor the innovation and in particular sustain the supply chain in the various contexts, it was hoped that the operation would have a sufficient demonstrating effect to raise the interest of third parties (including the line ministry, and even IFAD itself). These would then continue supporting the initiative for it to become sustainable. Quite encouraging results were obtained: the dissemination targets were indeed achieved, with interesting spin-offs in terms of nutrition improvement (Guatemala) and local manufacturing (India and Madagascar) – thus creating a significant momentum and awareness in each of the areas where the project was implemented. However, with no donor taking over, the project external funding ended after the initial three-year phase, causing a severe blow to these dynamics – although in both Madagascar and Guatemala the initiative managed to mutate and survive somehow. For instance, in Madagascar, at least one small company established during the project for assembling and disseminating low cost drip was still in business five years after the project ended, and some government projects continue to promote LCMIS. In Guatemala, other actors promoted drip in new areas.

Jean Payen

The *innovation systems hypothesis* conceives of technologies as being embedded – i.e. used, rejected, re-interpreted and shaped, within a sociocultural-economic system or network. Technology is not seen as having specific intrinsic characteristics that remain valid regardless of the social context; rather, it 'acquires its characteristics only through, and within, the network of institutions, discourses and practices that enact it' (Venot et al., 2014). Therefore, the focus should not be simply on improving the technology, doing more training, or creating more conducive markets to enable more efficient 'transfer' – though these are not irrelevant. Rather, technologies such as drip irrigation kits should be promoted by encouraging dialogues and mutual learning processes among the various actors with shared interests – including farmers. Garb and Friedlander (2014) use the metaphor 'technology translation' to describe this approach.

Garb and Friedlander (2014) compare the successful uptake of drip irrigation in Israel to the 'mostly failed' uptake in sub-Saharan Africa. They accept that in dry marginal lands, drip irrigation has 'great potential' – i.e. the technology itself is good. They attribute these different experiences to their very different contexts. In Israel, the uptake of drip irrigation occurred in a dynamic socio-technical context where the technology evolved in response to lessons learned within a system conducive to mutual sharing of lessons and adaptation of the technology and its use. In Africa, they argue that the technology has been transferred as a static physical device with no corresponding sociotechnical innovation and translation process: farmers are presented with the technology 'as is'.[10] Wanvoeke et al. (2016a) provide interesting examples from Burkina Faso: use of the drip kits requires changes in farmers' normal patterns of work, but there was no attempt to encourage innovative solutions, so farmers soon abandoned their use.

Attempts to boost adoption have focused on simplifying the technology to be 'system-free' (for example iDE), or in a few cases such as AMG, to bundle the kits with a minimum non-hardware support system – but still no re-interpretation and re-innovation process. Although AMG has shown some success compared to other programs, it is limited, and Garb and Friedlander argue that this is an artefact of the key personnel driving the program; therefore, it is not likely to be scaled up successfully. In western India, they claim there has been a sustained uptake and expansion of the use of drip irrigation systems, precisely because there is a pre-existing innovation system (though this uptake is of more robust drip systems by relatively prosperous farmers, not low-cost kits by poor farmers; see Namara et al., 2005). Attempts to transfer the same technology to African farmers have not met with sustained success because the alleged 'pre-disposing factors' are missing.

Garb and Friedlander (2014) accept the efficacy of drip irrigation technology; their recommendations focus on the need to support the emergence of innovation systems linking suppliers, farmers, markets and researchers, preferably through 'intermediary organisations' such as NGOs. They recommend promoting innovation platforms as a way to encourage shared learning of lessons, and possibly collective organisations to enable smallholders to purchase larger and more sophisticated systems. They propose that aid agencies move away from promoting technologies to more emphasis on capacity building and promoting learning organisations.

Why does promotion of drip irrigation kits continue?

The question addressed in this section is, why do donors, governments and NGOs continue to promote drip irrigation kits using similar strategies in successive projects? One possible answer to this question is a variation on the innovation systems hypothesis. This is a *contextual or 'vested interest'* hypothesis proposed as an explanation for continued support for the technology.

Based on a detailed case study in Burkina Faso, Wanvoeke et al. (2015) (see also Wanvoeke, 2015 and Chapter 13) argue that a coalition of key actors has developed a shared discourse and mutually reinforcing incentives and messages to sustain the flow of resources in support of drip irrigation. These include donors and policymakers who see drip irrigation investments as a way to modernise agriculture, alleviate poverty, save scarce water supplies, and promote economic growth; intermediaries such as NGOs, research organisations and extension services who serve as the conduits between donors and farmers and directly benefit from the flow of resources; drip kit providers, generally from the private sector, whose role is the manufacture and/or sale of drip kits – mainly to intermediaries; and 'users' of the kits – i.e. farmers who become involved in pilot testing the kits and as a result receive other benefits unrelated to drip irrigation. These actors form a network, with each one motivated to maintain the flow of resources justified by rhetoric emphasizing their shared goal of achieving development impacts. Wanvoeke (2015: 105) argues the actors continue to present – and believe – the positive image of the potential developmental impacts of drip irrigation kits 'through conscious strategic efforts' despite the lack of evidence of farmers' interest in using the kits in the absence of continued project support.

Wanvoeke and his colleagues do not use the term 'conspiracy' but that is what it appears to be: the actors may well have convinced themselves of the efficacy of what they are doing, and may well believe that just a few more pilot studies will lead to sustained take-off and game-changing transformations of smallholder agriculture. All the parties involved have developed strong vested interests in continuing these pilot programs. They seem to this author to resist commissioning credible systematic independent impact assessments.[11] One is reminded of the story of the emperor who is wearing no clothes but his subjects are reluctant to point this out to him until an innocent child does. Researchers ought to play the role of the innocent child but too often seem complicit in continuing a charade, not in some cynical calculating conscious manner, but based on their belief that somehow the problems can be solved and the technology will then take off. While attempting to explain the persistence of drip irrigation kit promotion projects, the 'vested interest' perspective does not offer a strategy for promoting drip irrigation kits more effectively – indeed it questions their efficacy in transforming smallholder agriculture.

To date, with the exception of the early studies in Zimbabwe cited above, there have been no systematic and independent assessments of the actual use, benefits and costs to farmers, and impacts of drip irrigation kits in sub-Saharan

Africa. Nevertheless, drip irrigation is widely considered a success (Wanvoeke et al., 2015), despite the absence of any evidence of sustained use by farmers. The question is, why is this so? The next section uses evidence from a specific program as a case study to address this question.

Feed the future – through drip irrigation?

USAID has been a major supporter of research on, and promotion of, drip irrigation kits. Its Office of Disaster Assistance (OFDA) financed NGOs to distribute free drip irrigation kits for drought relief in a number of countries, including Zimbabwe, as discussed above. USAID has over the years supported iDE and other NGOs as well as private firms such as Netafim to test and disseminate drip irrigation kits to smallholders. Many of its country offices have supported drip irrigation as part of their packages of agricultural development projects. USAID is also one of the donors that supported the drip kit promotion projects in Burkina Faso examined by Wanvoeke (2015) and his colleagues. Feed the Future (www.feedthefuture.gov/) is the American government's global hunger and food security program, under which a variety of projects are being supported, including several promoting drip irrigation kits.

To investigate further, the author obtained from the USAID documentation website[12] recent semi-annual and annual reports of Feed the Future's 'Partnering for Innovation' ('FTF-P4I') project, prepared by the implementing firm, Fintrac (2013, 2014, 2015a, 2015b). This project has invested more than US$1.6 million in drip irrigation products for smallholders in the past two years.[13] The project is designed to help the private sector scale up and market agricultural technologies for smallholders, and also supports a knowledge exchange website.[14] Drip irrigation kits are one of the nine technologies listed for commercialisation. Current or recent drip irrigation partners include Driptech to implement a project in India, Netafim to implement a project in Kenya, and iDE for a project in Zambia (Fintrac, 2015a). Each of these is promoting a specific drip irrigation kit in the designated country; performance is measured in the reports in terms of the number of kits sold (Driptech and IDE) or drip packages financed (Netafim). Under 'Progress Update', the number of kits sold is listed for Driptech and iDE, but the buyer is not clear – they may actually be intermediaries or distributors, not farmers per se (see for example Fintrac, 2015a: Annex II). Netafim has partnered with a Kenyan bank to finance drip packages to small commercial farmers; as of the end of March 2015, it claims US$165,000 from sales of 193 kits, against a target of US$10 million in sales to 4,600 farmers (Fintrac, 2015a: Annex II).

A search for 'irrigation' on the project's knowledge exchange website led to a number of testimonials to the claimed success of the project's drip irrigation investments from Driptech, Netafim and iDE. For example, iDE has partnered with Toro, a US agricultural equipment company, to promote a branded Toro drip kit in Zambia. The site has a 'success story' consisting of one Zambian farmer's experience. The use of interesting 'success stories' is a frequently used communication tool. Such cases are inspiring, but nearly always reflect the experiences of farmers with good connections receiving high levels of support from the project or NGO. Similarly, Driptech and Netafim are receiving grants to promote their particular drip kit in specific countries. The little information that is provided by Fintrac suggests that in each of the countries described, something like the coalition of vested interests analogous to that described for Burkina Faso has emerged.

The Fintrac reports are in a format, and make use of indicators, required by USAID (Feed the Future, 2013). These indicators are in terms that do not capture the actual impacts and outcomes of the investments beyond the number of kits sold (to farmers or to intermediaries?) or number of hectares irrigated (not broken down by type of irrigation). Are farmers actually using the kits after the project finishes? Are farmers buying kits even in the absence of the project? Do farmers continue to use the kits over time, or replace and upgrade them, and actually benefit? There is no evidence on these and other questions. In an email exchange with a USAID water management professional, it emerged that while there have been internal discussions of the need for impact evaluations, there is no agreement to make the needed investment – i.e. there is apparently internal resistance.[15]

A search of the USAID clearinghouse site was carried out using key words such as 'impact [also 'evaluation'] of drip irrigation kits'. The first document listed is the Southern Africa study discussed above (Merrey et al., 2008). This is the only independent impact assessment report based on actual fieldwork listed; most others list project progress reports in which the key word 'impact' is highlighted in terms of number of kits sold, or success stories of individual farmers receiving support from a USAID project. The only exception is a recent evaluation of a drip irrigation project in Honduras aimed at poor horticulturalists in the dry western part of that country (Hanif, 2015). That report emphasizes the potential for drip irrigation to lift poor farmers out of poverty, but even in a region with a relatively well-developed market-based support system compared to much of Africa, success will require considerable public support: 'there is no clear scaling pathway for drip irrigation for the poor and extremely poor' (Hanif, 2015: vi).

All three of the drip irrigation organisations being supported to promote drip irrigation kits – iDE, Driptech and Netafim – are social enterprises driven by charismatic leaders who are brilliant at communicating the potential of this technology to transform the lives of smallholder farmers. The paper by Venot (2016a) is in part based on interviews and public statements of the founders of iDE (Paul Polak) and Driptech (Peter Frykman). Although Netafim is a more classical private business based in Israel but operating in many countries, it too has a charismatic leader promoting drip kits to smallholders. He is Naty Barak, its Chief Sustainability Officer. He is a member of an Israeli Kibbutz where the modern version of drip irrigation was supposedly invented, and has spent forty years promoting it.[16] He is clearly highly committed, and like the other two social entrepreneurs, a 'true believer' in the efficacy of drip irrigation, even to achieve social change by empowering women. His 'vision is to create villages in Africa where the whole village uses drip irrigation.'[17] The AMG program in West Africa was similarly driven by a charismatic true believer (also Israeli) and used Netafim's technology (Wanvoeke et al., 2015).

Conclusions and recommendations

For over fifteen years, low-cost low-pressure drip irrigation kits have been promoted by numerous organisations to smallholder farmers in developing countries. In many cases, these organisations have perceived these kits as potential game changers – i.e. pathways out of poverty on a mass scale. The proponents of these kits have published papers arguing, based on small pilot studies and their own theories of development, that such kits can lead to huge reductions in rural poverty; examples are the literature on AMGs using Netafim kits (Woltering et al., 2011a, 2011b) and market-based promotion strategies using iDE kits (Magistro et al., 2007). Donors such as USAID, SDC and the World Bank have continued to support pilot projects to test the technology and strategies to promote their uptake. However, as demonstrated by the independent studies reviewed here, there is no evidence of widespread adoption of drip irrigation kits among African smallholders, and the available evidence suggests their use is rarely, if ever, profitable and sustained over time.

Several hypotheses have been suggested to explain this low rate of uptake and impact. Many studies start with the assumption that the technology is basically good and the bottlenecks are contextual: there is a need for more farmer education, further improvement of the technology, policy reforms to make investments more attractive or development of competitive markets for both inputs and outputs – or some combination of these interventions. A more recent hypothesis draws from innovation systems theory: what is needed is to create an effective network of key stakeholders in the technology to collaborate and learn together from experience, and co-create the conditions needed for the kits to be successful. All of these explanations share a strong faith in the efficacy of the technology itself.

An in-depth case study of the history of efforts to promote drip irrigation kits in Burkina Faso has addressed the question of why donors, NGOs, private firms and governments continue to promote low-pressure drip irrigation kits (Wanvoeke, 2015; Wanvoeke et al., 2015, 2016a, 2016b, 2016c). In essence, that study found that a coalition of participants in drip irrigation kit projects – often including researchers – have developed a shared discourse and set of messages that support their shared vested interests to continue making these investments. This chapter has used publicly available information to present a case study, USAID's Feed the Future initiative, to further examine the persistence of investments in drip irrigation kits even in the absence of evidence on their benefits and sustainability.

The Feed the Future technology projects, whether for drip irrigation or other products, each focuses on promotion of a specific technology whose adoption would allegedly improve food security. The widespread adoption of drip irrigation kits is supposed to achieve substantial water savings, reduce labour, improved food production, reduce malnutrition, increase farmers' incomes and even empower women. All of these projects are based on a linear theory of change: getting the context right, through improving the design and performance of the technology, providing more training and education in its use, improving access to output markets and in some cases credit to purchase the equipment, or some combination thereof. There is no evidence of major efforts to promote policy reforms to support adoption of these new technologies. More important, there is no evidence that the USAID projects are promoting a 'technology adaptation' approach – i.e. creating the conditions to strengthen innovation capacities in their target countries.

Based on the limited data available, it seems likely that the explanation of continued investment in drip irrigation kits proposed by Wanvoeke and his colleagues – the 'vested interest hypothesis' – applies to the Feed the Future programs as well. There are clearly shared vested interests among the parties involved: the donor wants to be seen as promoting environmentally and gender-friendly technologies that can improve nutrition and food security; the NGOs and private firms believe strongly in the efficacy of their technologies and are happy to use donors' funds to find ways to disseminate them widely; some researchers are also true believers and continue to pilot test the kits to overcome perceived bottlenecks in collaboration with the NGOs and manufacturers. In other words, the Feed the Future programs are replicating experiences already well-documented over the past decade, while there continues to be no evidence of widespread uptake or substantial positive impacts.

The vested interest hypothesis raises fundamental questions about donor-funded aid projects and NGO- and private firm-driven projects funded by donors. It is notable that the drip irrigation private firm involved is apparently not willing to use substantial shareholders' capital to develop markets for small low-cost drip irrigation kits. Another important point is that no one has financed proper independent impact evaluations of their investments in drip irrigation kits. One wonders whether there is active resistance to assessing their impacts. The absence of such impact assessments after over a decade of pilot testing and dissemination of free kits, combined with hints from colleagues working with and within donor programs, suggests this is a possibility.

The vested interest hypothesis offers important insights into why the coalition of interested parties continue to 'perform' drip irrigation projects in various countries even in the absence of evidence for real impacts. However, that analysis fails to offer suggestions on the way out of this developmental *cul de sac*. This is not surprising, as there are no clear entry points to change the current trajectory: the underlying assumption that specific technological innovations will lead to game-changing breakthroughs permeates not only development financing institutions but even many international agricultural research institutions. There is a growing academic literature criticizing the basic assumptions of top-down linear development programs and suggesting more complex learning-based approaches to development based on innovation systems concepts (e.g. Burns and Worsley, 2015).[18] But following the advice of such authors would require fundamental reform of development agencies supported by progressive politicians willing to accept that there will be no short-term silver bullets to justify the development budget. The 'technology as solution' perspective also reflects deeply held cultural values among donors and their constituents. *Navigating Complexity in International Development* (the title of the book by Burns and Worsley) would need to be preceded by navigating the complexity of the culture and politics of aid budgets.

Nevertheless, we researchers need to go beyond criticising failures to offering positive advice on how to move forward. We conclude with three recommendations.

1. A critical flaw in most of the drip irrigation kit projects reviewed is that project performance is assessed based on immediate outputs: number of kits distributed, number of farmers trained or amount of credit issued. Donors and their partners should invest in more rigorous assessments of actual impacts over time: how many farmers actually continue using the technology, disaggregated by gender, farm size and other variables; impact of use of these technologies on production, labour and income – including understanding who really benefits; and whether the systems are in place to support not only continued use of the technology but upgrading and adoption of new technologies over time.

2. Garb and Friedlander (2014) have an important insight that a network of interested parties focused on translation and integration of technologies and local realities can be an effective way to encourage innovation. However, they are overly focused on a single technology: programs to encourage local innovation should be open to multiple interested parties and to a menu of technologies and practices that farmers can test, mix and match, and adapt to their needs. There is a growing literature suggesting that investing in promotion of innovation platforms can catalyse agricultural change that is scalable (e.g. Klerkx et al., 2013; Cullen et al., 2014).

3. Finally, independent researchers can play a critical role but they must get away from doing research that focuses on testing single technologies such as drip kits with a view to validating their efficacy. Rather, providing support to innovation platforms through which multiple innovations are tested would be a more valuable contribution. As proposed by Merrey and Langan (2014), researchers should develop participatory diagnostic tools to enable farmers, extension agents and local firms to identify and prioritize problems and identify potential solutions.

Researchers should advocate for more independent evaluations and analyses of development initiatives, and where possible, carry out independent impact evaluations aimed at learning lessons and identifying what works and why. This would go far to solving the mystery of persistent development investments that make no developmental contributions.

Acknowledgments

The author wishes to acknowledge the useful advice and comments on an earlier draft of this paper received from Beverly McIntyre, an IWMI scientist based at USAID in Washington, D.C. He also benefited from discussions of this and other papers at a workshop to discuss some of the book chapters held at Wageningen University on 17 December 2015. He is grateful to Margreet Zwarteveen and several anonymous reviewers for their critical comments as well; these have helped to focus the paper. The author remains solely responsible for its contents.

Notes

1 See Merrey and Langan, 2014: Appendix 1 for information on the main commercial and non-profit suppliers.
2 The first author of the Shah and Keller paper is a senior scientist at IWMI; the second author was a prominent irrigation engineer who worked for iDE at the time the paper was written.
3 This author was a consultant for this study, contracted by ICRISAT.
4 Indeed, farmers sometimes 'supplement' the drip irrigation using buckets, as they perceive the kits do not apply sufficient water (Rohrbach et al., 2006; personal observation made in Ethiopia in 2015.
5 The AMG work was financed by Taiwan through a project on 'Affordable Micro Irrigation of Vegetables' (Woltering et al., 2011b).
6 Upadhyay et al. (2005) provide some evidence of positive impacts on women in Nepal.
7 According to a statement by Netafim on the USAID Feed the Future website, '[t]he primary barrier blocking smallholder access to drip irrigation is cost.' See http://agtech. partneringforinnovation.org/community/technologies/blog/2014/04/08/netafim-family-drip-system, accessed July 11, 2016.
8 Five of the seven co-authors of the Magistro et al. (2007) paper were iDE employees; the third author of the two Woltering et al. papers is associated with MASHAV and Netafim.
9 The project was funded by the private foundation COOPERNIC, entrusted to IFAD for fiduciary management and implemented by various operators : AVSF (a french NGO) in Madagascar, iDE India in Orissa, FUNCAFE – the social arm of the coffee producers association – in Guatemala.
10 As one reviewer of an earlier version of this paper noted, the differences between Israeli and African economies may be part of the explanation for the differences.
11 A point confirmed by email exchanges with colleagues working with a major donor.
12 This is a public website that acts as a repository of documents related to USAID programs, called Development Experience Clearinghouse. https://dec.usaid.gov/dec/home/Default.aspx, accessed July 11, 2016.
13 D. Hamilton, *What I learned from Naty Barak, chief sustainability officer, Netafim*, 2015. http://agtech.partneringforinnovation.org/community/technologies/blog/2015/10/13/what-i-learned-from-naty-barak-chief-sustainability-officer-netafim, accessed July 11, 2016.
14 http://agtech.partneringforinnovation.org/welcome, accessed April 13, 2016.
15 The US Congress recently (July 2016) passed the Foreign Aid Transparency Act which will require US government agencies to closely monitor and evaluate all foreign aid

programs based on their *outcomes* and improve transparency. This may lead to more impact evaluations of investments in future.

16 The roots of drip irrigation are usually traced to Germany in the 1860s.
17 D. Hamilton, *What I learned from Naty Barak, chief sustainability officer, Netafim*, 2015. http:// agtech.partneringforinnovation.org/community/technologies/blog/2015/10/13/ what-i-learned-from-naty-barak-chief-sustainability-officer-netafim, accessed July 11, 2016.
18 See blog on this recent book: www.ids.ac.uk/opinion/navigating-complexity-in-inter-national-development, accessed July 11, 2016.

References

Abric, S., Augeard, B., Onimus, F. Durlin, D., Ila, A., and Gadelle, F. (2011). *Lessons learned in the development of smallholder private irrigation for high value crops in West Africa.* Joint Discussion Paper 4. The World Bank, FAO, IFAD, Practica, and IWMI. Washington, DC: The World Bank. Available at www-wds.worldbank.org/external/default/WDSContent-Server/WDSP/IB/2012/02/20/000333038_20120220043308/Rendered/PDF/669170 WP00PUBL0st0Africa0Irrigation.pdf.

Belder, P., Rohrbach, D., Twomlow, S., and Senzanje, A. (2007). *Can drip irrigation improve the livelihoods of smallholders? Lessons learned from Zimbabwe.* Global Theme on Agroecosystems Report No. 33. Bulawayo, Zimbabwe: International Crops Research Institute for the Semi-Arid Tropics (ICRISAT). Available at http://ejournal.icrisat.org/volume5/aes/aes3.pdf.

Burns, D., and Worsley, S. (2015). *Navigating complexity in international development: Facilitating sustainable development at scale.* Warwickshire, UK: Practical Action Publishing.

Cullen, B., Tucker, J., Snyder, K., Lema, Z., and Duncan, A. (2014). An analysis of power dynamics within innovation platforms for natural resource management. *Innovation and Development* 4(2): 259–275.

Dittoh, S., Akuriba, M., Issaka, B., and Bhattarai, M. (2010). *Sustainable micro-irrigation systems for poverty alleviation in the Sahel: A case for "micro" public-private partnerships?* Poster presented at the Joint 3rd African Association of Agricultural Economists (AAAE) and 48th Agricultural Economists Association of South Africa (AEASA) Conference, Cape Town, South Africa, September 19–23, 2010. Unpublished. Available at www.researchgate. net/publication/254383659_Sustainable_Micro-Irrigation_Systems_for_Poverty_Allevia tion_in_The_Sahel_A_Case_for_Micro_Public-Private_Partnerships.

Feed the Future. (2013). *Feed the future indicator handbook: Definition sheets.* U.S. Government Working Document, updated October 18, 2013. Available at http://pdf.usaid.gov/ pdf_docs/pbaaa322.pdf.

Fintrac. (2013). *Feed the future: Partnering for innovation.* Annual Report No. 1, October 31, 2013. Available at http://pdf.usaid.gov/pdf_docs/pa00jhxc.pdf.

Fintrac. (2014). *Feed the future: Partnering for innovation.* Semi-annual Report No. 2, April 30, 2014. Available at https://dec.usaid.gov/dec/GetDoc.axd?ctID=ODVhZjk4NWQtM2Y yMi00YjRmLTkxNjktZTcxMjM2NDBmY2Uy&pID=NTYw&attchmnt=VHJ1ZQ= =&rID=MzY0NDIz.

Fintrac. (2015a). *Feed the future: Partnering for innovation.* Semi-annual Report No. 3, October 1, 2014 to March 31, 2015. Available at https://dec.usaid.gov/dec/GetDoc.axd?ctID =ODVhZjk4NWQtM2YyMi00YjRmLTkxNjktZTcxMjM2NDBmY2Uy&pID=NTY w&attchmnt=VHJ1ZQ==&rID=MzY0NDI0.

Fintrac. (2015b). *Feed the future: Partnering for innovation.* Annual Report No. 3, October 1, 2014 to September 30, 2015. Available at https://dec.usaid.gov/dec/GetDoc.axd?ctID=ODV

hZjk4NWQtM2YyMi00YjRmLTkxNjktZTcxMjM2NDBmY2Uy&pID=NTYw&attc
hmnt=VHJ1ZQ==&rID=MzY4MjM3.

Friedlander, L., Tal, A., and Lazorovich, N. (2013). Technical considerations affecting adoption of drip irrigation in sub-Saharan Africa. *Agricultural Water Management* 126: 125–132.

Garb, Y., and Friedlander, L. (2014). From transfer to translation: Using systemic understandings of technology to understand drip irrigation uptake. *Agricultural Systems* 128: 13–24.

Giordano, M., and de Fraiture, C. (2014). Small private irrigation: Enhancing benefits and managing trade-offs. *Agricultural Water Management* 31: 175–182.

Hanif, C. (2015). *Drip irrigation in Honduras: Findings and recommendations. Scaling up agricultural technologies from USAID's feed the future.* Arlington, VA: Management Systems International for E3 Analytics and Evaluation Project. Available at http://pdf.usaid.gov/pdf_docs/PA00KFQH.pdf.

Intermediate Technology Consultants (ITC), HR Wallingford, EDA Rural Systems Pvt. Ltd., India, International Development Enterprises, India, Birsa Agricultural University and Grammin Vikas Trust, India, Intermediate Technology Development Group, Zimbabwe and Hancock, I. (2003). *Low cost micro irrigation technologies for the poor.* DFID KAR R7392 Final Report. London: ITC.

IWMI-Southern Africa Regional Office. (2006). *Agricultural water management technologies for small scale farmers in Southern Africa: An inventory and assessment of experiences, good practices and costs.* Final Report produced by the International Water Management Institute (IWMI) Southern Africa Regional Office Pretoria, South Africa for Office of Foreign Disaster Assistance, Southern Africa Regional Office, United States Agency for International Development Order No. 674-O-05–05227–00 (USAID/OFDA/SARO); and the Investment Centre of the Food and Agriculture Organisation of the United Nations Letter of Agreement No. PR 32953. Pretoria, South Africa: IWMI.

Klerkx, L., Adjei-Nsiah, S., Adu-Acheampong, R., Saïdou, A., Zannou, E., Soumano, L., Sakyi-Dawson, O., Paassen, A. van, and Nederlof, S. (2013). Looking at agricultural innovation platforms through an innovation champion lens: An analysis of three cases in West Africa. *Outlook on* Agriculture 42(3): 185–192, September.

Magistro, J., Roberts, M., Haggblade, S., Kramer, F., Polak, P., Weight, E., and Yoder, R. (2007). A model for pro-poor wealth creation through small-plot irrigation and market linkages. *Irrigation and Drainage* 56: 321–334.

Merrey, D. (2013). *Water for prosperity: Investment guideline for smallholder agricultural water management.* IWMI Guideline. Available at http://imawesa.info/wp-content/uploads/2013/06/Water-for-prosperity_Douglas-J-Merrey2.pdf.

Merrey, D., and Langan, S. (2014). *Review paper on 'garden kits' in Africa: Lessons learned and the potential of improved water management.* IWMI Working Paper 162. Colombo: International Water Management Institute. Available at www.iwmi.cgiar.org/Publications/Working_Papers/working/wor162.pdf.

Merrey, D., and Sally, H. (2008). Micro-agricultural water management technologies for food security in southern Africa: Part of the solution or a red herring? *Water Policy* 10: 515–530.

Merrey, D., Sullivan, A., Mangisoni, J., Mugabe, F., and Simfukwe, M. (2008). *Evaluation of USAID/OFDA small-scale irrigation programs in Zimbabwe and Zambia 2003–2006: Lessons for future programs.* Report submitted by the Food Agriculture and Natural Resources Policy Analysis Network (FANRPAN) to USAID's Office of US Foreign Disaster Assistance, Southern Africa Regional Office (USAID/OFDA/SARO) in fulfilment of Contract 674-O-00–07127–00. Unpublished. Available at http://pdf.usaid.gov/pdf_docs/Pdacn691.pdf.

Moyo, R., Love, D., Mul, M., Mupangwa, W., and Twomlow, S. (2006). Impact and sustainability of low-head drip irrigation kits, in the semi-arid Gwanda and Beitbridge districts,

Mzingwane Catchment, Limpopo Basin, Zimbabwe. *Physics and Chemistry of the Earth, Parts A/B/C*, pp. 885–892.

Namara, R., Upadhyay, B., and Nagar, R. (2005). *Adoption and impacts of microirrigation technologies: Empirical results from selected localities of Maharashtra and Gujarat states of India*. IWMI Research Report 93. Colombo: International Water Management Institute. Available at www.iwmi.cgiar.org/Publications/IWMI_Research_Reports/PDF/pub093/RR93.pdf.

Postel, S., Polak, P., Gonzales, F., and Keller, J. (2001). Drip irrigation for small farmers. *Water International* 26(1): 3–13.

Rohrbach, D., Belder, P., Senzanje, A., Manzungu, E., and Merrey, D. (2006). *Final report on the contribution of micro irrigation to rural livelihoods in Zimbabwe*. Submitted by the International Crops Research Institute for the Semi-Arid Tropics (ICRISAT) to the Food and Agriculture Organisation of the United Nations (FAO). Unpublished.

Shah, T., and Keller, J. (2002). Micro-irrigation and the poor: A marketing challenge in smallholder irrigation development. In H. Sally and C. Abernethy (Eds.), *Private irrigation in sub-Saharan Africa: Regional seminar on private sector participation and irrigation expansion in sub-Saharan Africa*. Colombo: International Water Management Institute (IWMI); Food and Agriculture Organisation of the United Nations (FAO) and Technical Centre for Agricultural and Rural Cooperation (CTA), pp. 165–184. Available at http://publications.iwmi.org/pdf/H030864_TOCOA.pdf.

Sijali, I. (2001). *Drip irrigation: Options for smallholder farmers in eastern and southern Africa*. RELMA Technical Handbook Series 24. Nairobi, Kenya: Regional Land Management Unit (RELMA), Swedish International Development Cooperation Agency (Sida).

Upadhyay, B., Samad, M., and Giordano, M. (2005). *Livelihoods and gender roles in drip-irrigation technology: A case of Nepal*. Working Paper 87. Colombo: International Water Management Institute. Available at www.iwmi.cgiar.org/Publications/Working_Papers/working/WOR87.pdf.

Venot, J-P. (2016a). A success of some sort: Social enterprises and drip irrigation in developing countries. *World Development* 79: 69–81.

Venot, J-P., Zwarteveen, M., Kuper, M., Boesveld, H., Bossenbroek, L., Kooij, S. van der, Wanvoeke, J., Benouniche, M., Errahj, M., Fraiture, C. de, and Verma, S. (2014). Beyond the promises of technology: A review of the discourses and actors who make drip irrigation. *Irrigation and Drainage* 63(2): 186–194.

Wanvoeke, J. (2015). *Low cost drip irrigation in Burkina Faso: Unravelling actors, networks and practices*. Thesis submitted for degree of Doctor at Wageningen University, Wageningen, The Netherlands.

Wanvoeke, J., Venot, J-P., Fraiture, C. de, and Zwarteveen, M. (2016c). Smallholder drip irrigation in Burkina Faso: The role of development brokers. *Journal of Development Studies* 52(7): 1019–1033, July.

Wanvoeke, J., Venot, J.P., Zwarteveen, M., and Fraiture, C. de. (2015). Performing the success of an innovation: The case of smallholder drip irrigation in Burkina Faso. *Water International* 40(3): 432–445.

Wanvoeke, J., Venot, J-P., Zwarteveen, M., and Fraiture, C. de. (2016a). Farmers' logics in engaging with smallholder drip irrigation projects in Burkina Faso. *Society and Natural Resources* 29(9): 1095–1109, September.

Woltering, L., Ibrahim, A., Pasternak, D., and Ndjeunga, J. (2011a). The African Market Garden: The development of a low-pressure drip irrigation system for smallholders in the Sudano Sahel. *Irrigation and Drainage* 60: 613–621.

Woltering, L., Ibrahim, A., Pasternak, D., and Ndjeunga, J. (2011b). The economics of low pressure drip irrigation and hand watering for vegetable production in the Sahel. *Agricultural Water Management* 99: 67–73.

15 'Bricolage' as an everyday practice of contestation of smallholders engaging with drip irrigation

Marcel Kuper, Maya Benouniche, Mohamed Naouri and Margreet Zwarteveen

Introduction: a *'bricolage'* lens of innovation

Farmers in North Africa are often very proud to show their drip irrigation system, be it high-tech with a large basin and sophisticated headworks or a more modest version on a small plot. All are happy to be associated with a 'modern' and 'clean' technology of which the media, government agents, private companies, and fellow farmers speak so highly. When visiting a farm, the farmer will particularly insist on how he/she adapted drip irrigation to his/her requirements. Rather than a black-boxed standardized technology, one discovers a multitude of systems operated by a great diversity of farmers (see Chapters 3 and 11 on some of the adaptations farmers typically make on drip irrigation systems). When pushing the enquiry to the often 'grey' local support sector, one is struck by the industry of craftsmen, fitters, retailers, and intermediaries all working hard to make drip irrigation 'work' for individual farmers by designing new systems, selling or even manufacturing spare parts, carrying out repairs, and providing advice (Benouniche et al., 2016; see Chapter 17).

The expression *'bricolage'*[1] came up frequently in our discussions with farmers in the Saïss plain in Morocco, as they explained how they tinkered on drip irrigation systems until these systems fitted with their needs. Bricolage had a positive connotation for them, as it was understood as a creative process by which they adapted and thus appropriated an imported technology, and of which they were proud.[2] Smallholders also understood that they had been capable, through their own initiative, to gain access to a high-tech technology that was initially not necessarily meant for them by taking it from large-scale farms. Moreover, smallholders were not merely consumers or users of drip irrigation, but entered the spheres of production and distribution of drip irrigation (see Geels, 2004, for the different functions of innovation: production, distribution, utilisation). Engineers of irrigation companies were more circumspect about bricolage, as it went against the idea of engineering purity.[3] For them, it had a negative connotation of non-professional work and practice, of a process from which they were excluded.[4] It felt as if their competence, which had been so hard to acquire over many years of study and professional activity, was not recognized. At the same time, in the field they were engaged themselves in bricolage, 'working with local fitters, farmers and farm labourers to make technologically advanced international equipment fit with local markets and fields' (Benouniche et al., 2014). Bricolage can, therefore, be understood as contested practice but also a form of everyday contestation as local actors (farmers, local craftsmen, and retailers) effectively share responsibilities of the design process with engineers and shape drip irrigation systems in a different way that was envisioned by irrigation companies or the state.

This 'bricolage' lens of innovation draws attention to the 'distributed innovation agency of the drip irrigation community (farmers, fitters, welders, engineers)' that explains, in part, the success drip irrigation has had (and has) with smallholders in North Africa (Benouniche et al., 2014). Indeed, following our observations in the Saïss, we observed many instances of bricolage on drip irrigation in different settings in North Africa. Bricolage materializes in the different drip irrigation systems that can be observed in the field and that are competing with some of the more standardized systems proposed through the official subsidy procedures. In what follows, we will present three typical instances of bricolage in three different settings, showing how smallholders gained access to, adapted and appropriated drip irrigation. We will then discuss the implications of bricolage for the wider issues of social and technological change, and notably highlight that bricolage can be understood as an everyday form of contestation whereby farmers renegotiate their relationships with their elders, the state, or the market.

Three instances of bricolage

In this section we will show how smallholders 1) gained access to the technology and then continuously adapted the technology to their needs, 2) redesigned and re-engineered drip irrigation systems, and 3) renegotiated their relationships to the state and the market.

Saïss: trespassing of smallholders in the world of drip irrigation

We have documented elsewhere how drip irrigation, introduced from the mid-1980s onward on large-scale farms in the Saïss (Morocco) by irrigation companies as a single package, was transformed in three steps to a plurality of systems for different users and uses through bricolage (Benouniche et al., 2014). On these large-scale farms, farm managers and labourers first had to make an imported technology work without proximity support, as the irrigation companies were situated 200 km away. This included making small repairs, inventing new irrigation scheduling or maintenance routines. In a second step, farm managers and labourers adapted the drip irrigation systems to the particular conditions of the farms they worked in. For instance, they replaced imported filter systems by locally manufactured filters (Plate 4 in the color plate shows local welders manufacturing a filter system); and they removed certain components that did not function satisfactorily, such as the automatic chemical injection system. The labourers, being smallholders themselves, also adapted drip irrigation to the specific conditions of their own farms through a process of re-engineering, leading to a diversity of drip irrigation systems. Third, in close association with the 'grey' support sector, which by now had emerged, a hybrid system with imported and locally manufactured components became the local standard and was even considered to be eligible to the official subsidy scheme.

We will focus here on the instance when drip irrigation moved from large-scale farms to those of smallholders. Our observations suggest that this happened more or less simultaneously through different (distributed) processes. There is a similar pattern to these processes (Ameur et al., 2013; Benouniche et al., 2016): 1) labourers on large farms participated in the installation of the drip irrigation system by irrigation companies; they were impressed by the technology and observed keenly the configuration and functioning of the system; 2) they operated and maintained the drip irrigation system on a daily basis on these farms, and entered in a process of bricolage to 'make it work', thus demystifying the technology; 3) they obtained some equipment, either from the large farm (secondhand) or from a nearby retailer, and tried it out on their own small plots while adapting the system; 4) once it 'worked' for them, they invited neighbours and friends to come and see their system, positioning themselves as 'experts' within their social network; 5) they became innovation intermediaries, by offering advice and services to fellow farmers; drip irrigation became a (part-time) business opportunity (see also Chapter 17); 6) irrigation companies realised the potential of this business niche and started to provide the equipment smallholders required; and 7) large-scale farmers connected to the networks of smallholders, as some innovations were also of interest to them.

The trespassing of smallholders in the world of drip irrigation can be illustrated by the case of Driss (interviewed in 2016), who was a labourer on a large farm and a smallholder (see Benouniche et al., 2016):

> I was impressed with drip irrigation the first time, when I participated in its installation. I was in charge of the maintenance of the system and I discovered that finally it was not very complicated. I wanted to have it myself, but it was expensive and my father did not believe it would work in our farm. I waited until the day the owner [of the large farm] asked me to install drip irrigation in his small garden. That is when I discovered the system with the small valves. I stole some of them and showed them to a friend who is a retailer of drip irrigation nearby. He did not know these valves either, so I had to go to Casablanca, where a retailer explained the functioning of these valves to me. I designed in my mind the drip irrigation system with the small valves that we called *roubiniyatte*. I bought the equipment and I came back to my village. And then I organised a big party, everybody came to see my system. It was a big day for me. I explained the advantages of the system: it is a simple and inexpensive system, as you don't need all the components, it is adapted to small farms, particularly for those who do not have much water or have a hillside plot. Two or three days later,

several farmers went to see the local retailers to purchase the equipment. It was not available then, but the retailers contacted the irrigation companies in Casablanca to provide them with the required equipment.

In the Saïss, drip irrigation is now widely used by a large diversity of farmers for different crops. Ameur et al. (2013) estimated, for instance, that in two particular villages about 25% of the farmers used drip irrigation on 27 to 44% of the surface area.

Biskra: diffusion of smallholder innovations to large farms

In Chapter 16, it is shown how, in the Algerian Sahara, smallholders stripped the standard drip irrigation system that was proposed in the region (in particular for date palms) in the framework of a public subsidy scheme through a process of *bricolage*, and re-engineered it to adapt it to their situation. Local farmers usually use the Arabic words *Khdamt* (I repaired) and *Sanaet* (I fabricated) when talking about these adaptations. This refers to a process of self-repair and fabrication. This farmer-led innovation process was at the basis of a swift diffusion of drip irrigation to smallholders and accompanied the rapid extension of greenhouse horticulture since the early 2000s. Farmer-led innovation prompted equipment suppliers to adapt their supplies to the emerging demand, thus tailing this innovation process but also trying to influence it, by proposing technological improvements.

A double testimony (from field interviews in March 2015) of a local farmer and a development engineer of an irrigation company illustrates this power game in the innovation process. Mustapha a 56-year-old landowner wanted to invest in date palms in Biskra:

> In 2000 the government launched a subsidy program to develop date palms and promote new irrigation systems. The government gave us the drip irrigation equipment but I didn't know how to use it. I threw half of this equipment (central fertigation units, filters, . . .) in the garage. A few years later, when I visited retailers, I noticed that farmers were looking for drip irrigation lines and sharecroppers were asking for it too. Yacine, who was a young sharecropper in my farm that year, told me: 'Uncle Mustapha, you already have half of the system; all I am asking for is new drip lines and I will show you the results' . . . so I bought imported drip lines from the local retailer. We made a combination of the system that I had with the new drip lines. The results were amazing that year. We had less weeds and sharecroppers had more time to do other tasks and take care of the crops. The year after, the sharecroppers fabricated small fertigation units, using small jerry cans, for each greenhouse. The information was spreading fast and local retailers were recommending the use of these fertigation units to feed the plant directly through irrigation water . . . Now all farmers in Biskra are using this irrigation system.
>
> (Mustapha, 56 years old, farmer)

Maybe farmers cannot see it but we were improving the system for them too. By being in touch with the retailers we were listening to the farmers' demands. We concentrated on the drip lines. In 2003, we proposed to farmers the complete range of drip lines (drip spacing, flow rate, . . .). Then based on farmers' feedback, we chose the more adapted ones for them. While improving the drip lines, farmers developed a small fertigation system at a low cost. We lost that part of the system but we still sell the drip lines. In the last few years, the competition is about providing a better flow with a low pressure. We are working with international manufacturers to improve this to stay competitive in this market.

(Younes, 39 years old, development engineer and local retailer)

As was the case in the Saïss, local innovation systems are firmly connected with more global innovation networks. However, what is important here is the balance of power among the different protagonists. Smallholders gained influence precisely because they had shown the capacity to re-engineer the drip irrigation systems, and could thus impose to the market (and the retailers) the type of (low-cost) equipment they required instead of being supplied by standard kits that would be too costly or not adapted to their needs.

What is also interesting in this process of farmer-led innovation is that some particular innovations of smallholders, for instance the micro-fertigation units used to inject chemicals in the system, made their way to the large farms. This was possible because these smallholders were active as sharecroppers on these large farms:

I installed two big Canarian greenhouses with a big central station for fertigation. . . . The supply market was not ready for such equipment and I didn't have a technician to make it work. I was blocked . . . then the sharecroppers who had worked before to the west of Biskra proposed to adapt their small fertigation units to these big greenhouses . . . they made a fertigation tank of 200 litres for each greenhouse . . . and it works very well, just like the small ones.

(Rachid, 43 years old, farmer, interviewed in 2015)

This is why Naouri et al. (see Chapter 16) insist on the fact that not only the technical innovations travel, but – and perhaps foremost – the innovators and their capacity to innovate do so as well. This means that these innovators can adapt their innovations to new conditions and to a constantly evolving context. It is this dynamic capacity and interest to engage in *bricolage* that keeps the local innovation systems alive.

Tadla: smallholders negotiating autonomy in a state-managed irrigation system

In large-scale irrigation schemes, smallholders depend on the state for water supply and are part of a larger project of planned rural development. They cope with bureaucracy through 'informal adjustments, evasive actions that circumvent formal rules or procedures' (Lees, 1986). We showed elsewhere that in the Tadla irrigation scheme (Morocco), farmers were looking for more autonomy from 'state' water in an irrigation scheme where even cropping patterns were imposed on them in the past (Kuper et al., 2009). They did so through a myriad informal adjustments, of which one of the most important was the installation of private tube-wells, providing a complementary source of water. At the same time, the relationship with the state, in particular the irrigation administration, was never severed. Farmers were rather looking to renegotiate their dependence on the state (Kuper et al., 2009).

This double hypothesis on the relation smallholders have with the state, consisting of a simultaneous quest for autonomy and a renegotiation of their dependence, appeared clearly when the state implemented a collective project to convert 10,235 ha to drip irrigation (see Box 10.1 in Chapter 10). On the one hand, farmers accepted the project[5] (they actually were required to do so formally under the agreement of the state with the international donors financing this project) and had a standardised drip irrigation system installed, even though not all farmers were enthusiastic about the project: 'Today, drip irrigation is like the iPhone, you are obliged to have it even if you don't want it, to be up to date and fashionable, especially when they impose it on you' (a smallholder, interviewed in 2015). The discourse of the irrigation administration goes in the same sense: 'We chose a favourable site [where pressurized water can be delivered without any pumped energy] where the farmers are easy to convince' (agent of the irrigation administration, interviewed in 2016). In this view, farmers need to adopt drip irrigation, as part of a nationwide project: 'Today, farmers are obliged to modernise their irrigation system. And thanks to this project, they can do so. I will do everything I can to guarantee the success of this project, it will serve as a model for other regions'.

On the other hand, farmers negotiated that the system be provided with a 'free valve', meaning that, if needed, they could provide water to crops through gravity irrigation, alongside drip irrigation. This valve was not used much by farmers after the implementation of drip irrigation, but it provided farmers with a certain feeling of freedom. More importantly, farmers informally linked their private tube-wells to the state's drip irrigation system, thus reducing their (future) dependence on the irrigation service. Farmers gave two reasons for doing this. First, farmers made the link with the price of water, as they were afraid that the price for pressurized water would be higher than the water they were used to receive. The tube-well was, therefore, a credible alternative for 'state' water: 'I plugged in the drip irrigation system to the tube-well, because tomorrow the irrigation administration may increase the price of irrigation water. Drip irrigation is there now, I can't remove it, but I can irrigate without water from the administration' (a smallholder, interviewed in 2015). In other words, linking up the tube-wells was part of a larger negotiation process about the price of water. Second, farmers understood that their dependence on the efficiency of the irrigation administration had increased, as drip irrigation – unlike gravity irrigation – maintains the soil water content within a narrow range, as irrigation is supposed to be applied more frequently (every two to three days instead of the fortnightly irrigation applications under gravity irrigation). Drip irrigation thus requires a much more efficient irrigation service than gravity irrigation:

> I do not believe much in this project, I am sure that tomorrow there will be a lot of breakdowns in the system because the water comes from far. I took my precautions and I asked the company to plug in the drip irrigation system to the tube-well. It was done outside of the company's working hours. So I am fine and I am independent, I irrigate when I want to. A day of breakdown equals a day of irrigation lost, and even if the price of irrigation water increases, I am fine.
>
> (a smallholder, interviewed in 2015)

In this case, *bricolage* of smallholders served to maintain a certain autonomy from the state, while formally acknowledging the state's important role in irrigation management.

Conclusion: bricolage as an everyday practice of contestation

We showed how smallholders in North Africa embraced enthusiastically drip irrigation through a process they refer to as *bricolage*, thereby achieving a distributed innovation agency across different kinds of local actors (Garud and Karnøe, 2003).

Smallholders went through a lot of trouble to gain access to the technology. They then engaged in a tedious process of experimentation to adapt the technology to their situation, by stripping out all the options they deemed unnecessary, by redesigning the system or by re-engineering certain components, with the help of local craftsmen and retailers. Their engagement with the technology was often contested, in particular by engineers of irrigation companies and of the agricultural administration (see Benouniche et al., 2014). Smallholders then popularised and disseminated new drip irrigation systems, based on local standards, and made sure that the market would supply 'their' systems and not some black-boxed foreign system.

So what explains the drive and enthusiasm of smallholders in this long battle to create, adopt, distribute, and use the technology? Of course, smallholders gained a lot in the process. There were always sound reasons for adapting the technology in each situation we studied, including ease of use, reducing the cost, extending the irrigated area, saving water, saving labour, etc. Smallholders, often supported by local craftsmen, designed simple and low-cost systems that were practical to use. Those smallholders who were at the forefront of drip irrigation innovation also positioned themselves as 'experts' and innovation intermediaries, thus making money in the drip irrigation hype they contributed to create (Benouniche et al., 2016). There was a lot of learning and capacity building that occurred in the process. Farmers, and especially their young sons, enjoyed engaging with such a creative process, which also provided them with social status.

However, we postulate that the impressive drive of smallholders to engage with *bricolage* on drip irrigation can only fully be understood by considering *bricolage* as an everyday form or practice of contestation. In the Saïss, smallholders struggled to gain access to a technology that was not meant for them in the first place. They literally trespassed on the domain of large-scale farms for the purpose of taking the technology for their own, which is of course the definition of poaching (or *braconnage*; see de Certeau, 1984), and reminds us of the 'homers' on the factory floor described by Anteby (2003). Especially young men were fascinated by the technology and gained legitimacy and authority with regard to their parents, by handling drip irrigation and aspiring to modernise the family landholding. In doing so, they thus also challenged indirectly the patriarchal family hierarchies (Bossenbroek et al., 2015; see also Box 6.1 in Chapter 6). In Biskra, smallholders challenged the configuration of the (black-boxed) drip irrigation system that was subsidized by the state and marketed by irrigation companies. They designed a low-cost system of their own that became the dominant model in the market even though it was not subsidised. This is no mean feat, as thousands of smallholders thus reversed a power balance with respect to the market of drip irrigation equipment and with respect to the state. In the Tadla, smallholders accepted the collective drip irrigation project, but opposed the increased dependence on 'state' water, by plugging their private and often illicit tube-wells into the drip irrigation system. This reflects the complicated and long-standing relationship they have with the irrigation bureaucracy and their quest to continually renegotiate their dependence on the state (see Kuper et al., 2009).

We, therefore, conclude that *bricolage* on drip irrigation systems by local actors can be seen as contested practice, and at the same time as an everyday practice of contestation bringing about a role reversal, where such actors effectively share responsibilities of the design process with engineers, and renegotiate their relationships with their elders, the state, or the market. This draws attention to the implications of bricolage for some of the wider issues of social and technological change, in particular in terms of power (re)distribution and public responsibility. It also shows that caution is needed (in research and interventions) in too rapidly assuming a direct causal relation between the presence or use of drip irrigation and some larger phenomena such as water saving or green modernisation. Original intentions and designs get muddled in processes of bricolage, altering the nature and direction of processes of change in often surprising and contingent ways. Bricolage can be seen as an everyday form of 'contestation by appropriation': a small form of and performance of mimicry, which – through its enactment – subverts the real (and pure) drip irrigation technology.

Acknowledgments

The research was conducted in the framework of the Groundwater Arena Project (ANR CEP S 09/11) and the project Drip Irrigation Realities in Perspective (NWO-MVI).

Notes

1   According to the Merriam Webster dictionary and thesaurus, *bricolage* is the 'construction (as of a sculpture or a structure of ideas) achieved by using whatever comes to hand'. *Bricolage* is a so-called loanword. It 'made its way from French to English during the 1960s, and it is now used for everything from the creative uses of leftovers ('culinary bricolage') to the cobbling together of disparate computer parts ('technical bricolage')'. As we discovered in the field, the term *bricolage* had also been adopted in the Moroccan dialect of Arabic.

2   Interestingly, when we went to the field in 2016 the enthusiasm about the conquest of drip irrigation seemed to have waned as the technology had been domesticated and adapted by farmers: 'drip irrigation today is like the laptop computer, everybody has it from the simplest to the most complicated system; it is like a phone, from the iPhone to the simplest mobile phone. Drip irrigation no longer impresses me' (a smallholder in the Saïss, interviewed in 2015). This phenomenon may relate to what Pinch and Bijker (1984) call closure in technology – i.e. the stabilisation of the artefact and the fact that the *'relevant social groups see the problem as solved'*.

3   We showed elsewhere how the discussions we had with farmers resonated well with the more scholarly debates in innovation studies on *bricolage* (Benouniche et al., 2014). Most of these studies are inspired by the metaphor proposed by Lévi-Strauss (1964) who distinguished two modes of thinking, that of the *bricoleur* and that of the engineer (see Keck, 2004). This metaphor was taken up by Paul Pascon, head of the social sciences department in an agricultural engineering school in Morocco, in his educational project for Moroccan engineers (Pascon, 1980). He opposed dialectically the two antagonistic options an engineer has when working on the design of, for instance, an irrigation system. Either the engineer sticks to official design procedures and norms, creating sophisticated systems that may not work but gain the respect of his peers. Or the engineer adapts the ideal and scholarly models he was taught to field realities, by making do with whatever is available locally and designing systems that work. In other words, he engages in a process that can be called *bricolage*. In this case, the system may be well received by end users, but may be rejected as *bricolage* by his peers. Cleaver (2012) adapted, more recently, the concept to explain processes of institutional formation and development, thereby disputing the model of 'development by design'. Instead, she argues, 'institutions are formed through the uneven patching together of old practices and accepted norms with new arrangements' or in other words through institutional *bricolage*.

4   More generally, *bricolage* has a somewhat ambiguous connotation in French. On the one hand, the Larousse dictionary defines it as a nonprofessional activity, or as work that is not serious. On the other hand, there is a serious, official and booming 'bricolage' sector in France, presenting itself as the leading sector equipping contemporary households. It is estimated that 80% of the households engage in bricolage in house and garden.

5   In a similar way, smallholders in sub-Saharan Africa 'accepted' drip irrigation, as it came as part of a large project (Wanvoeke et al., 2016).

References

Ameur, F., Hamamouche, M.F., Kuper, M., and Benouniche, M. (2013). La domestication d'une innovation technique: la diffusion de l'irrigation au goutte-à-goutte dans deux douars au Maroc. *Cahiers Agricultures* 22(4): 311–318.

Anteby, M. (2003). The 'moralities' of poaching: Manufacturing personal artifacts on the factory floor. *Ethnography* 4(2): 217–239.

Benouniche, M., Errahj, M., and Kuper, M. (2016). The seductive power of an innovation: Enrolling non-conventional actors in a drip irrigation community in Morocco. *The Journal of Agricultural Education and Extension* 22(1): 61–79.

Benouniche, M., Zwarteveen, M., and Kuper, M. (2014). Bricolage as innovation: Opening the black box of drip irrigation systems. *Irrigation and Drainage* 63(5): 651–658.

Bossenbroek, L., Ploeg, J.D. van der, and Zwarteveen, M. (2015). Broken dreams? Youth experiences of agrarian change in Morocco's Saïss region. *Cahiers Agricultures* 24(6): 342–348.

Certeau, M.D. (1984). *The practice of everyday life.* Trans. S. Rendall. Berkeley, CA: University of California Press.

Cleaver, F. (2012). *Development through bricolage: Rethinking institutions for natural resource management.* London and New York: Routledge.

Garud, R., and Karnøe, P. (2003). Bricolage versus breakthrough: Distributed and embedded agency in technology entrepreneurship. *Research Policy* 32(2): 277–300.

Geels, F.W. (2004). From sectoral systems of innovation to socio-technical systems: Insights about dynamics and change from sociology and institutional theory. *Research Policy* 33(6): 897–920.

Keck, F. (2004). *Lévi-Strauss et la pensée sauvage.* Paris: Presses universitaires de France.

Kuper, M., Errahj, M., Faysse, N., Caron, P., Djebbara, M., and Kemmoun, H. (2009). Autonomie et dépendance des irrigants en grande hydraulique: observations de l'action organisée au Maroc et en Algérie. *Natures Sciences Sociétés* 17(3): 248–256.

Lees, S.H. (1986). Coping with bureaucracy: Survival strategies in irrigated agriculture. *American Anthropologist* 88(3): 610–622.

Lévi-Strauss, C. (1964). *La Pensée sauvage.* Paris : Plon.

Pascon, P. (1980). Le technicien entre les bavures et le bricolage. *Etudes rurales. Idées et enquêtes sur la campagne marocaine. Société marocaine des éditeurs réunis,* 3–12.

Pinch, T.J., and Bijker, W.E. (1984). The social construction of facts and artefacts: Or how the sociology of science and the sociology of technology might benefit each other. *Social Studies of Science* 14(3): 399–441.

Wanvoeke, J., Venot, J.P., Zwarteveen, M., and Fraiture, C. de. (2016). Farmers' logics in engaging with projects promoting drip irrigation kits in Burkina Faso. *Society & Natural Resources* 29(9): 1095–1109.

16 The 'innovation factory'

User-led incremental innovation of drip irrigation systems in the Algerian Sahara

Mohamed Naouri, Tarik Hartani and Marcel Kuper

Introduction

Drip irrigation is increasingly promoted for smallholder agriculture in Asia and Africa and the adoption of low-cost drip irrigation equipment has recently been the subject of increased attention in the literature (Namara et al., 2007). Most of this literature focused on the promotion of low-cost drip irrigation kits and their potential contribution to poverty alleviation (Polak et al., 1997). However, the use of the kits, which is generally based on linear innovation processes, has met many problems (Kulecho and Weatherhead, 2006; Wanvoeke et al., 2015): 'the failed uptake of drip irrigation in many sub-Saharan African countries can be viewed as a consequence of the transfer of static physical artifacts into new contexts lacking similar local systems into which these could be absorbed and evolve (re-innovated)' (Garb and Friedlander, 2014). Similarly, Verma et al. (2004) explained the 'limited growth' of drip irrigation among smallholders in India by the fact that a 'package' was provided, which did not match the farmers' needs. As a counter example, these authors show that low-cost drip irrigation systems can spread rapidly. Elsewhere, there is also increasing evidence for local innovation processes catering successfully to small-scale farmers installing and running drip irrigation. In Morocco, for example, there has been rapid diffusion of drip irrigation to smallholders, who collectively 'domesticated' a technology that was initially not meant for them (Ameur et al., 2013). Such farmer-led innovation favoured and was supported by the emergence of an informal support sector, able to adapt drip irrigation systems to the specific and evolving needs of smallholders (Benouniche et al., 2014). Despite the fact that drip irrigation is now being adopted by an ever-wider range of smallholders for a diversity of cropping systems, and in different bio-physical and socio-economic contexts (Venot et al., 2014), these local innovation processes, like the ones documented in India and Morocco, have not received much attention.

In this chapter, it is postulated that user-led incremental innovation can be particularly suited to meeting the dynamic and diverse requirements of smallholders. Innovation is defined as the interactive process of developing and implementing a new idea, technique, know-how or institution (Olivier de

Sardan, 1995; van de Ven et al., 1999). Smallholders play an active role in steering this process in close interaction with other innovative local actors who can respond to diversified demand by proposing creative composite irrigation systems that incorporate the evolving needs of different farmers, and which can be modified at any stage. This is characteristic of a local innovation environment where multiple actors interact and adapt systems to specific conditions (Chesbrough, 2003). Farmer-led innovation, or more generally user-led innovation, thus relates to the idea that innovation is being 'democratized', and that 'users of products and services – both firms and individual consumers – are increasingly able to innovate for themselves' (Von Hippel, 2005). 'Farmer-led innovation' acknowledges the local knowledge and know-how of farmers (Scoones and Thompson, 2009), and the capacities of local actors to develop locally specific options and rapidly adapt to changing conditions (Waters-Bayer et al., 2009). The 'innovation factory' is thus a process mobilizing local actors and their capacities to gradually adapt the technology through re-engineering, and to disseminate it through networks of users and innovation intermediaries.

The specific objective of this chapter is to analyse the different innovations smallholders may make to drip irrigation systems in a context of rapidly changing farming systems. Our hypothesis is that user-led innovation enables the diffusion of drip irrigation to a wide range of farmers, as local actors *translate* international drip irrigation systems to local situations.

Study area and research approach

The study was conducted in the Biskra region, situated in the arid Algerian Sahara (see Figure 16.1). In a context of favourable agricultural markets and state support, a groundwater-based agricultural boom has taken place over the past 30 years with the rapid expansion of palm groves and greenhouse horticulture (Côte, 2002). Mostly irrigated with drip irrigation, greenhouse horticulture increased rapidly from 1,370 ha in 2000 to 5,165 ha in 2014 (this is about 130,000 greenhouses, each measuring 8 m by 50 m). The study focused on three agricultural municipalities, with a total area of 174 km², reputed for their high horticulture greenhouse activity. The first, called El Ghrouss, is located 60 km to the west of Biskra and is characterised by a mixed system of date palms and greenhouse horticulture irrigated by drip irrigation. The other two municipalities, M'ziraa and Ain naga, are located 30–40 km to the east of Biskra, and are specialized in greenhouse horticulture and field crops also irrigated by drip irrigation.

The surveys were carried out between December 2012 and June 2013. Forty-two interviews were conducted with a range of different farmers (landowners, lessees and sharecroppers). The interviews focused on identifying: 1) the farming system, including the farmers relation with the land (owner, lessee or sharecropper), the cropping pattern (mono-cropped, multiple crops), the number of greenhouses, the source of capital (personal, credit, subsidies); 2) the mode of access to groundwater: collective or individual, which generally results

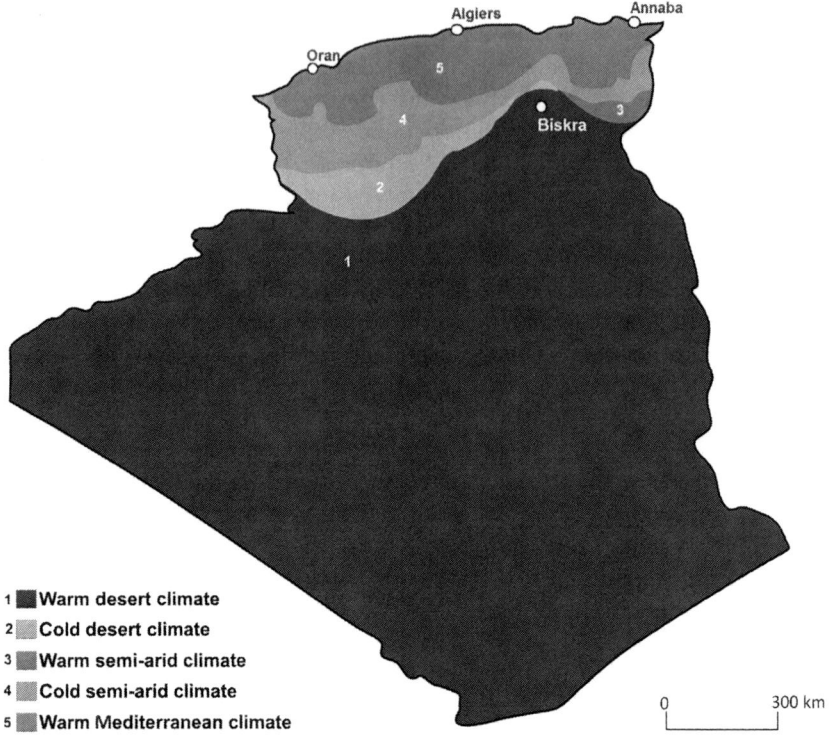

Figure 16.1 Situation of the study area: Biskra, Algeria.

Source: the authors, adapted under a creative commons license, based on the updated world map of Koppen-Geiger Climate Classification by Peel et al. (2007).

in a situation of, respectively, limited or full access to irrigation; and 3) the modifications made to the components of the drip irrigation systems.

Results

Transitional 'pooled' farming systems requiring constant innovation

Before the introduction of drip irrigation, crops in greenhouses in Biskra were irrigated by gravity (*Al Amla*). According to the farmers, this system required large volumes of water, was labour intensive and meant that there was considerable growth of weeds inside the greenhouse. This resulted in an increase in the cost of irrigation and inputs, and limited the irrigated area that could be served by one tube-well. For these reasons, since 2010, all the farmers in our sample and the vast majority of greenhouse horticultural farmers in the region have adopted drip irrigation. In the case of rented lands, a landowner has to provide land and water to lessees, and the more efficient lessees are in their use of water, the more land the landowner can rent out. In this context, landowners are requiring the lessees to use drip irrigation because it enables them to rent out more land.

In the mid-1990s, a few large landowners introduced drip irrigation systems, which catered to 10 greenhouses in the area. After these first tests, other farmers tried to adopt this mode of irrigation but failed. They had difficulty with the irrigation practices, the equipment was not always suited to the local context (rigid drip lines, for instance, did not suit the farming systems which involve changing location), and drip irrigation was very costly. In 1998, companies supplying agricultural inputs again introduced drip irrigation in the region, this time providing drip irrigation tubing with in-line emitters. They provided this equipment to some local greenhouse farmers, who were well known in the area for their horticultural performance. These experiments showed that drip irrigation was of potential interest for greenhouse horticulture, but the farmers required a lot of follow-up as they still had many technical difficulties.

In the early 2000s, the government initiated the National Agricultural Development Program (PNDA), which allowed farmers to gain access to generous subsidies (up to 100%) for tube-wells, plantations and drip irrigation systems. Even though these subsidies mostly targeted and benefited palm groves, this government programme enabled farmers to gain experience in drip irrigation, and encouraged companies selling inputs (seeds, fertilisers and pesticides) to include drip irrigation systems in their catalogue.

These three consecutive complementary initiatives paved the way for the adoption of drip irrigation by a considerable number of farmers, including smallholders. It was probably the arrival of lessees and sharecroppers from the north of Algeria from the early 2000s onward that spearheaded the later boom in greenhouse horticulture associated with the use of drip irrigation. The new arrivals were interested in the land and water resources available in Biskra, as well as the possibility of producing early-season tomatoes which would fetch higher prices. They had experience in tomato production in greenhouses associated with drip irrigation. Because drip irrigation had already been tested in the area, they were able to find everything they needed and brought in their expertise. Greenhouse horticulture consequently spread rapidly throughout the study area together with drip irrigation, confirming the new agricultural image of a region that was previously mainly reputed for the production of dates.

These rapidly evolving dynamics are based on a complex farm structure with many different actors on the same farm, including landowners, lessees, sharecroppers and labourers working together and pooling different resources together (capital, land, water, labour; see Figure 16.2). Naouri et al. (2015) identified three different horticultural farming systems in the area. The 'mobile frontier' farming system is the classical greenhouse farming system in the area based on the mobility of greenhouses which move to new unexploited land every three years as this intensive farming system exhausts the soils. This system accounts for two thirds of the area under greenhouses. Greenhouse 'tunnels' (8 by 50 m) associated with drip irrigation enable the production of tomatoes,

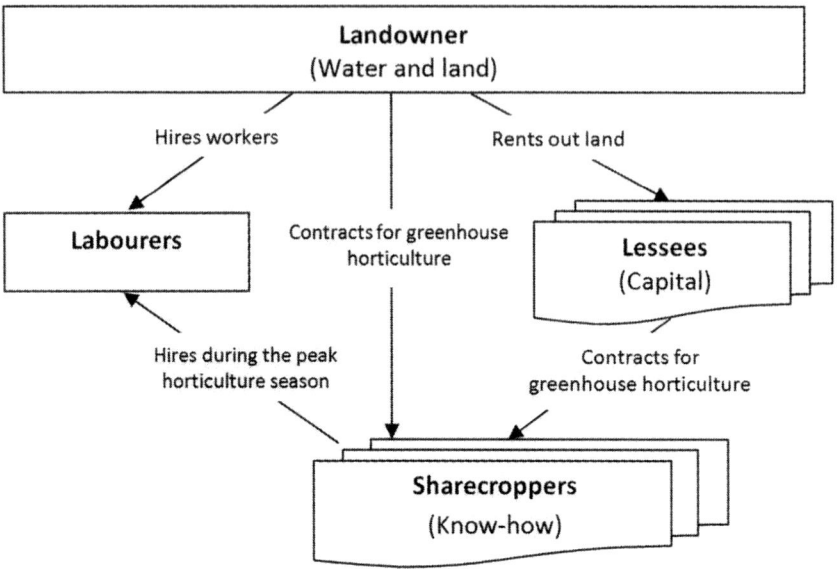

Figure 16.2 Relations between different actors on the same farm.
Source: the authors.

bell peppers, aubergines and watermelons. Usually, the landowner clears and levels the land and installs a tube-well. The landowner then contracts with one or more lessees, who rent in the land including access to water. The lessees set up the greenhouses and contract with sellers to obtain seeds, fertilisers and pesticides, using credit facilities. These lessees then engage sharecroppers who generally take responsibility for 5–10 greenhouses. Typically, 30–40 sharecroppers are present on the same farm, and have collective access to the tube-well. They use the land without crop rotation. After three years of cultivation, the land is abandoned due to declining fertility. Lessees and sharecroppers move on and colonise new unexploited land.

During this time, the landowner accumulates capital and may decide to invest more permanently in the land by planting palm trees. Horticultural land is thus progressively transformed into palm groves. The 'mobile frontier' is transformed into an intensive 'fixed' farming system, which currently applies to about one third of the total number of greenhouses in Biskra. In that second system, the greenhouses, drip irrigation systems and agricultural practices are similar to those in the 'mobile frontier' farming system, but the landowner's objective is to gradually replace the greenhouses by date palms after the third year (in El Ghrouss). In the east (M'ziraa and Ain naga), there is no mobile frontier farming system. Landowners prefer fixed farming and invest directly in greenhouse horticulture in association to field crops like beans or barley to continue greenhouse horticulture beyond three years on the farm (from 30–50% of the farm). If the landowner rents out the land, only one lessee is generally involved. The farms are managed directly by the landowner or by a single lessee with the help of sharecroppers (not more than 10 greenhouses per sharecropper). In all cases, the landowner is much more present on the farm than in the case of the mobile frontier farming system.

The third farming system is a high-tech hyper-intensive farming system using Canarian and Almeria greenhouses (each greenhouse covers 1–6 ha) with a lifespan of 30 years. Characterised by the full exploitation of the covered area both horizontally (by using the space otherwise lost between tunnel greenhouses) and vertically (tomato plants can grow to a height of 4 m rather than only 2 m in tunnel greenhouses) and intensive horticultural production, this farming system is used by large-scale farmers, who hire sharecroppers to do the work. These greenhouses are resistant to the strong winds in the area. This farming system requires precise agricultural practices, and the greenhouses are equipped with sophisticated drip irrigation systems. Only a few farmers use such greenhouses, but the area concerned is increasing rapidly. In 2009, there were only two greenhouses, each covering 1 ha, while in 2014 more than 90 ha were covered with these greenhouses, and in 2016, 250 ha.

These three farming systems enable 1) continuous access to resources (land, water and capital) for new young sharecroppers, who often come from hundreds of kilometres away and have no financial capital of their own; 2) these young farmers to acquire financial capital and technical skills and continue their upward professional mobility (Naouri et al., 2015) (see Figure 16.2). Indeed, it is this socio-professional mobility along with relatively high profits that encourages these young farmers to get involved in these farming systems.

The systems rely on distributed organisation characterised by decentralised management and high mobility of actors (lessees and sharecroppers). The lessees move their greenhouses every three years because soil fertility drops after the third season (in the frontier system). The sharecroppers are hired for only one season, and can thus change to another lessee or, if they have enough capital, become lessees themselves. The number of greenhouses is increasing rapidly and the actors constantly prospect for new land and associations with other actors. Finally, farmers continually adapt their short-cycle cropping systems to the agricultural markets. For instance, when tomato prices drop toward the end of the season, farmers may dig up their tomato plants and plant melons instead. Biskra has two national markets for agricultural products. On these national markets, the farmers' margins are generally high because they sell their produce directly to traders who come from all over the country, and thus no intermediaries are involved. It is indeed the only region in Algeria which provides vegetables to 40 million domestic consumers during the winter.

Re-engineering drip irrigation systems to adapt them to decentralised mobile farming

The farming systems described above are characterised by multilayered and distributed management, by the high geographical and social mobility of its actors and by strong connections to agricultural markets. The drip irrigation systems had to adapt to these farming systems, as the classical configuration (single owner, centralized command, etc.) of drip irrigation systems was not appropriate (see Figure 16.3). First, many actors are organised around a single tube-well, and use the same irrigation system. The actors on the farm share the investment in the drip irrigation system. The landowner owns the tube-well, water tower, main lines and main valves. The lessee owns the sub-main lines, the fertigation units and the drip lines. To the east of Biskra, the lessees even share the cost of the drip lines with the sharecroppers. These actors made many adaptations. They removed and replaced some parts and accessories, such as the accumulation tank and filters, which they deemed unnecessary.

Second, the hierarchical management of the system materialises the sharing of responsibilities among actors. This means that each actor has a specific domain of intervention; for example, a landowner is responsible for operating the tube-well and delivering water to the main lines. The landowner, therefore, does not intervene at lower levels in the irrigation system. Within a given level of organisation, specific actors will be active and all of them experiment to improve the functioning of the system under their responsibility for their

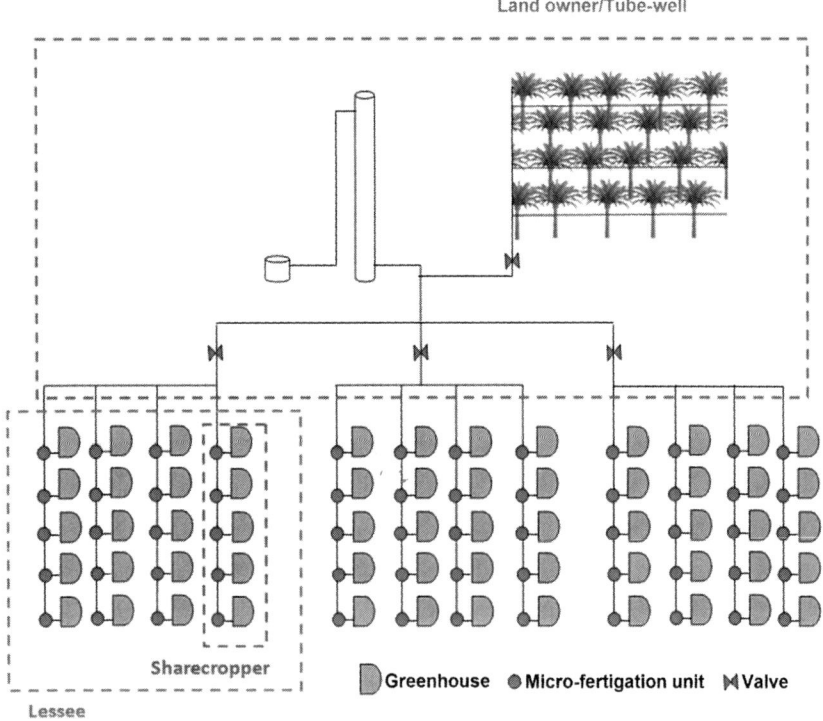

Figure 16.3 The hierarchical organisation of the drip irrigation system used in farming systems in Source: the authors.

specific requirements. Notably, the drip system had to make it possible to manage the water and associated fertilisers at the level of each greenhouse rather than the more classical centralised control system. Third, due to the high mobility of the actors, some parts of the irrigation system were made mobile (main lines) and others are disposable (drip lines).

Two different local innovations are presented in further detail below: the water tower, used as a regulation system, and the micro-fertigation units for decentralised management of fertigation.

Multiple advantages of one adaptation: water tower for regulation
and water management

The water tower (as the farmers call it), is one of the most notable innovations in the area. The water tower is made of an iron pipe up to 50 cm in diameter and around 10 m in height that farmers obtain secondhand from petrol companies (see Figure 16.4).

The water tower has an overflow opening at the top to indicate any dysfunction of the system – i.e. when the water pressure exceeds the height of 10 m

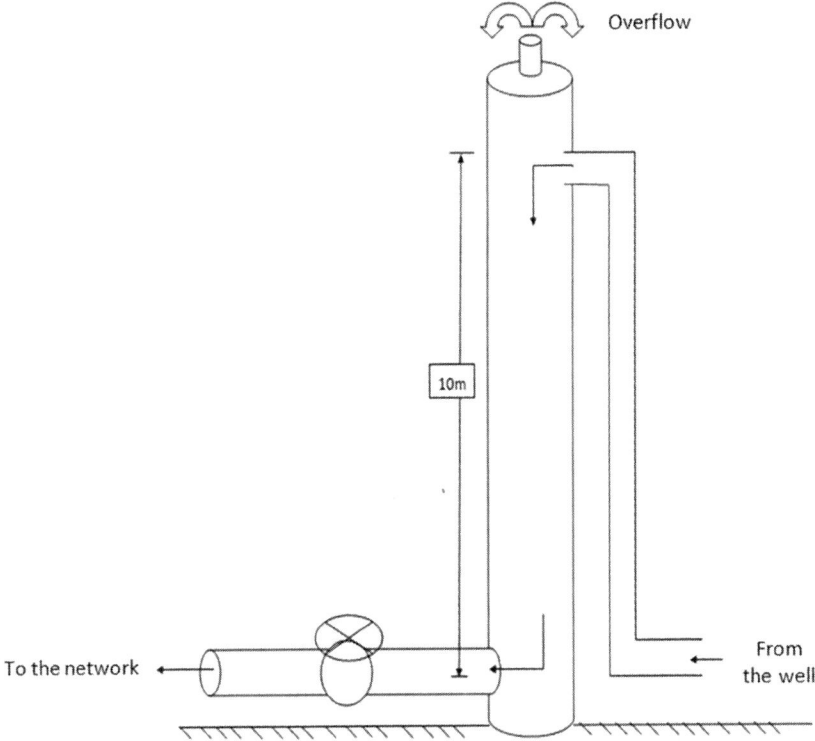

Figure 16.4 Schematic diagram of a water tower.

Source: the authors.

(1 bar), the overflow can be seen from a distance. The water tower is placed next to the tube-well and connected to it by an external pipe to the top of the tower. The water is delivered from an opening at the bottom of the tower to the piped system that transports the water to the greenhouses. The water tower simplifies irrigation management in a complex farming system with multiple actors. After the calibration of the system and the drawing of an irrigation programme, which are done at the beginning of the season in a meeting with the owner and the lessees, the latter know their water turn. The rules are clear, every greenhouse has two irrigation rotations of three hours per week and the irrigation system is switched off for maintenance one day a week. The land-owner stipulates that the system has to be shut down between 5 pm and 11 pm due to the high electricity rate.

During the course of this meeting, the lessees test the water pressure to see how many greenhouses can be irrigated simultaneously. The sharecroppers check if there is enough pressure in the drip lines. Often two to three share-croppers have their water turn at the same time. If there is an overflow at the top of the tower, a sharecropper who has his water turn can open additional valves. If the pressure is low in the drip lines, he can close some valves without having to go to the tube-well to increase the flow rate. The water tower plays the role of an indicator and encourages the users to respect the rules. If the water tower overflows, it means that someone forgot to open the valves for his greenhouses, and if a sharecropper observes the pressure has dropped in the drip line, he can check by closing the valves and watching the water tower. If it does not overflow it means that someone else is using the system and if it overflows that means that there is a leak somewhere on his own drip system.

Water towers like this appeared in the mid-1990s, even before the introduc-tion of drip irrigation systems. At the time, they were used in gravity irrigation to deliver water to the different farm plots. The first water towers were made of stacked oil barrels, and their maximum height was 4 to 5 m. Later, when farm-ers started using drip irrigation, they continued to use water towers. However, they increased the height to 10 m to provide sufficient pressure for all the drip lines, and they replaced the barrels by tubes. In this way, landowners were also able to use the water tower as a substitute for the horizontal pump which is generally used to provide sufficient pressure for drip irrigation.

The water tower is both a pressure regulator and a management device. It protects the submersible pump in the tube-well against high pressure (more than 1 bar) by overflowing. At the same time, it provides water at a constant pressure to the drip irrigation system, which is thus also protected against vari-ations in pressure that are often due to problems in the energy supply or to irri-gation operations by the multiple water users (lessees and sharecroppers) who may make mistakes when transmitting the water turn. Most tube-wells serve 10–40 farmers (in general lessees or sharecroppers) with 100–400 greenhouses. As a management device, the water tower has made it possible to decentralize the management of irrigation water, and to introduce a 'water turn' (similar to the way water is distributed in gravity irrigation systems).

Fertigation systems: adaptations and transfer between small-scale and large-scale farms

Starting in 2000, farmers replaced their gravity irrigation systems with drip irrigation. In the first few years, they continued to apply fertiliser manually. It was only in 2005 that soluble fertilisers became widely available, prompting farmers to start fertigation. Lessees and sharecroppers had observed fertigation in palm groves and were eager to use it for greenhouse horticulture. However, the fertigation units in palm groves had been designed for centralized fertigation management, whereas greenhouse sharecroppers wanted to be able to manage the fertigation for each greenhouse separately.

Sharecroppers consequently adapted small jerry cans to be used as micro-units of fertigation. The jerry cans are located at the entrance of each 400 m² greenhouse (see Figure 16.3). The flow of water within the canister mixes the fertiliser to be introduced in the irrigation network in the greenhouse. The new system works like a conventional fertigation tank. The partial closure of the valve located in the middle (see Figure 16.5) increases the flow rate of water in the pipe, thus causing a partial vacuum injection of the solution in the canister.

This facilitates fertigation, as sharecroppers often cultivate different crops with different needs, in different greenhouses. The system was thus adopted for technical and agronomic reasons, but also for financial reasons as the cost of fertilisers for each greenhouse and for each sharecropper can be calculated at the end of the season. The micro-fertigation units were developed, tested and

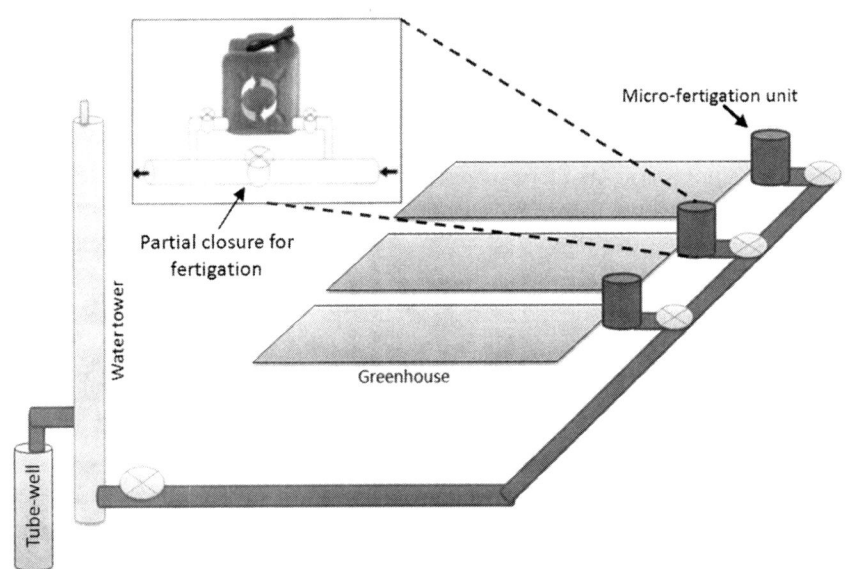

Figure 16.5 Micro fertigation units.

Source: the authors.

improved by the sharecroppers, falling within what can be called their innovation perimeter inside the shared farming systems.

The sharecroppers had also an important role in the diffusion of this technique. Thanks to their mobility, knowledge of the model spread throughout the entire region of Biskra through a process of learning and knowledge application. Interestingly, this innovation was not only used for small greenhouses. Recently, the big farms in the east also adopted the technique for their large-scale Canarian greenhouses, when the large-scale farmers hired sharecroppers who often came from the region west of Biskra. The first Canarian greenhouse was installed in the Biskra region in 2009. It was imported from Morocco as a complete sophisticated package, and was deconstructed by the local actors. In this system, fertigation was meant to be managed by a conventional system, composed of fertigation tanks associated with pumps that measure the required amounts of fertiliser. Those who used this type of greenhouse considered that the fertigation system was expensive and too complicated for local sharecroppers. In addition, the market for fertigation inputs was not suited to this technology, as premixed fertigation elements (NPK) made the dosing pumps unnecessary. Sharecroppers adapted the system by installing a 100 litre fertigation unit for each hectare under greenhouse using the same technique they had used for the micro fertigation units commanding a single greenhouse. The sharecroppers thus contributed their knowledge of small-scale green housing horticulture to these large-scale sophisticated farming systems.

When other actors enable local innovations systems to function

Retailers and large input companies stimulated farmer-led innovation

While the local adaptations of drip irrigation can be considered a farmer-led innovation, the role of the other actors should not be underestimated (see also Box 16.1 on how market can influence on farm irrigation practices). The dynamic farming systems created a demand for technical and organisational support from sellers of inputs and private engineers and technicians. There are many actors in the (competitive) market for inputs who support the different farming systems. Two types of actors in particular should be mentioned.

First, local retailers provide inputs to land owners and lessees (60 retailers for about 130,000 greenhouses). These retailers also provide credit facilities and they play an important role in the financing of greenhouse horticulture. In addition, they give advice to farmers and may visit the greenhouses in the case of crop diseases and pests or other problems. Being at the interface between farmers and (inter)national inputs companies, these retailers are nodes in this innovation network, connecting actors and transmitting new ideas. In the case of new seeds, products or technologies proposed by companies, the retailers are the best way to promote them, as they are in direct contact with the farmers. For example, when a new drip line was introduced, in order to promote them, retailers provided better credit facilities, whereas previously, the farmers had to pay cash for the drip lines. The farmers also have confidence in the local retailers. When a farmer tries out new practices or adapts his equipment, he will share the information with his retailer. Inversely, when a problem arises, the retailer is the second person a farmer will ask for a solution (after his neighbour). The retailer helps the farmer find a solution or asks other retailers in other regions for help, or, if this is the first time the problems have arisen, they ask the company to find a solution.

Second, larger companies provide inputs and advice to local retailers, but also have development agents, mainly young engineers or technicians, in the field for direct contact with the farmers. Around 25 agricultural engineers ensure the permanent local presence of these companies in a two-way approach: (1) they are aware of the problems farmers face in the field, without the information being filtered by the retailers, and (2) they can experiment their own inventions. They organise technical training days for both farmers and retailers, and visit greenhouses at the request of retailers. These engineers not only sell products but also give practical advice to farmers. This advice is free of charge, but is part of a strategy to market the products of the companies, which also conduct experiments in farmers' greenhouses. Farmers who have a good reputation for the performance of their horticulture and are well-known among fellow farmers are thus approached to serve as 'pilots'. The experiments may be the introduction of new seeds, the application of plant protection products or the implementation of techniques such as drip irrigation.

Large companies originally supplied conventional drip irrigation systems to farmers that were more suited to individual large-scale farms. Smaller scale input retailers allowed small-scale horticultural farmers, especially lessees, to access the different components of the drip irrigation systems marketed by large companies. Indeed, they sold the components separately, so the farmers had the choice of creating or adapting their own systems. The companies carefully monitored the adaptations to the irrigation systems made by the farmers thanks to feedback from the retailers who worked closely with the farmers to solve the different problems that arose with drip irrigation. For example, the drip line was progressively adapted to the needs of the farmers by providing more discharge with less operating pressure, closer spacing of the emitters, and emitters better suited to situations where there is a lot of sand in the irrigation water. As the companies are in competition today, there is a wide range of drip lines to respond to the different situations encountered in the fields.

Box 16.1   How agricultural markets determine drip irrigation performance at field level

In a context of increasing water scarcity, the performance of irrigation systems is a growing concern. Engineers classically evaluate irrigation performance by measuring (1) the irrigation efficiency, i.e. how much water is beneficially used by the plant (including water necessary for leaching salts) as a percentage of the total irrigation volume applied, and (2) the irrigation uniformity, i.e. how uniformly the irrigation water is distributed to the crop. As far as drip irrigation is concerned, theoretical reference values of 90% in terms of efficiency and uniformity are frequently quoted. However, these figures generally refer to new irrigation equipment and to measurements in experimental stations, but do not account for farmers' practices and constraints.

A study was conducted on the performance of drip irrigation systems used for horticulture (mainly tomato) in 25 greenhouses (8 m x 50 m) belonging to 13 farmers in field conditions in Algeria's Sahara. Irrigation practices were monitored and the discharges of individual drippers were measured along the crop cycle to determine the irrigation efficiency and the irrigation uniformity. Here, only the results on distribution uniformity (DU) are presented. DU values at field level changed during the cropping season as a result of multiple factors at farm and regional levels, including groundwater management, competition between crops and market prices. The results show that DU values at the field level are comprised between 50% and 70%. The lack of filter systems and the use of soluble fertilizers led to clogging of drip emitters, explaining a quick degradation of DU values along the crop cycle. Interestingly, in some greenhouses an improvement of DU values was observed. Farmers perceived the impact of clogging on the growth of plants and renewed the disposable drip irrigation lines (sometimes twice a year), thereby increasing the DU values.

At the farm level, a considerable difference in DU values was observed for farmers having an individual or a collective access to groundwater. In case of a collective tube-well, farmers carefully manage irrigation and take a keen interest in the state of their equipment. These farmers obtained higher values of DU, typically around 70%. Another important factor determining irrigation practices is the presence of competing crops on the farm. For example, during the early and mid-season of tomatoes, there was much pressure on water resources, as everybody wanted to irrigate. Towards the end of the season, the pressure on water resources is much less as only a few farmers plant a second crop (e.g. melons). This means that the irrigation performance is not a crucial issue anymore for farmers and DU values typically decrease. Finally, at the regional scale, the prices of tomatoes on agricultural markets determined whether farmers wanted to accelerate or slow down the ripening process. Irrigation supplies play a determinant role in this. When prices were very low, farmers even stopped irrigating their crops towards the end of the season, as they had already gained sufficient money with earlier harvests.

Irrigation performance is a dynamic process with large variations along the cropping season and influenced by many factors beyond strict crop water requirements, such as the organization of the farming systems and market prices. It is recommended not to take the theoretical figures of irrigation performance for granted, but rather to carefully analyse the factors determining irrigation performance in the field to determine opportunities for water saving in agriculture.

Khalil Laib, Tarik Hartani and Sami Bouarfa

The role of the state enabling innovation through subsidies, experiments and infrastructure

At first sight, the state appears to be absent from the greenhouse dynamic, which seems to be mainly driven by private initiatives. However, the state has frequently played a role in initiating and accompanying agricultural development, even though the official actors such as agricultural extension agents are not directly involved in the innovation process. The state, therefore, played more of an enabling than a direct role. The state allowed access to land on the basis of the well-known slogan that the land belongs to 'those who work the land' (in particular the 1983 Law on the Access to Agricultural Land Ownership). It also facilitated access to groundwater through subsidizing deep tube-wells in the early 2000s. The state also supported the development of agricultural markets, enabling the marketing of greenhouse horticulture. In addition, it was involved in funding development projects and promoting new technologies including drip irrigation. From 2000 onward, public subsidies for drip irrigation encouraged private actors to settle in Biskra to sell drip irrigation systems (as well as to provide advice and support). Finally, the government provided facilities to these new agricultural regions, which had become of national importance, by developing basic infrastructure, including roads and electricity networks. Due to the strong agricultural dynamics around greenhouse horticulture, the state is increasingly acknowledging and promoting the private dynamics of developing this type of horticulture.

Discussion and conclusion

A farmer-led distributed innovation system mobilizing 18,000 to 26,000 potential innovators

Farmers seek to improve the efficiency of their production systems but also their working conditions by progressively adapting the components of greenhouses, in our case the drip irrigation system, and adopting both local and external improvements. They experimented in stages and made adjustments as needed to seize opportunities which arose or to deal with constraints. According to the agricultural administration, in 2014, there were around 130,000 greenhouses in Biskra. Based on an average of five to seven greenhouses per sharecropper and counting landowners and lessees, this means that there were around 18,000 to 26,000 actors potentially able to make changes in the study area.

These farmers can be considered as local innovators. They conducted small-scale experiments to gradually improve the technology, evaluated their consequences, and reorganised their farm operations whenever necessary, thereby producing local references for greenhouse techniques. In addition, local farmers developed learning skills and acquired know-how in the process. They were able to develop new agricultural farming systems like the intensive frontier farming system with collective access to water resources and decentralised management of drip irrigation systems. These adaptations or innovations were created in a distributed manner, with each actor attempting to face the constraints of everyday farming within his perimeter. Landowners were concerned by the functioning of the tube-well and the regulation system, which prompted them to install the water tower. The lessees and sharecroppers were concerned by water distribution and fertigation, which led to the invention and application of micro fertigation units.

Farmer-led innovation versus low-cost prefabricated kits: what makes the difference?

In the literature, there has been a wide call for cheaper (low-cost) equipment for small-scale farmers, and different authors have reported on the success or failure in the introduction of such equipment (e.g. Kulecho and Weatherhead, 2016; Garb and Friedlander, 2014). However, these were generally centrally designed standardized drip irrigation kits, which were then proposed to farmers despite recent evidence of many failures, and non-adoption, for instance in West Africa (Garb and Friedlander, 2014). This was also the case in our study area where, at first, standardized drip irrigation systems were promoted. Their failure to be adopted by farmers was because they were not appropriate to the physical context and existing farming systems, and also because there was no support sector and not much know-how and experience. By contrast, this study showed how farmers themselves (landowners, lessees and sharecroppers) in the Algerian Sahara, with the support of a dynamic support sector, adapted conventional drip irrigation systems to a wide range of uses and users in the context of a rapidly growing greenhouse horticultural sector. These adaptations enabled farmers to acquire cheaper drip irrigation systems, but were also related to specific technical and management constraints in a context of pooled access to productive resources with multiple and highly mobile actors, intensive high-value farming systems with frequent interactions with the support sector, and shared water management.

However, smallholder-led innovation in drip irrigation only took off once drip irrigation had been adopted by large-scale date palm growers and the supply sector had developed sufficiently to provide the hardware necessary for the development of suitable drip irrigation systems. The users put together their own drip irrigation systems using reusable parts (dripper lines) and got rid of certain components (e.g. filters and basins). These systems were more flexible and mobile, which suited the lessees and sharecroppers. Drip irrigation in itself turned out to be a good investment for all actors. The landowner could rent out more land through more efficient irrigation, thus providing opportunities for more lessees. In addition, drip irrigation reduced the labour requirements both in terms of agricultural practices (e.g. weeding) and irrigation. The sharecroppers were thus able to manage more greenhouses. This generated financial capital for all the actors involved in these hierarchically organised farming systems. Interestingly, the support sector readily accepted the important role of farmers in local innovation systems and adapted their commercial range of products to the changing needs of farmers. In comparison, the literature on small-scale and low-cost kits proposed by NGOs in Africa generally report on prefabricated rigid systems, which prevent the user from participating in the development of adapted and dynamic systems.

Innovation perimeters and mobility of actors

The innovation process of drip irrigation in our case study mimicked the distributed organisation of the farming systems (see Figure 3.1), which confined each actor to a specific perimeter of innovation. The sharecroppers, for example, intervened only from the secondary drip lines to the emitters, as the other parts of the system were under the responsibility of landowners or lessees. However, the innovations travelled with these sharecroppers to other locations and other farming systems. This is illustrated by the example of the fertigation units, which were developed by sharecroppers for small-scale greenhouses (400 m²) but also travelled to the large-scale Canarian greenhouses located 80–120 km away where the sharecroppers also went to find work (Naouri et al., 2015). On the other hand, the regulation system (water tower), developed by landowners to the west of Biskra, did not travel to the east. Landowners to the east of Biskra were aware of this regulation system, but they did not think they required these water towers because their farming systems are less distributed and involve fewer actors given their 'fixed' nature. While this appears to be an objective reason why this innovation did not travel, its non-diffusion may also be linked to the fact that the landowners in the west – the innovators – did not farm in the east.

Interestingly, the high mobility of actors – and hence of their innovations – over hundreds of kilometres challenges the usual observation that farmer-led innovation in one locality can seldom be replicated elsewhere (Waters-Bayer et al., 2009). In our case, some innovations, such as the fertigation systems, travelled over a large distance within the same farming systems, and were also adopted in other farming systems, thanks to the geographical mobility of, for instance, sharecroppers. In addition, there was also a rapid upward shift in the social mobility of the actors. The high profitability of the farming systems enabled sharecroppers to become lessees in a period of only a few years and even landowners a few years later (as they acquired the means to invest in a tubewell). Socio-professional mobility allowed today's sharecroppers more scope for the application of their technical capital and also enabled the entry of new generations of innovators. Socio-professional mobility facilitated the shared understanding of the different actors on the requirements of, for example, the water management system. When sharecroppers become landowners, they already understand the needs of sharecroppers and lessees, thereby extending their perimeter of innovation.

The incremental innovation factory: building innovation and innovators

Adapting drip irrigation to local situations was shown to have gone through many steps. Farmers have a sense of observation, replication and re-engineering. They stripped existing high-tech equipment to keep only the vital minimum. In the beginning, they used drip irrigation as a distribution system to deliver water from the tube-well to plants. In a second step, different actors experimented with different options, some taken from conventional systems such as the fertigation units used in palm groves, and implemented them within their perimeter of innovation. The adaptations were tested and gradually improved through re-engineering. In the next step, selected innovations that were considered to be efficient were used in places which were not those where they originated and by other users, who might, in turn, tinker with them. These innovations were not restricted to specific farming systems, but travelled farther due to the high geographical and socio-professional mobility of actors. These innovations were thus put through different improvement tests. In the transfer process, only what 'worked' and was applicable was kept. For example, fertigation units grew in scale and the water tower failed to find a place in the new farming systems such as the hyperintensive farming system relying on Canarian greenhouses.

No innovation is permanent: a solution to a given problem does not remain valid forever and the conditions farmers face are constantly changing (Waters-Bayer et al., 2009). When networks of users are involved in the development of a technological package and when they are not considered mere users of black-boxed technology, the technology is continuously adapted. We have shown that the dynamics of greenhouse horticulture in the Algerian Sahara, involving a wide range of dynamic actors who interact intensively over rather large areas (several hundreds of square kilometres), provide fertile ground for farmer-led innovation. The social and geographical mobility of actors turned lessees into landowners and sharecroppers into lessees looking for new sharecroppers or trained labourers. During their careers, they thus built a capacity of adaptation and comprehension of technologies and of their use and functionalities. This means that greenhouse dynamics not only produced technical innovations but also allowed and were made possible by the emergence of actors with specific and transferable know-how and organisational skills, who are capable of innovating and of adapting technologies in new situations when they move on.

Acknowledgments

This study was conducted with the support of the ANR project 'Groundwater Arena' (CEP S/11–09). We thank the anonymous reviewers for their insightful comments.

References

Ameur, F., Hamamouche, M.F., Kuper, M., and Benouniche, M. (2013). La domestication d'une innovation technique: la diffusion de l'irrigation au goutte-à-goutte dans deux douars au Maroc, *Cahiers Agricultures* 22(4): 311–318.

Benouniche, M., Zwarteveen, M., and Kuper, M. (2014). 'Bricolage' as innovation: Opening the black box of drip irrigation systems. *Irrigation and Drainage* 63(5): 651–658. doi:10.1002/ird.1854.

Chesbrough, H.W. (2003). *Open innovation: The new imperative for creating and profiting from technology*. Cambridge, MA: Harvard Business School Publishing.

Côte, M. (2002). Des oasis aux zones de mises en valeur: l'étonnant renouveau de l'agriculture saharienne. *Méditerranée* 99(3): 5–14.

Garb, Y., and Friedlander, L. (2014). From transfer to translation: Using systemic understandings of technology to understand drip irrigation uptake. *Agricultural Systems* 128: 13–24.

Kulecho, I.K., and Weatherhead, E.K. (2006). Adoption and experience of low-cost drip irrigation in Kenya. *Irrigation and Drainage* 55(4): 435–444. doi:10.1002/ird.261

Namara, R.E., Nagar, R.K., and Upadhyay, B. (2007). Economics, adoption determinants, and impacts of micro-irrigation technologies: Empirical results from India. *Irrigation Science* 25(3): 283–297.

Naouri, M., Hartani, T., and Kuper, M. (2015). Mobilités des jeunes ruraux pour intégrer les nouvelles agricultures sahariennes (Biskra, Algérie). *Cahiers Agricultures* 24(6): 379–386.

Olivier de Sardan, J.P. (1995). Anthropologie et Développement. Essai en socio-anthropologie du changement social. Marseille: APAD; Paris: Karthala.

Peel, M.C., Finlayson, B.L., and McMahon, T.A. (2007). Updated world map of the Köppen-Geiger climate classification. *Hydrology and Earth System Sciences Discussions Discussions* 4(2): 439–473.

Polak, P., Nanes, B., and Adhikari, D. (1997). A low cost drip irrigation system for small farmers in developing countries. *Journal of the American Water Resources Association* 33(1): 119–124.

Scoones, I., and Thompson, J. (2009). *Farmer first revisited: Innovation for agricultural research and development*. Rugby: Practical Action Publishing.

Van de Ven, A., Polley, D.E., Garud, R., Venkataraman, S (1999). *The Innovation Journey*. New Yord: Oxford University Press.

Venot, J.P., Zwarteveen, M., Kuper, M., Boesveld, H., Bossenbroek, L., Kooij, S. van der, Wanvoeke, J., Benouniche, M., Errahj, M., Fraiture, C. de, and Verma, S. (2014). Beyond the promises of technology: A review of the discourses and actors who make drip irrigation. *Irrigation and Drainage* 63(2): 186–194.

Verma, S., Tsephal, S., and Jose, T. (2004). Pepsee systems: Grassroots innovation under groundwater stress. *Water Policy* 6(4): 303–318.

Von Hippel, E. (2005). Democratizing innovation: The evolving phenomenon of user innovation. *Journal für Betriebswirtschaft* 55(1): 63–78.

Wanvoeke, J., Venot, J.P., Zwarteveen, M., and De Fraiture, C. (2015). Performing the success of an innovation: The case of smallholder drip irrigation in Burkina Faso. *Water International* 40(3): 432–445.

Waters-Bayer, A., Veldhuizen, L. van, Wongtschowski, M., and Wettasinha, C. (2009). Recognising and enhancing processes of local innovation. In P.C. Sanginga, A. Waters-Bayer, S. Kaaria, J. Njuki and C. Wettasinha (Eds.), *Innovation Africa: Enriching farmers' livelihoods*. London: Earthscan, p. 409.

17 Intermediaries in drip irrigation innovation systems

A focus on retailers in the Saïss region in Morocco

Caroline Lejars and Jean-Philippe Venot

Introduction

Drip irrigation has long been promoted as a promising way to use water more efficiently and to meet today's world water and food challenges (Venot et al., 2014). Its introduction and development have been supported and in some cases funded by multiple actors (government, private companies and NGOs alike) as a technological way of using water more efficiently, notably in arid or semi-arid regions where the expansion of irrigation is often underpinned by increasing abstraction of groundwater (Siebert et al.,2010; Shah, 2009).

The use of drip irrigation, especially when associated with private groundwater abstraction, gave rise to different types of farming systems based on intensification and new crops. It also led to the development of new markets, both for irrigated crop productions and for inputs. A case in point is that of North Africa, where rapid changes in groundwater abstraction and in the structure of the agricultural sector occurred from the 1980s onward and have accelerated since the 2000s. Today, groundwater is delivered through hundreds of thousands of mostly private tube-wells to more than 500,000 farm holdings in Algeria, Morocco and Tunisia, Together, these cover more than 1.75 million hectares and this trend defines new irrigation frontiers every day (Kuper et al., 2016; see also Chapter 16 in this volume). From the mid-1990s onward, the increasing use of groundwater was often accompanied by pressurized piped irrigation (particularly drip irrigation) in mutually reinforcing processes toward intensification and modernization. More recently, the development of groundwater-based drip irrigation has also been partly linked with agricultural policies that promote 'modern and intensive' agricultural models and goes hand in hand with subsidy programs (see Chapter 10).

Recent research showed that the emergence of groundwater-based drip irrigation and the expansion of intensive cropping systems affect both the organization of the food processing sector and the manufacturing and supply of inputs and equipment, eventually leading to the development of new markets and to the reconfiguration of 'supply chains'[1] (Llamas, 2010; Bouarfa et al., 2011, Lejars et al., 2012; Lejars and Courilleau, 2015). Multiple actors including local manufacturers and/or suppliers of irrigation equipment (pumps, drip irrigation and engines) play active roles in these reconfigurations, which can also involve the re-engineering of the hardware itself, as shown by Benouniche et al. (2014) in Morocco.

In this chapter, we focus on local retailers of irrigation equipment in the groundwater-based irrigated area of the Saïss in Morocco. We draw from the innovation system literature and conceptualize local retailers of drip irrigation equipment as 'innovation intermediaries'. As identified by Stewart and Hyysalo (2008), retailers are often a type of intermediaries with the most direct link-ages between use and the supply side of products. This is partly because they exercise competence and power over multiple technologies and the factors that influence their uptake such as pricing, distribution channels, marketing, branding, feedback from other intermediaries and end users. They also often have connections with the administration, which puts them in an ideal position to facilitate access to subsidy schemes for instance. Like other 'intermediaries', they can provide information, knowledge, advice and funding (Howells, 2006; Moss et al., 2009, 2009b), thus enabling the 'translation' (Garb and Friedlander, 2014) or 'domestication' (Ameur et al., 2013) of a somehow 'standard' technol-ogy to fit local circumstances. They may facilitate the supply and dissemination of specific pieces of equipment, and may disseminate and provide information concerning these innovations. This puts retailers in a privileged position to shape and support innovation processes, and explains why we chose them as an entry point for our study.

Another reason for 'entering' the innovation system through these local retailers is that they remain largely 'invisible' though they play a key role in supporting the expansion of drip irrigation. Because 'intermediation' is partly informal and usually seen as a side activity (Klerkx and Leeuwis, 2009), the role of intermediaries in supporting innovation has often been neglected especially in public policy (see Schlager, 2007 for a generic argument on the invisibility of intermediaries in public policy and Poncet et al., 2010 for a Moroccan case study on the topic). Starting with retailers is also a pragmatic choice: they may be the easiest innovation intermediaries to identify because they have a shop and at least some of their activities are formal.

Finally, the innovation system literature highlights the networked and dynamic dimension of innovation and the multilateral and systemic nature of 'intermediation' (Klerkx and Leeuwis, 2009). This led us to look 'beyond' the sole retailers and to understand the relations that retailers nurture with other actors (among which are farmers, public agencies and importers) and how these have evolved with time as drip irrigation use extended in the region.

In the section that follows we describe the study area and our methods of enquiry. We then present our results, focusing (1) on the characteristics and the roles played by local retailers of drip irrigation equipment and (2) on their place in the broader network of intermediaries involved in the drip irrigation innovation system of the Saïss region. We then discuss how the role of local retailers evolved with the development of the drip irrigation market and the transformation of drip irrigation equipment.

Study area and methods

Study area: tube-wells and drip irrigation development in the Saïss Plain

The Saïss is a rich agricultural plain in Morocco, located south of the imperial cities of Fez and Meknes (see Figure 17.1). Covering about 220,000 hectares, until the 1970s, the Saïss was mainly known for rain-fed crops (cereals, vineyards and olive trees). Since the droughts of the early 1980s, along with the liberalization of the agricultural sector, there was a tremendous increase in the groundwater-based irrigated area. Technical innovation associated with a decrease in the cost of irrigation equipment allowed the exploitation of deep aquifers by tube-wells and the rapid diffusion of drip irrigation systems. Farmers first tapped the phreatic aquifer through wells (15 to 50 meters deep) and – from 2000 onward – the Lias confined aquifer through deep tube-wells 120–200 m deep. Secure irrigation enabled the development of irrigated high value crops (onions and potatoes) and led to new and more intensive farming systems, based on orchards (olives, plums, peaches and apples), horticulture (onions and potatoes), and fodder crops.

More recently, the new Moroccan agricultural strategy called the Green Morocco Plan (GMP), launched in 2008, accelerated this trend. The GMP promotes cultivation of high value added crops (fruit trees and vegetables), notably by subsidizing private investment in the agricultural sector. In the irrigation sub-sector, one major incentive for investment takes the form of subsidies for drip irrigation. The latter can reach 100% of the total investment costs for farms less than 5 ha in size, and 80% if more than 5 ha are equipped. Because

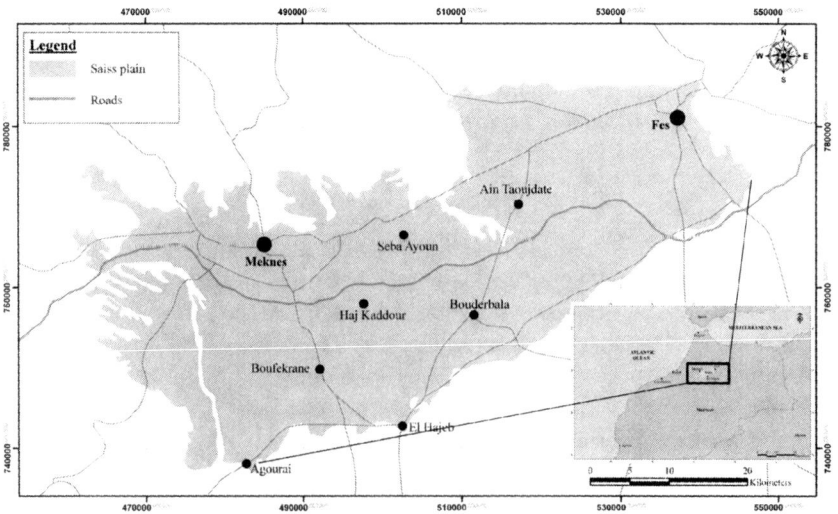

Figure 17.1 The Saïss Plain in Morocco.

Source: Fatah Ameur.

the GMP supports the development of agro-investment, the sharp increase in the production of irrigated fruit crops (apples, grapes and plums) and vegetables (onions and potatoes) also led to congested domestic markets (Sellika and Faysse, 2015; Lejars and Courilleau, 2015).

Surveys and data

As drip irrigation developed in a context of rapid change, we conducted two series of surveys. The first, conducted in 2012, was an exhaustive survey of retailers of drip irrigation equipment in the Saïss region (21 retailers were interviewed); which forms the basis of this study. The 2012 survey was updated in 2015, when 32 retailers were interviewed including 18 retailers already interviewed in 2012 and 14 who had started their business since then (another three had ceased their activity). Two surveys three years apart allowed analyzing the history and trajectory of retailers, their marketing strategy, the type of suppliers and customers they dealt with, the type and origin of the equipment they sold, and their roles in facilitating access to credit and subsidies and in disseminating information.

We also identified the innovation network in which local retailers were embedded and how it had evolved over time. We analysed the local and national relations between the local retailers of drip irrigation equipment and other types of actors involved in drip irrigation, including fitters of pumps and irrigation equipment, drillers, craftsmen, and also national equipment suppliers and importers. In 2012, we interviewed 18 actors of this network; in 2015 we interviewed an additional 10.

Results: characteristics, role and evolution of the drip irrigation intermediaries

Evolution in the number and profiles of retailers in line with drip irrigation development

Retailers are now scattered over the entire Saïss plain. The highest concentration of retailers is in Meknes (the main town in this region) and Boufekrane (in the latter town, all the retailers are located along the main road that goes through the town). Many retailers also settled in smaller towns, where they are more easily accessible by farmers. (See Table 17.1.)

Table 17.1 Spatial repartition and total number of retailers in the Saïss Plain

Number of retailers	Fès	Meknès	Ain Taoujdate	Bouderbala	Haj Kaddour	Boufekrane	El Hajeb	Total
In 2005	0	3	2	1	2	3	1	**12**
In 2015	3	6	3	4	5	7	4	**32**

Source: the authors.

Since 1994, when the first retailer settled and opened a shop in Meknes, the number of retailers has continuously increased (see Figure 17.2). Their profile, background and clientele has also evolved significantly (see Figures 17.3 and 17.4). We identified four distinct periods, which mimic the drip irrigation development trend described by Ameur et al. (2013) and Benouniche et al. (2016).

The first period, before 2000, corresponds to the progressive introduction of drip irrigation by large-scale farmers mainly for fruit trees cultivation. Until 2000, farmers who installed drip irrigation were mainly large-scale farmers and sourced their equipment directly in Casablanca or even Agadir (the major entry points into Morocco for imported merchandise), without going through local suppliers. Only two retailers settled in the area during this period (Figure 17.2). The first retailer to settle in Meknes was a private broker[2] who sourced drip irrigation equipment from Casablanca and Agadir. He progressively specialized in irrigation, conducted studies and fitted drip-irrigation systems for medium scale farms (i.e. between 5 ha and 20 ha). The second one was a teacher of agriculture from Fez, who, in 1996, invested in a shop in Ain Taoujdate. For both of them, customers were large-scale farmers and a majority of medium-scale farmers, who were willing to test drip irrigation.

During the second period, between 2000 and 2005, large-scale farmers extended drip irrigation to other plots, a few medium-scale farmers also tested drip irrigation on their farms, and their laborers installed secondhand equipment on their own small farms, mostly for horticultural crops (potatoes and

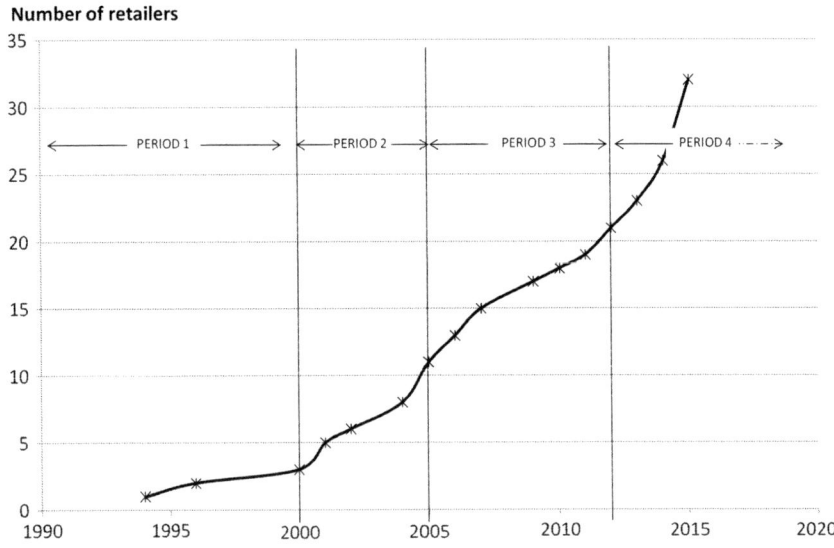

Figure 17.2 Changes in the number of retailers between 1994 and 2015.

Source: this study, interviews with retailers.

onions). During this period, engineers and agricultural technicians who had installed drip irrigation systems on large-scale farms started setting up as retailers (see Figure 17.3). Some farmers also diversified their activities and/or sold their land and set up as small-scale local retailers. In 2001, the first farmer/retailer settled in Boufekrane. He was a farmer from the Saïss region who had installed drip irrigation in his own farm in 1998 and decided to invest in the retailing business because, as he said, 'I had installed drip irrigation on my farm, and I spent most of my time helping my friends and neighbors' (Ahmed, Boufekrane). Finally, some traders and brokers decided to specialize in the sale of drip irrigation equipment.

During the third period, from 2006 to 2012, drip irrigation spread fast, and was installed on all types of farm, including small-scale horticultural farms. The increase in the sales price of onions and state subsidies promoting water-saving techniques (accounting for up to 100% of the investment cost of drip irrigation) encouraged both the extension of drip irrigation for horticulture and

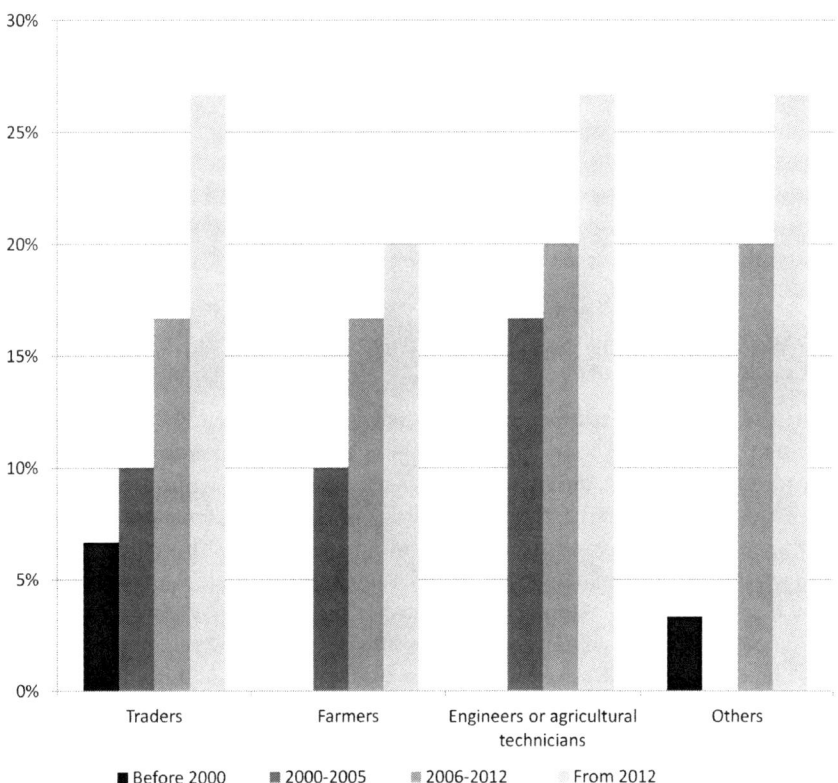

Figure 17.3 Differences in retailers' backgrounds (in percentage of the total number interviewed) depending on the date of their installation.

Source: this study, interviews with retailers.

the development of retailing. Farmers also had better knowledge of how to make drip irrigation work, and were backed up by increasingly solid networks of support services. During this period, a new category of retailers emerged: whereas in the beginning, retailers were mainly traders extending their business, after 2000, they came from other professions or/and with a university degree (teachers, professors, new but unemployed graduates), sometimes even outside the agricultural sector (see Figure 17.3).

Since 2012, drip irrigation is widely used in the area. As the number of retailers has boomed, competition increased. Three retailers quit, others managed to continue by sticking to 'classic' intermediation functions, while many new retailers with different profiles settled in the area (the number of new farmers who have settled as retailers is now lower than between 2006 and 2012). The demand for drip equipment and repair services has increased, and the demand for design studies has gone down.

Customer base and services provided: adaptation to new conditions

Changes in the type of customer base from large-scale farmers to smallholders

The retailers who established their business before 2006 obtained most of their income from large- and medium-scale farmers. Those who settled between 2006 and 2012 started targeting smallholders and conducted most of their business with the latter. Those who settled after 2012 clearly targeted smallholders. In 2015, irrespective of when they settled, most retailers do most of their business with smallholders (see Figure 17.4).

Changes in the type of equipment: imports, local production
and secondhand equipment

Retailers of drip irrigation equipment acted as intermediaries between demand and supply, translating users' cost preferences and technical requirements.

Indeed, the type of equipment retailers sell has changed significantly over the last 15 years. Before 2012, the vast majority of the equipment sold by retailers in the Saïss was imported from Italy and Spain. In 2015, the equipment was imported from a wider range of countries, including China, Turkey and Egypt, which are known for manufacturing cheaper products (see Figure 17.5).

According to the retailers we interviewed, this rapid change is mainly explained by the need to reduce the cost of drip irrigation equipment, in relation to an increasing demand by smallholders. Even if up to 100% of the cost of drip irrigation can be subsidized, the majority of smallholders do not benefit from subsidies, because the administrative procedure to obtain a subsidy can be long and complex and also because they would then be obliged to declare their boreholes (most were dug without authorization – required since the creation of the River Basin Agencies in Morocco). (See Venot et al., 2014 for the challenges smallholders face to access public subsidy schemes on drip

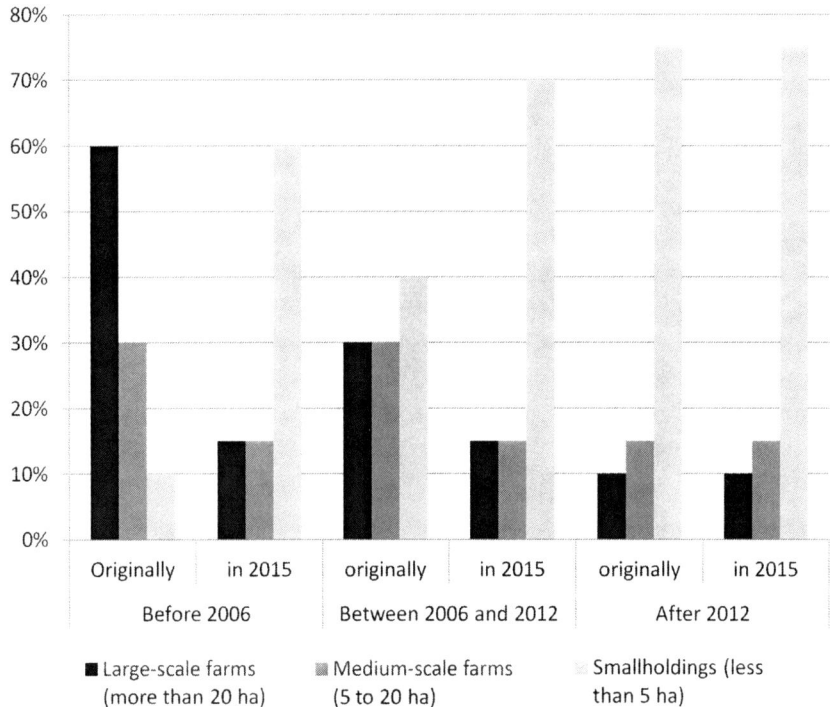

Figure 17.4 Differences in retailers' consumer base (in percentage of the total number interviewed), depending on the date they established their business.

Source: this study, interviews with retailers.

irrigation.) Finally, retailers hardly sell secondhand material equipment, which is commonly sold from farmers to farmers.

Providing technical support and credits

Formal support services for agriculture are weakly developed in the study area. Despite the diversity and the dynamism of the agricultural sector, organizations such as banks, insurance companies and agricultural extension agencies are lacking. Moreover, smallholders generally prefer calling on individuals 'who speak their language' and understand their problems (Poncet et al., 2010), which is the case of the local retailers. In addition to their core activities, local retailers thus provide many ancillary services to smallholders such as access to credit, facilitating application procedures for subsidies, and technical and installation support.

Payment facilities offered by retailers are closely linked to the public subsidy scheme. To obtain subsidies, farmers have to have the equipment and the installation technically approved by authorized bodies. Some retailers are certified

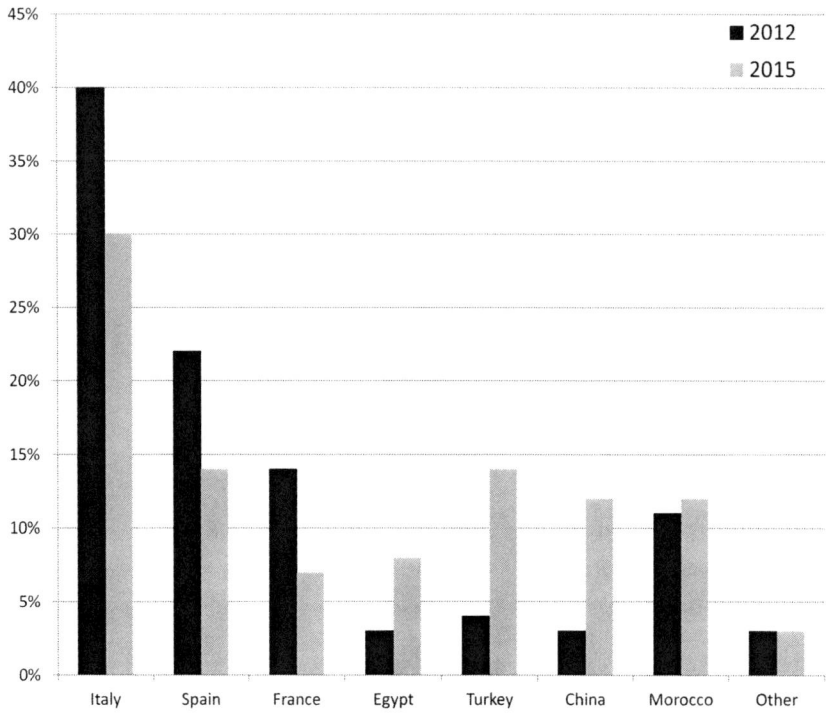

Figure 17.5 Changes in the origin of the imported equipment (in volume sold by retailers) and changes between 2012 and 2015.

Source: interviews with retailers.

and have the right to provide this technical approval; others fill in the application forms for the subsidy on behalf of the farmers and provide credit while waiting for approval of the subsidies from the government. These retailers have relationships with the farmers and close links with state representatives, which are both needed to facilitate what is a time-consuming and cumbersome process. In this case, the price of the equipment is on average 10% to 15% more expensive than if the retailer does not apply for the subsidy on behalf of the farmer. This difference in price is linked to the fact that only certified equipment (mainly imported) is eligible for a subsidy and that the installation has to respect certain standards. This difference in price leads to some 'grey market' transactions. For instance, the imported hydro-cyclones used for filtering water are subsidized while local ones are not; some farmers who benefit from subsidies resell imported hydro-cyclones and buy cheaper local ones, making a profit on the difference.

In 2012, half of the retailers provided facilities to obtain subsidies to 30% of their clients. In 2015, 80% of retailers provided these services but the percentage of farmers requesting it had decreased to 20%. According to the retailers, this is because farmers now have a better knowledge of subsidy application procedures, which they can handle on their own or can ask for help from other farmers (see Table 17.2). The decrease in the share of farmers requesting this service from any single retailer may also be due to the increase in the number of retailers who provide the service, hence diminishing the customer base. This may also be linked to the fact that farmers now renew part of their system instead of installing full new systems and prefer buying and installing non-subsidized equipment at a lower price.

The 80% of retailers who provided credit and certification also conducted technical studies and installed drip irrigation systems for farmers on their farms. The income provided by these ancillary services (technical studies and installation) dropped by half between 2012 and 2015 (see Table 17.3). According to retailers, drip irrigation techniques are now well known by farmers, meaning many of them install the system themselves. Those who still need such services are the new investors who rely on technicians to work their plots or farmers requesting subsidies and credits – as subsidies depend on whether specific designs, which have been elaborated according to engineering standards, are respected.

Providing ancillary services is a way to attract and keep clients, which may explain why an ever increasing number of retailers do it. In 2015, only 20% of the retailers limited themselves to selling drip irrigation equipment and did not offer other ancillary services. Nevertheless, we observed a decrease in the demand for ancillary services per retailer (see above).

With the generalization of drip irrigation and with farmers progressively acquiring the knowledge needed to install and manage drip irrigation systems, retailers were obliged to adapt to farmers' demand and to diversify their services to stay in business. The provision of ancillary services, the capacity to become a facilitator in obtaining subsidies and to serve as a link between the state and

Table 17.2 Evolution in the ancillary services provided by retailers to farmers

	2012	2015
% of retailers providing facilities to obtain subsidies (credit, follow-up procedures)	50%	80%
% of clients who benefits from credit and follow-up procedures	30%	10%

Source: the authors.

Table 17.3 Importance of ancillary services for retailers' business

	2012	2015
% of retailers providing installation, studies and technical advice	70%	80%
% of revenue provided by clients who benefit from 'the whole package'	60%	30%

Source: the authors.

farmers have become the main ways for retailers to remain key actors in an innovation system that changes quickly. However, maintaining their place in the system requires an ability to adapt rapidly, to be aware of new innovations, and also to invest and be able to ensure sufficient funding, which is unlikely to be possible for all retailers currently operating in the region. The examples of the three retailers who stopped their activities between 2012 and 2015 may herald further consolidation of the sector with some actors assuming an ever increasing share of the transactions of drip irrigation equipment, although these will need to remain close to the farmers, which is the main reason why they assumed a key role in supporting the spread of drip irrigation in the region in the first place.

Beyond retailers: a changing local network with links to national actors

At the regional level (i.e. the Saïss plain), local retailers are major actors supporting the expansion of drip irrigation. They act both as a 'hub' through which drip irrigation equipment is supplied to farmers (Kilelu et al., 2013) and as 'intermediaries' who play a key role not only in technical innovation (Benouniche et al., 2016) but also in providing financial and knowledge support to farmers. They also connect different actors and, as such, they constitute a key node in a regional drip irrigation innovation system that is rapidly evolving with (1) changes in the types of actors involved and (2) differentiation of the relationships that local retailers have with bigger manufacturing and import companies that operate at national level (see Figure 17.6). In the following paragraphs, we describe these transformations.

An evolving local network of intermediaries

As highlighted in the introduction, innovation systems are 'systemic' and dynamic. These involve a multiplicity of ever changing actors. Since the early 2000, several types of innovation intermediaries have played a key role in supporting the development of the drip irrigation innovation system of the Saïss Plain, in addition to, and even in association, with retailers: 'local craftsmen' and 'fitters' who re-engineered irrigation equipment to match farmers' needs and demands as these evolved, and drillers of tube-wells.

Two main types of craftsmen were identified: local welders who build hydrocyclone filters from old butane gas bottles and local mechanics who adapt old automobile engines so they can be used with butane – which is highly subsidized, in principle for domestic purposes – to pump groundwater. These newly engineered (and cheap) products are supplied directly by the local craftsmen to the farmers or through retailers. Craftsmen act as 're-engineering intermediaries' and play a key role in the innovation process and in the extension of groundwater use, partly because of the wide range of artifacts needed to make a drip irrigation system work in practice (Benouniche et al., 2014). Some retailers serve as 'marketing intermediaries', between farmers and these 're-engineering intermediaries'. In most cases, the retailers who sell this kind of locally made equipment are those (i) who established their business after 2012, and who targeted smallholders, or (ii) who are not certified to provide the technical approval necessary to obtain subsidies.

Other 're-engineering intermediaries' who also supported the development of drip irrigation use, notably in the early stages of development, are 'fitters' (of pump and irrigation equipment). Fitters are individuals who adjust the original design of a drip irrigation system to real-world conditions on farms (in other words, they transform a 'paper' drip irrigation system in a 'material' system on the farm; see also Chapter 3 for a detailed description of the practices of designing and installing drip irrigation systems). Fitters can be farmers, laborers or technicians who learned to install drip irrigation systems on the large-scale farms in which they were employed or on their own farm (Ameur et al., 2013; Benouniche et al., 2016). Some fitters may be full-time independent workers, others may be employed by regional or national companies, or by local retailers. At the early stage in the development of drip irrigation, all retailers provided fitters or mediated between independent fitters and farmers. However, the situation gradually changed from 2010 onward, as demand dropped. As one retailer told us, 'We used to work with two fitters. But, today everyone knows how to install drip; nobody needs a fitter' (Ali, Haj Kaddour). The real changes in the number of fitters are difficult to evaluate, but, according to retailers, there is less and less demand from farmers for installation and technical studies (see preceding section, 'Providing technical support and credits'), which herald the progressive disappearance of fitters as a specific category of actor in the innovation system (fitting is still needed but is done by other actors, notably by the farmers themselves; see also Benouniche et al., 2016).

While some 're-engineering intermediaries' tend to disappear, 'pure marketing intermediaries' have emerged in the last five years. These are drip irrigation 'brokers' (*courtiers* in French, see endnote 2 for the meaning attributed to this word in the Moroccan context). Brokers do not have a local office and may not even live in the area; they buy equipment directly from national firms, and resell it to farmers cheaper than local retailers. According to local retailers, 'They just have a phone and a car. Their practices are abusive and this competition is unfair. Some of them work in the fields with the farmers, but others just wait for them in the front of our shops' (Mostafa, retailer, Boufekrane). Brokers come closest to being 'pure marketing intermediaries' – i.e., they mostly buy and sell, without providing (i) other ancillary services, such as credits or advice, which still is a key aspect of retailers' activities or (ii) direct feedback to companies on the type of adjustments that specific pieces of equipment require to better match farmers' real needs. Nevertheless, they provide indirect feedback through the increase of specific component demands. As far as 'pure marketing' intermediaries are concerned, some suppliers of fertilizers and crop protection products have also started selling drip irrigation equipment. Two such agricultural merchants started in 2014. As one of them said, 'I don't know anything about drip irrigation, but many clients asked me to supply them with drip irrigation equipment. They wanted to buy spare parts. So I started providing this service' (Ahmed, Boufekrane).

Finally, we observed that, in addition to the above actors who play a direct role in the development of drip irrigation, actors 'dealing with' wells and boreholes (e.g. drillers and drilling companies) have played and continue to play an indirect role in supporting the development of drip irrigation. Drillers may be small companies that own a few drilling machines and employ a few local specialized technicians or they may be individual owners of drilling machines, operating at regional or national level. A few retailers, who deal with large-scale farmers and provide a complete drip irrigation installation 'package' mediate between farmers and drillers. But, most often, the drillers and retailers act independently. 'I know them of course, but I do not want to mediate between farmers and them. The issue is too sensitive. I don't want to be held responsible if the driller does not find water!' (Rachid, Bouderbala).

A changing supply network

The regional innovation system (i.e. in the Saïss plain) does not function in a vacuum; local actors have multiple links to national manufacturers and importers. Importers are mainly based in Casablanca (40% of the companies), Agadir (21%) and Tangier (11%), the main ports where manufactured goods arrive in Morocco. Drip irrigation equipment is imported from a wide range of countries. According to the foreign exchange bureau, between 2005 and 2010, 32 countries exported irrigation equipment to Morocco (Italy, Spain, France, China, Turkey and Egypt are the main ones). Other manufacturers of drip irrigation equipment have factories in Morocco (in 2015, 45 factories existed in Morocco). These are usually owned by foreign investors installed in Morocco, and who have two comparative advantages over importers (i) knowledge of the Moroccan market (they started their business in the mid-1990s and have acquired knowledge about its specificities and the adaptations required to meet the needs of Moroccan farmers) and (ii) a competitive price, as costs of production and transports in Morocco are lower than in other countries.

Before 2006, national companies introduced drip irrigation to selected large-scale farmers. These companies imported equipment mostly from European countries and served as official representatives of foreign international firms. They 'incubated' local retailers to sell their products (see also Benouniche et al., 2016):

> when I started as a retailer (in 2004), my only supplier was an importer in Casablanca. He rented my sales room and I sold his equipment. But I had to diversify my suppliers and the type of equipment to respond to the farmers' demand, they were looking for cheaper equipment. As soon as I purchased my own salesroom (in 2010), I diversified my suppliers, and I now sell equipment made in Turkey, in China as well as produced locally.
>
> (Mostafa, Ain Taoujdate)

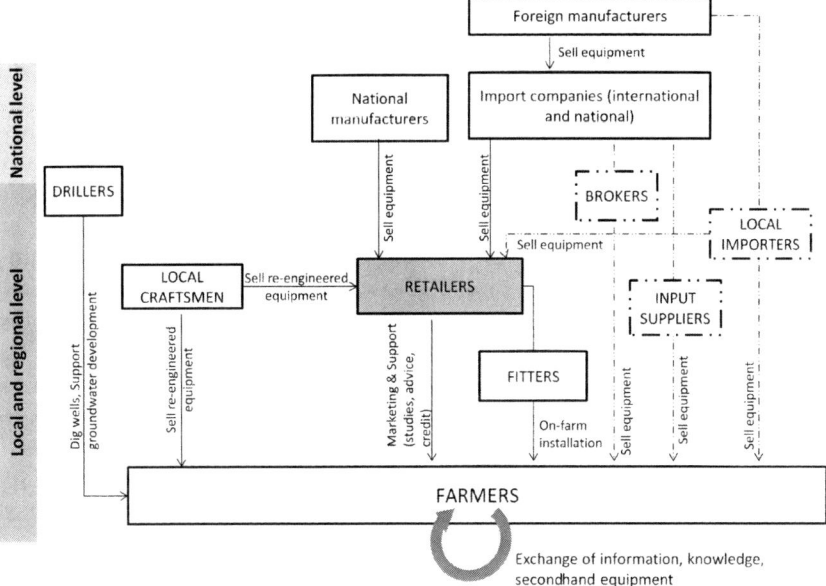

Figure 17.6 A complex and ever changing network of actors.

Source: this study; Note: Actors in dotted box appeared after 2012.

This comment illustrates how 'early retailers', who were often traders, technicians or engineers who had privileged relationships with these big firms (and may even have been employed by them at some point), progressively became independent and built up their own network of providers to be able to cater to the change in demand. Retailers were familiar with local conditions and local users while drip irrigation companies were a source of information about new technologies and subsidy procedures. These actors used each other in their quest to extend their business.

Finally, in 2015, an importer opened an office in Meknes. He works with some local retailers but deals mainly directly with farmers. He can sell his imported products at a lower price to farmers. The arrival of this 'retailer-importer' can be seen as a sign of the evolving role of retailers but also as a tendency toward fewer intermediaries in the drip irrigation equipment supply network.

Discussion

The development of drip irrigation: a temporary opportunity?

The rapid increase in the number of actors involved in the drip irrigation innovation system (local retailers, craftsmen, importers, etc.) is evidence that drip irrigation was a spectacular economic opportunity for actors in the region.

Most actors involved in the drip irrigation innovation system have two concerns: (1) increasing competition among themselves and (2) changing demand, because farmers now need to renew their equipment rather than install new systems. As mentioned by one of the older retailers in the area, 'I was the first to sell and install drip irrigation in Boufekrane, now there are seven of us providing the same kind of services. The golden age of drip irrigation is coming to an end!' (Ahmed, Haj Kaddour). Increased competition among local drip irrigation retailers led to a reduction in the cost, production and diffusion of innovation but also raises questions about their own future.

Like other intermediaries, some retailers are looking for new opportunities in the agricultural sector and/or in other sectors. Several retailers explicitly told us they were saving money to start an alternative activity, not in the irrigation sector, in the medium term. They felt the profitability of retailing drip irrigation equipment will decrease due to stronger competition – including from 'brokers' who incur little financial cost. Others try to target quality, new equipment and to diversify their services. 'Farmers will always need to irrigate, they need someone they can trust, who can supply them with quality equipment' (Mohammed, Ain Taoudjate). With the generalization of drip irrigation and the diffusion of knowledge related to it, one may wonder who will remain in the system and what their role will be. Will retailers remain specialized in irrigation, proposing more and more ancillary services, or will they diversify the products they sell and include inputs, for instance?

In the Kairouan plain in Tunisia and in the Algerian Sahara, local retailers are not specialized in drip irrigation equipment, which is simply another product they sell (Lejars et al., submitted). There, the farmers have the knowledge they need to install and manage the systems themselves. What about the Saïss plain? Is market segmentation a stage in the evolution of the local drip irrigation equipment market or a more structural characteristic? Most of the retailers have had to diversify their services and adapt to farmers' demand. The provision of ancillary services, the capacity to become a facilitator to obtain subsidies and to be a link between the state and farmers have played a fundamental role in retailers' ability to become and remain key actors in an innovation system that changes quickly. Preserving their place in the system requires the capacity to adapt, to be aware of new innovations, and also to invest in – and have access to – sufficient funds (for instance to provide credits to farmers), which is unlikely to be possible for some of the actors currently operating in the region.

An 'ecology of intermediaries': a booming drip irrigation sector?

Studies conducted in sub-Saharan Africa (Villholth et al., 2013) showed that the lack of ancillary services such as financial or extension services but also the lack of dynamic networks distributing equipment and disseminating technology are huge constraints limiting groundwater irrigation development. Many actors promoting drip irrigation in sub-Saharan Africa also attribute the low uptake of this technology to such contextual factors (see for instance Kulecho and Weatherhead, 2005; Friedlander et al., 2013; Chapters 12, 13 and 14 in this volume provide alternative explanations to this state of affairs by focusing on the political economy of drip irrigation promotion in sub-Saharan Africa).

In the Saïss, the dynamic and evolving network of intermediaries around local retailers and the capacity of this network to translate the technology for farmers is certainly one of the reasons of the rapid extension of drip irrigation in this region. The progressive structuring of the innovation system is based, of course, on the ability to supply drip irrigation equipment, on the capacity to adapt and produce ancillary services, but also on the capacity of retailers to connect different fields of knowledge, most notably the 're-engineering knowledge' of craftsmen and fitters and the 'marketing knowledge' of importers and manufacturers of drip irrigation equipment. The capacity to create connections between local farmers and international firms, between local craftsmen and farmers, is an activity in itself and has been a key factor in explaining why some retailers were successful and others less so. As different bodies of knowledge spread, the role that different actors play in the network is deemed to change. As drip irrigation use extends, farmers are building and developing their own connections; notably they appear increasingly able to design and install drip irrigation systems and apply for public subsidies on their own. The role of re-engineering intermediaries, like fitters, tends to become obsolete and to progressively disappear. New intermediaries (local importers and brokers) are, on the other hand, playing an increasingly important role in marketing to farmers who know more and more clearly what kind of products they want. The emergence of brokers, working directly with national firms that are not physically present in the Saïss plain, or of importers who set up local offices, might be early signs of large-scale changes to come in the structure of the drip irrigation innovation network.

As noticed by Stewart and Hyysalo (2008), when it comes to innovative new products, no established functioning chain of intermediaries can be assumed to exist. What happens is a gradual adjustment and learning about suitable forms of technologies, the actors to be involved in their elaboration and dissemination, and the complementary services and support to be supplied. There is thus an evolving 'ecology of intermediaries' in and between supply and use. 'Re-engineering intermediaries', like fitters and craftsmen, configure usage and knowledge about drip irrigation innovation systems and 'marketing intermediaries', such as brokers, facilitate its deployment, both helping users to domesticate it and helping suppliers to respond to actual uses. However, when the benefits or the appeal of being an intermediary wane or when the innovation tends to be 'domesticated', these people are likely to shift location or to change their role.

Conclusion

The development of irrigation and irrigated production was a significant economic opportunity for many actors in the Saïss region of Morocco. This chapter analyzed the roles played by different actors in supporting the massive adoption and 'translation' of the drip irrigation technological package by small-holders. We notably stressed the key role played by local retailers in supporting the innovation process over the last 15 years. They acted as 'marketing interme-diaries' facilitating the supply of new technologies and new equipment but also as 'knowledge intermediaries' providing information and services to farmers as well as facilitating their access to credit and subsidies, thus helping to reduce market risks. Their proximity to farmers meant they could constantly adapt to the needs of the latter and to evolving demand. This was possible because of the relations they established (1) with national and international companies who manufacture and import equipment and (2) with local craftsmen and with fit-ters who 're-engineer' the technology.

What this chapter also shows is that the 'ecology of intermediaries' is con-tinuously evolving: the profile of local retailers evolved with time, the type of services they provided and the network in which they are embedded as well. They, however, continue to act as catalysts of the drip irrigation innovation sys-tem because they are both close to farmers and connected to formal policy and economic spheres. Their dynamic and informal networks allow them to adapt to fast changes in the physical or socio-economic environment.

The key role of 'intermediaries' in supporting innovation has long been rec-ognized, but the latter are still rarely involved in water management or agricul-tural development discussions and policies. Our analysis suggests that it could be useful to involve local innovation intermediaries (such as retailers) in the implementation of agricultural and/or water management policies. Making them visible is the first step to giving them a part to play in the dissemination of information and innovation. They may also have a role to play, not only in supplying drip irrigation innovation but also in bringing manufacturers closer to local demand. Drip irrigation equipment is continually being adapted by 're-engineering intermediaries' as well as by the farmers themselves. Identifying and making farmers' needs and preferences visible at the national or global level could happen through retailers' networks. To this end, further research is required on how local retailers build and develop their connections with national and global networks of suppliers and how these networks evolve with time.

Acknowledgments

We thank IRD as well as the SICMED program for their financial support to the study as well as two students, Imane Rais and Abdel Aziz Khalifa, who collected the data upon which this analysis is based. The authors also benefited from the financial and scientific support of the Arena Groundwater Project. We also thank the anonymous reviewers who helped us improving the quality of this chapter.

Notes

1   The term 'supply chain' refers to the full range of activities required to bring a product or service through the different stages of production (including processing and the input of various producers and services) in response to consumer demand (Beamon, 1998).

2 The term 'broker' is widely used in the innovation system and the anthropology of development literature to designate individuals or organizations who 'mediate' or link actors who are otherwise not connected (i.e. they act as 'nodes' in a social network). In this chapter, the term 'broker' is intended in a more narrow sense. In Morocco, the term 'courtier' (broker) is used to define an individual whose sole job is to put different actors in touch (in our case, mostly farmers and retailers). They do not have any other activities: they mostly rely on an extended social network and a mobile phone to conduct their job.

References

Ameur, F., Hamamouche, M.F., Kuper, M., and Benouniche, M. (2013). La domestication d'une innovation technique: la diffusion de l'irrigation au goutte-à-goutte dans deux douars au Maroc. *Cahiers Agricultures* 22(4): 311–318. doi:10.1684/agr.2013.0644

Beamon, B.M. (1998). Supply chain design and analysis: Models and methods. *International Journal of Production Economics* 55(3): 281–294.

Benouniche, M., Errahj, M., and Kuper, M. (2016). The seductive power of an innovation: Enrolling non-conventional actors in a drip irrigation community in Morocco. *The Journal of Agricultural Education and Extension* 22(1): 61–79. doi:10.1080/1389224X.2014.977307

Benouniche, M., Zwarteveen, M., and Kuper, M. (2014). Bricolage as innovation: Opening the black box of drip irrigation systems. *Irrigation and Drainage* 63(5): 651–658.

Bouarfa, S., Brunel, L., Granier, J., Mailhol, J.C., Morardet, S., and Ruelle, P. (2011). Évaluation en partenariat des stratégies d'irrigation en cas de restriction des prélèvements dans la nappe de Beauce (France). *Cahiers Agricultures* 20(1–2): 124–129.

Friedlander, L., Tal, A., and Lazarovitch, N. (2013). Technical considerations affecting adoption of drip irrigation in sub-Saharan Africa. *Agricultural Water Management* 126: 125–132.

Garb, Y., and Friedlander, L. (2014). From transfer to translation: Using systemic understandings of technology to understand drip irrigation uptake. *Agricultural Systems* 128: 13–24.

Howells, J. (2006). Intermediation and the role of intermediaries in innovation. *Research Policy* 35(5): 715 728. doi:10.1016/j.respol.2006.03.005

Kilelu, C.W., Klerkx, L., and Leeuwis, C. (2013). Unravelling the role of innovation platforms in supporting coevolution of innovation: Contributions and tensions in a smallholder dairy development programme. *Agricultural Systems* 118: 65–77.

Klerkx, L., and Leeuwis, C. (2009). Establishment and embedding of innovation brokers at different innovation system levels: Insights from the Dutch agricultural sector. *Technological Forecasting and Social Change* 76(6): 849–860.

Kulecho, I.K., and Weatherhead, E.K. (2005). Reasons for smallholder farmers discontinuing with low-cost micro-irrigation: A case study from Kenya. *Irrigation and Drainage Systems* 19(2): 179–188.

Kuper, M., Faysse, N., Hammani, A., Hartani, T., Hamamouche, M.F., and Ameur, F. (2016). Liberation or anarchy? The Janus nature of groundwater use on North Africa's new irrigation frontiers. In T. Jakeman, O. Barreteau, R. Hunt, J.D. Rinaudo, and A. Ross (Eds.), *Integrated groundwater management*. Dordrecht, The Netherlands: Springer, pp. 583–615.

Lejars, C., and Courilleau, S. (2015). Impact du développement de l'accès à l'eau souterraine sur la dynamique d'une filière irriguée Le cas de l'oignon d'été dans le Saïs au Maroc. *Cahiers Agriculture* 24(1): 1–10.

Lejars, C., Daoudi, A., and Amichi, H. (submitted). The key role of supply chain actors in groundwater irrigation development in North Africa. *Hydrogeology Journal*.

Lejars, C., Fusillier, J.L., Bouarfa, S., Coutant, C., Brunel, L., and Rucheton, G. (2012). Limitation of agricultural water uses in Beauce (France): What are the impacts on farms and on the food processing sector? *Irrigation and Drainage* 61: 54–64.

Llamas, M. (2010). Foreword. In L. Martínez-Cortina, A. Garrido, and E. López-Gunn (Eds.), *Re-thinking water and food security*. London: Taylor & Francis, pp. vii–x.

Moss, T. (2009). Intermediaries and the governance of sociotechnical networks in transition. *Environment and Planning A* 41: 1480–1495.

Moss, T., Medd, W., Guy, S., and Marvin, S (2009a). Organising water: The hidden role of intermediary work. *Water Alternatives* 2: 16–33.

Poncet, J., Kuper, M., and Chiche, J. (2010). Wandering off the paths of planned innovation: The role of formal and informal intermediaries in a large scale irrigation scheme in Morocco. *Agricultural Systems* 103: 171–179.

Schlager, E. (2007). Community management of groundwater. In M. Giordano and K.G. Villholth (Eds.), *The agricultural groundwater revolution: Opportunities and threats to development*. Oxford: CABI Head Office, pp. 131–152.

Sellika, I., and Faysse, N. (2015). Perspectives de production et de commercialisation de la pomme au Maroc à l'horizon 2025. *Alternatives Rurales* 3: 1–17.

Shah, T. (2009). *Taming the anarchy: Groundwater governance in South Asia*. Washington, DC: Resources for the Future Press, 310 pp.

Siebert, S., Burke, J., Faures, J.M., Frenken, K., Hoogeveen, J., Döll, P., and Portmann, F.T. (2010). Groundwater use for irrigation – A global inventory. *Hydrology and Earth System Sciences* 14: 1863–1880. doi:10.5194/hess-14-1863-2010

Stewart, J., and Hyysalo, S. (2008). Intermediaries, users and social learning in technological innovation. *International Journal of Innovation Management* 12: 295–325.

Venot, J.P., Zwarteveen, M., Kuper, M., Boesveld, H., Bossenbroek, L., Kooij, S. van der, Wanvoeke, J., Benouniche, M., Errahj, M., Fraiture, C. de, and Verma, S. (2014). Beyond the promises of technology: A review of the discourses and actors who make drip irrigation. *Irrigation and Drainage* 63(2): 186–194.

Villholth, K.G., Ganeshamoorthy, J., Rundblad, C.M., and Knudsen, T.S. (2013). Smallholder groundwater irrigation in sub-Saharan Africa: An interdisciplinary framework applied to the Usangu plains, Tanzania. *Hydrogeology Journal* 21(4): 863–885.

18 Drip irrigation and state subsidies in India

Understanding the success of the Gujarat Green Revolution Company

Janwillem Liebrand

Introduction

In India, as elsewhere in the world, drip irrigation is seen as a technology with great potential benefits. Ever since it was introduced in the country in the early 1970s, the technology has received ample attention in policy and research debates on water and agriculture. Government reports and researchers in India have stated that drip irrigation can substantially increase crop productivity and simultaneously save water in agriculture, up to 70% as a result of irrigation use efficiencies of more than 90% (GOI, 1990, 2006, 2010, 2014; INCID, 1994; Sivanappan, 1994; Narayanamoorthy, 1997a, 1997b, 2004; Jalajakshi et al., 2006; Kumar and Palanisami, 2011). In spite of the acclaimed qualities of the technology, adoption rates of micro-irrigation (MI) – the term most frequently used in India to denote drip and sprinkler irrigation systems – are deemed to be low (Palanisami et al., 2011; Pullabhotla et al., 2012). Drip irrigation in India has long been perceived in policy and research circles as the remit of large farmers; as a costly technology that mainly appeals to so-called gentlemen farmers (Shah and Keller, 2002). To also make the technology accessible and attractive to small farmers, various initiatives have been developed. These include those by iDE/India, focusing on the KB drip kits (KB stands for *krishak bandu*, meaning farmers' friend) and other forms of low-cost drip, such as Pepsee systems.[1] Yet, also for those customized systems, adoption rates generally are low (Shah and Keller, 2002; Verma et al., 2004). Poor users of drip irrigation face specific challenges, such as insufficient access to markets, and they often find out that low cost is still too expensive.

The low adoption rates of drip in India, however, have not stopped the government from promoting MI technologies. In 2005, with the objective to promote efficient use of water in agriculture, the government of India launched a new centrally sponsored scheme (CSS) on MI. In this scheme, high capital costs of especially standardized drip irrigation have been identified as the major barrier for widespread adoption.[2] Hereby, other factors known to hamper adoption were implicitly discounted, such as a lack of access to water or insufficient market access for smallholders (GOI, 2006; Phansalkar and Verma, 2008). Under the new scheme, the central government provides subsidy for 40% of the capital costs of standardized MI systems, and state governments contribute a minimum of 10%. The policy also stipulates that an additional 10% subsidy is provided by the central government when recipients are classified as marginal (< 1 ha) or small farmers (1–2 ha), or belong to scheduled castes or scheduled tribes. Throughout India, farmers can thus rely on a minimum of 50% to 60% subsidy for the installation of standardized MI systems on their land.[3] In 2010, the scheme was renamed as the National Mission on Micro Irrigation (GOI, 2010). As per 15 March 2012, a total amount of about US$847 million had been released for subsidies by the central government (Pullabhotla et al., 2012; Indiastat, 15–3–2012).[4] This level of subsidy release in India suggests that the spreading of MI technologies under farmers is (now) taking place at some scale.

The recent public policies aiming at supporting the diffusion and adoption of drip have attracted positive attention among policymakers and researchers in India, especially in the states of Andhra Pradesh and Gujarat. For these states, the current scheme for MI subsidies is described in government evaluations as a particularly successful one, especially in terms of implementation, area coverage, and achieved physical and financial targets (NABCONS, 2009; GOI, 2014). In this chapter, I focus my analysis on Gujarat and I critically analyse the context for such positive assessments of the subsidy scheme through a description of how it works and for whom. My objective is to develop a critical understanding of some of the knowledge and policy claims that surround drip technology in India, and to contribute to a debate on the meaning and use of drip beyond its acclaimed technical and economic advantages.

With the start of the CSS scheme on MI, and to avoid shortcomings of earlier federal and state schemes, the central government advised states to adopt 'greater flexibility' in the design of institutional mechanisms and implementation structures (Pullabhotla et al., 2012: 2). The idea was to promote independent authorities and circumvent classic government services, to achieve efficient subsidy disbursement. Two states in particular, Andhra Pradesh and Gujarat, have taken this advice seriously, establishing separate implementation agencies as Special Purpose Vehicles in the government administration. The Andhra Pradesh Micro Irrigation Project (APMIP) was already established in 2003, under the Directorate of Horticulture, two years before the start of the CSS on MI. The Gujarat Green Revolution Company (GGRC) was set up in 2005 in direct response to the new scheme on MI. The two agencies differ in character. The APMIP can be considered to operate as a normal government department, while the GGRC has the form of a public company. Both organisations have achieved high levels of subsidy disbursements and they are seen by the central government in India as the most successful models in terms of their 'capacity and quality' of implementation (NABCONS, 2009: 3). More specifically, APMIP is seen as a replicable model in terms of 'achieving physical and financial targets', and the GGRC in terms of having 'a multi-stage, multi-level monitoring process' in place (GOI, 2014: 88). The GGRC also is positively assessed by researchers in India. In a recent review on the spread of MI in nine states in India, Palanisami et al. (2011) for instance state that '[a] special purpose vehicle such as the Gujarat Green Revolution Company in each state *should be* created to handle MI implementation' (emphasis added) (p. 86).

Against the background of the positive reputation of the GGRC and the high levels of subsidy disbursement in India, I use the chapter for a case study of the GGRC, treating it as one significant episode in a larger history of planned efforts to promote drip and sprinkler irrigation in India. I am particularly interested in describing and tracing how and in what context GGRC obtained its praise in policy and research circles. My study relies on a review of academic and policy literature and interviews with key actors in Gujarat, conducted in November 2015, respectively with the GGRC, manufacturers and distributors of MI technologies, the Aga Khan Rural Support Programme (AKRSP) and a few farmer beneficiaries. A limitation of the analysis is that I have not collected detailed information on the beneficiaries of subsidized drip in Gujarat. Instead, I rely on impressions and personal interpretations of respondents' views, and a few direct field observations to scrutinize claims of success in the world of drip.

The remainder of this chapter is divided into four sections. First, a brief background is presented on the use and spreading of drip irrigation in India. Then, the subsidy disbursement model of the GGRC is discussed, followed by an analysis of (policy) claims that are made in relation to the GGRC and drip in Gujarat. The chapter ends with conclusions and a discussion.

Spreading of drip irrigation in India

Drip irrigation technology was first tested in India in the early 1970s, at the Agricultural College of Coimbatore, Tamil Nadu (Narayanamoorthy, 2005). The main rationale for state-led introduction of drip irrigation at the time was similar to that of other new technologies: to raise agricultural productivity and trigger a green revolution. We now know, with hindsight, that drip irrigation, unlike improved seed varieties, fertilisers, pesticides and tractors never played a significant role in these aims (Phansalkar and Verma, 2008). Throughout the 1970s and 1980s, the area under drip irrigation in India remained small (see Figure 18.1). Against this background, it is salient to observe that drip technology is now again promoted by policymakers in India, including the GGRC, as a catalyst for a second green revolution. The name itself of the GGRC is an apt illustration for this ambition, promoting the development of agriculture with a new focus on increasing productivity, sustainability and the efficient use of available resources.

Little is known about how exactly drip technology found its way into farmers' fields, or about who have been the main actors involved in this dissemination. It is clear, however, that farmers in Tamil Nadu, Andhra Pradesh and Maharashtra were the first to adopt drip irrigation (Verma et al., 2004). In Tamil Nadu, the technology was mainly used in banana and coconut production, and in Andhra Pradesh and especially Maharasthra, it was used for grapes and citrus fruits, banana, mango and pomegranate (Narayanamoorthy, 2005). Until deep in the 1990s, talking about drip in India was basically synonymous with talking

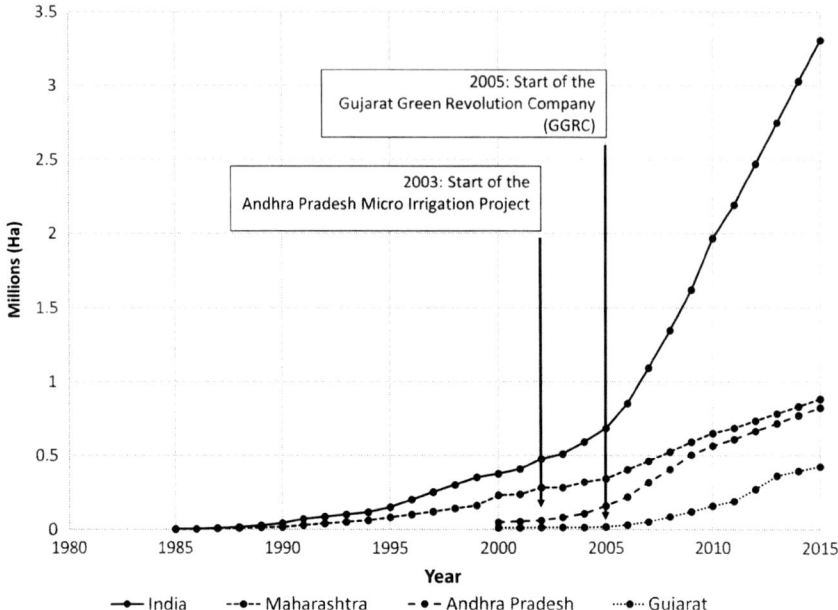

Figure 18.1 Drip irrigation in India, Maharashtra, Andhra Pradesh and Gujarat (1985–2015).

Source: Author's estimate, using numbers of Indian Statistics, 2001–2015 as well as Narayanamoorthy (2005); Pullabhotla et al. (2012); GGGI (2015) and GGRC (2015b).

about drip in Maharashtra. In 2000, over 60% of the total area under drip irrigation in India was located in this state (see Figure 18.1).

In 1981, the National Committee on the Use of Plastic in Agriculture was established and 17 experimental research stations (Plasticulture Development Centres) were set up to test drip among other (plastic) technologies. The first CSS scheme on MI started in 1982–1983 and was channelled through the Ministry of Water Resources, under the Minor Irrigation Division. A central feature of the early scheme was the focus on marginal (< 1 ha) and small (1–2 ha) farmers. Presumably, bigger farmers or areas were not eligible for subsidy because they were expected to manage themselves. Actual subsidy disbursements, however, were very low. One of the reasons was that application procedures for subsidies were cumbersome. By 1991–1992, the central government had only released about US$10 million (see Phansalkar and Verma, 2008: 40). The exception in India was Maharashtra. There, the government, using its own state resources, disbursed about US$126 million for subsidies on MI between 1986 and 1993 (Narayanamoorthy, 2005). There is, however, to my knowledge, no convincing explanation in policy and research literature about why subsidy delivery for drip and sprinkler adoption in Maharashtra was more successful than elsewhere in India.

In the 1990s, the central government continued its subsidy scheme on MI for small farmers, and so did states like Maharashtra, albeit on a smaller scale, but without much effect. Overall, progress was below expectations. By 2000, in total, 375,000 ha was reported to be under drip irrigation in the whole of India (Verma et al., 2004), while the total potential area for the technology in India was thought to be about 20 million ha (GOI, 2006).

In spite of the low adoption rates, or perhaps because of it, the government continued to promote standardized MI technologies. While productivity goals had been in the foreground at the time of introducing the technology, over the years and on the waves of ever more widely heard proclamations of a national crisis of water scarcity the focus shifted decisively to the potential water-saving capacities of drip. It is telling in this regard that the guidelines for a new CSS on MI, prepared in 2003–2004 by a special high-level taskforce on MI, start by saying there is a need for more 'judicious use of available water' in agriculture, explaining that 'conventional methods of (. . .) irrigation [are] highly inefficient', leading to a 'wastage of water' (GOI, 2006: 2) (see Box 18.1 on the difficulties to assess the water-saving potential of drip irrigation, given the diversity of Indian irrigation systems and ecosystems). The rationale for promoting drip thus changed, but the identification of high investment costs as being a major barrier for its adoption did not. This justified a new CSS on MI.

In this particular problem framing, which posits that high capital costs are the major impediment to adoption and proposes state (price) subsidies as the solution, commercial companies have always played a lobbying role in India. They are exerting pressure on the government to achieve that standardized MI technologies become affordable for smallholders, aiming to secure market prospects. Private sector involvement in drip irrigation goes back to the late 1980s in India, when companies such as Jain Irrigation Systems started to test and promote the technology (Singh, 2010), presumably working initially with bigger farmers. An important strategy of private companies has always been to lobby for what some have called a systems approach (Narayanamoorthy, 2005: see p. 12). Their argumentation was that drip is most effective when whole systems and standardized components are adopted. An emphasis on whole systems and standardized components clearly was a strategy of big market players to secure business prospects. Their success shows in how subsidies for MI in India have always been reserved for standardized equipment; this effectively prevented market actors in (and users of) non-standardized drip equipment to benefit from state support. Put differently, government actors buy in on (public) service delivery, economic growth and water saving, while (big) private sector actors of standardized equipment (manufacturers, wholesalers and dealers) buy in on a growing and yet unsaturated market in which profits are more or less ensured by a continuous flow of price subsidies, supplied by one powerful buyer (the government). In this regard, the state subsidy schemes on MI have always represented a strong alliance of interests, and this also has to be expected with the latest CSS for MI – the National Mission on Micro Irrigation – in states such as Gujarat, with public agencies such as the GGRC.

Before turning to an analysis of Gujarat, it is useful to reflect on available numbers on the spreading of drip in India. Amidst the various views on drip that exist (what it is or what it should be; who is using it or who should use it), it is difficult to establish how much actual use of drip there is in India. Government figures on area (ha) covered by drip systems, such as presented in Indian statistics, primarily reflect progress made in the subsidy programmes. Some of these figures represent actual progress of implementation, but others reflect planned areas in relation to targeted levels of subsidy disbursement. None of these figures may thus reveal actual use of drip irrigation by farmers in a given year, or in a given season. Furthermore, the figures ignore areas that are covered by non-standardized or low-cost drip systems, such as promoted by iDE/India. This NGO reported, as of June 2005, a total number of 150,000 users in its programmes (cited in Phansalkar and Verma, 2008: 41).

Taking these dynamics into account, I have made an estimate of the total use and spreading of drip irrigation in India, cross-checking and weighing available numbers (see Figure 18.1). In total, over 3 million ha is now (believed to be) under drip irrigation in India (at least for one season in a given year). It shows that Maharastha is still important, even though its percentage share has come down to about 25% of the total area in India (from about 60% in 2000). The figure also shows a spectacular growth (after 2000) in Andhra Pradesh, and especially in Gujarat. There, up to 2005, drip irrigation was practically non-existent, but a decade later, in 2015, over 400,000 ha was (stated to be) covered by it. It is generally assumed that this growth has been facilitated in important ways by the GGRC and its subsidy programme.

Box 18.1   The deep waters of determining whether drip irrigation saves water in India

India has the largest drip-irrigated area in the world (nearly 3 million ha) though it is only less than 4% of the country's net irrigated area. In spite of major advances in the science of hydrology and in crop sciences, the questions pertaining to the impact of drip irrigation in terms of water saving per unit area and at basin level remain largely unanswered. This is due to an issue of scale: the extent of water saving achieved through the use of efficient irrigation devices such as drip irrigation can reduce as the analysis moves from the field to the farm on to the watershed.

Assessing the extent of water saving requires an in-depth understanding of the different elements of the water balance. First, water that is 'lost' from the cropped field through field runoff can be available to the nearest plot in the farm, and the water which is lost in deep percolation recharges the shallow aquifer, which can in turn be picked up by farmers using pumps and wells. As for evaporation from the soil profile, it depends on the crops (for many distantly spaced crops such as fruit trees, cotton, castor, and fennel soil evaporation can be significant; for field crops, soil evaporation evolves with cropping stages and the extent to which the crop canopy covers the soil). Finally, evaporation is also closely linked to weather conditions: it is lower in humid and sub-humid climate than in to arid and hyper arid regions. Also key to assessing the extent of water saving linked to using drip irrigation, is understanding the 'rebound effect' (Berbel et al., 2015), which highlights that farmers who might save water per unit of land through the use of drip irrigation can use the 'saved' water to expand the area, hence resulting in no real water saving.

The issue of water saving through drip irrigation is especially difficult to assess in India, given the diverse geo-hydrological and climatic situations and irrigation types existing in the country and the complex soil-water-atmosphere-plant interactions that determine consumptive water use by crops. If drip irrigation systems are adopted in areas with high aridity and deep water table condition such as alluvial north Gujarat, there could be some real water saving as drip irrigation could prevent non-beneficial evaporation from the top soil strata (1–3 meters), non-recoverable deep percolation, or high content of hygroscopic water in the unsaturated zone of the soils. This could also be the case in the well-irrigated areas of semi-arid and arid peninsular India where top soil evaporation and non-recoverable deep percolation are high.

If the water table is very shallow and climate is humid, like in the alluvial areas of West Bengal, almost the entire water from deep percolation ends up in the aquifer. Hence, using drip irrigation would lead to 'dry water savings'. Though individual farmers would gain from the use of drip irrigation systems (for instance in terms of lower pumping cost), it would not save water in real term if the analysis focuses on the aquifer. The water balance, however, would altogether change if the area has saline aquifers like in parts of South Western Punjab, as in such a case, deep percolation water would not be available for reuse without being treated – which is expensive.

In the canal command areas of hot and semi-arid parts of peninsular India, particularly in Andhra Pradesh, Karnataka, Telangana and Maharashtra, return flow to shallow groundwater from gravity-irrigated fields is likely to be significant, as water table is generally shallow in such areas. Here as well, there may not be any real water saving through the use of drip irrigation system as farmers often tap in the shallow groundwater with wells and pumps. In such contexts, it is important to raise caution about over-estimating the benefits of drip irrigation in terms of water saving. Return flows from canal irrigation at times produce much greater benefits than what the same water would have generated if supplied through drip irrigation system. One other example of this is the canal irrigated areas of north eastern Punjab where paddy and wheat are irrigated through flooding method. The return flows augment recharge to groundwater in these alluvial plains, which is pumped out through wells to provide a much need supplementary irrigation to crops.

As regard to the rebound effect, the key factor to understand is that of land availability. Land may not be a constraint in many semi-arid and arid regions of peninsular India (Telangana, Rayalaseema region of AP, Vidarbha and Marathwada regions of Maharashtra, Karnataka and Tamil Nadu) and farmers could extend their cropped area. In other regions (central Punjab and the alluvial north of Gujarat as well as the Indus basin irrigation system in Pakistan for instance) pressure on irrigated land (and the groundwater resources) is high, hence de facto limiting the extent to which farmers can extent their cultivated area. In these regions using drip irrigation system is more likely to yield 'real water savings'.

It is important to raise caution about the extent to which drip irrigation can yield water saving given the huge public expenditures that are being made in the form of subsidies. The on-going debate about the impact of drip irrigation can be settled if we invest in high quality empirical research involving field measurements, and compare these with modeling results in different climatic and geo-hydrologic environments and cropping conditions.

Dinesh Kumar

The Gujarat Green Revolution Company

The GGRC was set up under the Modi administration (chief minister of Gujarat between 2001 and 2014, now prime minister of India). His term in Gujarat is generally praised as exemplary of a particular development model, one that is based on the active pursuit of business-based economic reform, the reduction of bureaucracy and the encouragement of foreign direct investment. Its success is associated with bold decision making and strong political will. The GGRC in many ways illustrates and embodies this. In consonance with the agricultural policy of Gujarat Vision 2010, aiming to save water and improve agricultural productivity through the promotion of MI systems (Singh, 2010), Modi's administration renamed a dormant government enterprise, the Gujarat Agricultural Processing Company, to create a new semi-autonomous state corporation: the GGRC (Pullabhotla et al., 2012). The ownership of the organisation remained vested in the same three public companies, respectively Gujarat State Fertilizers and Chemicals (46% shareholder), Gujarat Narmada Valley Fertilizers & Chemicals (46% shareholder), and Gujarat Agro Industries Corporation (8% shareholder).

Furthermore, the state irrigation department was appointed as the (overseeing) governing body of the GGRC (Singh, 2010), its office was established on the premises of Gujarat State Fertilizers and Chemicals, and management staff was brought in on deputation from other government departments. The idea was to adopt a business-based approach for a public agency, to implement the CSS scheme on MI in Gujarat, cut red tape and create a single window for beneficiaries (GGRC, 2015a).

Other than designing the GGRC as a Special Purpose Vehicle, Modi is said to have done two other important things. First, he pledged to allocate more than US$340 million to the GGRC for the implementation of drip and sprinkler irrigation (GOG, 2005; Singh, 2010). Promises for new subsidies for drip irrigation, most of which were to be provided by the central government, are nothing new in India, but such a pledge for a seamless access to financial resources certainly was. In 2005, this figure was more than the total subsidy amount for MI in the whole of India in the previous two decades that had been released by the central government. Second, Modi appointed a senior officer of the prestigious Indian Foreign Service to lead the GGRC, giving him security of tenure for four to five years. Reportedly, Modi made it clear (to him and others) that he did not want to be tied to the central government, giving him a long rope to design a new organisation and ensure flexibility in operation.

In this context, the GGRC created an organisational model for subsidy disbursement for MI in Gujarat that was new and unique in India. Based on a pilot project on 30 ha of land, involving 56 farmers in Vadodara and Banaskantha districts, respectively a well-watered district in central Gujarat and a dry district in north Gujarat, a standard business modality was developed for the agency (Singh, 2010: see p. 153). This model is shown in Figure 18.2.

The model works as follows. First, farmers who are interested in drip (or sprinkler) have to approach a local dealer (distributor) of a MI company of their choice – not the government as elsewhere in India. Then, the MI company undertakes a survey to provide a cost estimate and submits – with signatures of the farmer – an online application for subsidy to the GGRC. The farmer deposits his or her full share of the cost estimate (usually 50%) around the same time at a nearby bank, on the account of the GGRC.[5] The GGRC follows up by making a tri-party agreement between the applicant (the farmer), the MI company and the GGRC, and releases a work order for the MI company with a first instalment (25% of the cost estimate). Then, the MI company (often through a dealer) installs the system and a third-party inspection (consultants) is called in to witness a trial run of the equipment (by the MI company). At the occasion of the trial run, the farmer is requested to acknowledge that it works well. Once he has confirmed that and after the third-party inspection, an insurance certificate is issued for the GGRC. Finally, the MI company submits a final bill and the GGRC releases the remaining 75% of the funds.

Next to this procedure, the organisation of the GGRC has three other characteristics that make its design unique. First, the GGRC provides, on average, a 50% subsidy on the capital costs of MI systems, the lowest amount of subsidy for MI technologies in India (NABCONS, 2009). Second, to ease the complexities involved in getting subsidies, the GGRC does not impose a ceiling

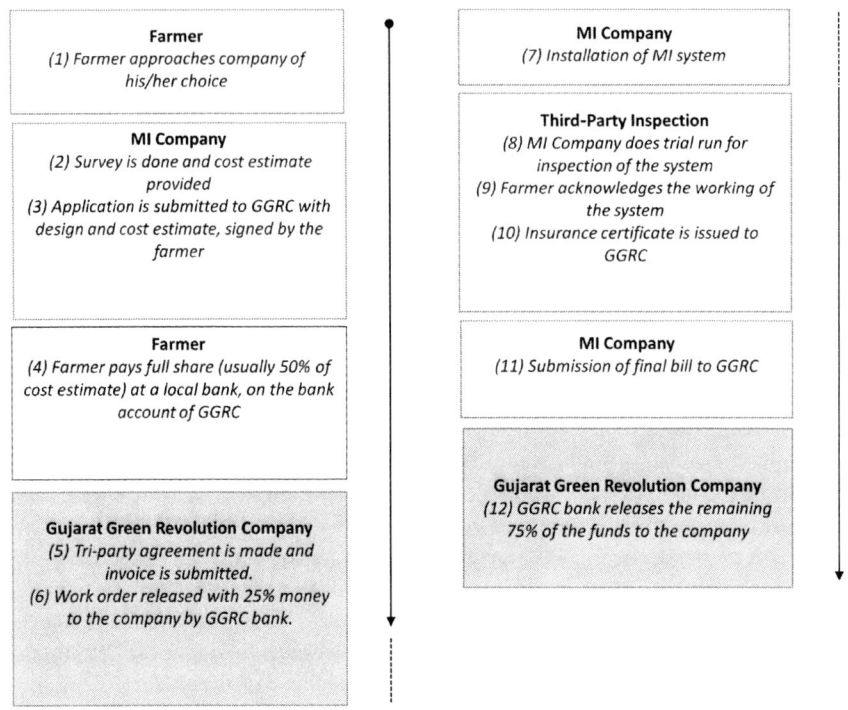

Figure 18.2 Subsidy disbursement under the GGRC in Gujarat.

Source: the authors, based on interviews as well as Singh (2010) and Pullabhotla et al. (2012).

of 5 ha in its scheme, nor does it stipulate crop choices or a maximum area for sprinkler coverage, as elsewhere in India (NABCONS, 2009; GGRC, 2015a). And third, the GGRC pays subsidies directly to MI companies – not to farmers (Singh, 2010). This means that farmers only need to make arrangements for 50% of the capital costs of MI systems, unlike elsewhere in India, where they are expected to advance 100% of the investment costs.

The model has been designed in such a way that most of the transaction cost for subsidy application, as well as some of the financial risks in getting the subsidies, are borne by MI companies and third-party inspectors, rather than by farmers and the government as was previously the case. The basic objectives in the design of the model are: (a) to ensure service delivery by the market, (b) get private sector actors involved in the paperwork for application, and (c) relieve the government of personnel liabilities, intensive extension work, and cumbersome field and technical inspections. Measured against these objectives, the model of the GGRC can be considered to work well. To illustrate the dynamic, as per November 2015, the GGRC had registered and was dealing with 65 MI companies in Gujarat and 36 consultancy firms for third-party inspections. A senior manager of the GGRC estimated that the MI companies were employing a total workforce of about 1000 field personnel (e.g. engineers, agriculturalists, extension workers), while the agencies for third-party inspections were making use of a total number of 200 inspectors. In addition, there are new dealer networks of MI companies. Singh (2010) reports a dealers density in the state of one dealer per 350 ha on average, ranging from 145 ha to 570 ha, based on a survey among 14 dealers of MI companies in Gujarat (p. 176). Jain Irrigation Systems, for instance, had already appointed 150 dealers for Gujarat by 2010 (p. 162) and Netafim 45 dealers (p. 165). These numbers reveal that big MI companies have high stakes in the GGRC, enabling the government of Gujarat to execute its MI subsidy programme through placing dealers in the state and selling them products.

These conditions – the reliance of GGRC on business actors – in turn have allowed GGRC to operate as a small and lean organisation, focusing on processing applications and releasing subsidies. As elsewhere in India, the application procedure is still cumbersome: in total 14 to 18 pieces of paperwork need to be completed, but GGRC manages it quickly. It deals with the whole process, from new applications to payment of the final bill for the MI company, within 14 months and often faster (Singh, 2010; GGRC, 2015a). Those acquainted with offices of the Indian bureaucracy may find the working atmosphere of the GGRC a revelation. In the offices, stacks of paper folders are still there, but applications are submitted and can be followed online (by farmers, MI companies and dealers). Every application is given a unique identification number, and in processing applications, GGRC adheres to the principle of FIFO (first in, first out). Farmers can deposit their share of the payment at a local bank, on the GGRC account (similar as with school fees in India). They, therefore, do not need to travel to the GGRC office. Locations of farmers' fields are verified with GPS coordinates, linking them to Google Earth images. This in turn, allows third-party inspectors to use a mobile phone application to verify the installation of new drip and sprinkler systems, using a QR code. For verification, third-party inspectors only need to upload GPS coordinates and photos of the installed system, with the farmer holding a signboard with the identification number of the application.

Through this procedure, from 2005–2006 to 2015–2016, GGRC reached more than 800,000 beneficiaries, covering an area of approximately 1.3 million ha under MI, of which about 495,000 ha is under drip irrigation (Indian Statistics; GGRC, 2015b, 2015c, 2016a, 2016b). As of November 2015, GGRC was receiving applications at a rate of 500 per day and was claiming to implement MI systems at a rate of 600 ha per day. According to my estimates, based on data made public by the GGRC in 2016 and using some simple assumptions (a drip systems costs about INR 125,000/ha; a sprinkler system about INR 19,000/ ha, see Singh, 2010: 152), the GGRC had released a total amount of subsidy of approximately US$686 million during the last decade; about US$550 million for drip systems and about US$136 million for sprinkler systems (of which 80% had been paid by the central government and

20% by the government of Gujarat).[6] In its most basic form, the GGRC thus represents a political pledge for spending public funds on MI technology. That this pledge has actually been fulfilled is new for MI in India. The Gujarat government now plans to also use the GGRC for bringing the entire canal command area of the Sardar Sarovar Project (1,792,000 ha) under MI technologies, with the goal to modernize agriculture, and many states in India, respectively Rajasthan, Madya Pradesh, Uttar Pradesh, Chattisgarh, Tamil Nadu and Karnataka, have expressed an interest to establish GGRC type agencies (Singh, 2010; GGGI, 2015). Not without coincidence, during a visit to the GGRC office in November 2015, the manager director of the GGRC was absent, away on a trip to Haryana state, to explain the state government how the GGRC model could be replicated.

Understanding the success of the GGRC and drip in Gujarat

As described above, GGRC's methods of disbursing subsidies for promoting drip and sprinkler irrigation are widely praised in policy and research circles. However, the fact that subsidies have been effectively delivered by the GGRC and that the government in Gujarat has lived up to its political promises does not tell much about whether and how drip irrigation systems are being used by farmers. To date, little research has been done on the actual use of subsidized drip irrigation in the field in Gujarat. Also, there is little known about who is gaining (most) from the GGRC and its state subsidies for drip systems. To understand the context in which the GGRC is seen as a success, especially in relation to an observed absence of studies on the actual use of drip irrigation in the field, I focus my analysis on two groups of actors who stand to gain (directly) from the subsidies. These are farmers and private sector actors (MI companies and their dealers). I focus on them because they are portrayed by the GGRC and the central government, in addition to Modi and the GGRC itself, as the main actors of and behind the success of the GGRC, respectively as 'beneficiaries' and 'system suppliers' (GOI, 2010; GGRC, 2015a). Other actors who (may) have stakes in promoting a positive image of the GGRC, such as government officials at central and state levels, politicians, consultants and staff of NGOs, are not portrayed by the GGRC and the central government as key to the success of MI adoption in Gujarat. In what follows, I first analyse which farmers have benefitted from the GGRC and then, I zoom in on private sector actors that have engaged with the GGRC.

Focus on farmers

As noted, I did not collect detailed information on beneficiaries of the GGRC. Instead, I use figures on beneficiaries presented in public information and promotional material of the GGRC, showing the achievements of its subsidy programme. For instance, a recent promotion leaflet of the GGRC presents MI on the front page as a technology capable of 'yielding green gold, anytime, every time for lifetime' (GGRC, 2015b). The leaflet shows a young, green seedling growing on a bed of gold coins, flanked by the logo of the GGRC. Here, the term 'green gold' refers to the capacity of MI systems, especially drip irrigation, to increase crop yields (gold) and save water (green). It thus conveys the message that MI can offer farmers a golden future in a sustainable or green way: hence, green gold.

The figures that are presented in public information of the GGRC do not allow for detailed analysis, but they give an idea who have been the main beneficiaries of the subsidies. It presents beneficiaries in classes of landholding, showing that it mainly is medium (2–10 ha) and large (> 10 ha) landholders in Gujarat who have benefited from MI subsidies, respectively 56% and 4% of all the beneficiaries of the GGRC (GGRC, 2016a). Conservatively estimated, based on the assumption that farmers of all classes obtain drip and sprinkler equipment in equal proportion, it implies that at least 75% of all subsidies in monetary terms went to medium and large landholders (more than US$514 million). To clarify, these numbers on Gujarat should not come as a surprise. Access to irrigated land is skewed in Gujarat in favour of larger landholders (Prakash, 2005) and they also tend to be the farmers in the state who mostly grow horticulture – and cash crops like cotton and groundnut. These type of crops are (very) suitable for MI technologies, especially for drip irrigation. In contrast, marginal (< 1 ha) and small (1–2 ha) landholders in Gujarat disproportionally have less access to irrigated land and many of them rely on rains to cultivate their land. Put differently, it would be wrong to expect that subsidies for drip and sprinkler irrigation will (or can) mainly benefit small and marginal farmers in Gujarat, because many of them have limited access to water for irrigation.

Notwithstanding, it implies that only a relatively small part of the total farmers' population in Gujarat has benefitted from the GGRC subsidies for MI. According to the Agricultural Census of 2010/11, there are close to 4.9 million landholdings in Gujarat. In total, 37% of these landholdings belong to marginal farmers and 29% to small farmers; only 33% and 1% of the landholdings belong respectively to medium and large farmers. In contrast, the information provided by the GGRC portrays an almost complete reversal of this division: just 10% and 30% of the beneficiaries are respectively marginal and small farmers, and the majority consists of medium and large farmers (total 60%) (GGRC, 2016a). The information of the GGRC does not specify which type of farmers have obtained subsidy for which type of MI technology (drip or sprinkler), but I assume that small and marginal farmers have mainly obtained subsidies for certified sprinkler equipment, knowing that those systems on average are about five times less costly than certified drip systems. Again, this is not particularly surprising information, but it reveals some of the context of the success of the GGRC. Only 4% of all marginal farmers in Gujarat and only 17% of all small farmers in the state have benefitted from MI subsidies, compared to 29% of all medium farmers and 71% of all large farmers in Gujarat.

These numbers and observations corroborate with a survey presented in Singh (2010) among 34 farmers who were using MI systems, mostly for drip irrigation, in the north of Gujarat (Sabarkantha, Banaskantha and Kachchh districts). In total, 69% of the sampled farmers had benefitted from GGRC subsidies most of whom were land owners, who had tube-wells and electric motors at their disposal to ensure reliable irrigation supply for cash crop production. Hence, Singh (2010) reported that most of the sampled farmers were cultivating almost their entire area of the farm (on average 90%), using drip and sprinkler irrigation for cash crops like cotton, groundnut, castor, potato and other vegetables, and for mango and papaya.

The numbers of the GGRC and the characteristics of farmer beneficiaries described by Singh (2010) also match my impressions from a field visit. I visited the village of Ode, Anand district, central Gujarat, a place with a lot of drip users and beneficiaries of GGRC subsidies. There, mainly Patel farmers, a land-owning class and upper caste in Gujarat, were using the latest drip in-line technology of Netafim for the production of high value crops, such as capsicum and chili peppers. In addition to some early adopters, it was reported to me that most farmers in the area had recently started using drip irrigation, using subsidies of the GGRC to install new systems. The farmers had good access to water. The village area falls in the Mahi Canal command and canal water supply ensured a regular recharge of groundwater aquifers. My respondents told how Patel farmers had invested massively in tube-wells, claiming that there are now 500 wells in the area. All farmers who were using drip irrigation in the area were reported to cultivate land of up to 10 ha and have access to at least one or two tube-wells. Drip irrigation enabled these farmers to increase productivity and maximize returns per unit of (pumped) water and per unit of land, using the technology to increase water use productivity, literally producing more crop per drop. However, the farmers did not seem pre-occupied with using less water per se; they had good access to water, it was available for them and they aimed to increase their income.

The farmers of Ode may not be the prototypical beneficiary of GGRC subsidies; farmers who are located in the Banaskantha district, for instance, in areas of severe groundwater over-exploitation, in so-called dark zone areas, are dealing with a very different situation, but their characteristics fit an image of who tend to be the main beneficiaries of the GGRC. It is farmers who focus on commercial production and have good access to (ground) water, credit, information and markets. For this type of farmers, raising the necessary credit to apply for the subsidies (50% of the total capital cost) is not a problem. This explains why, as of November 2015, GGRC reported that just 3% of the applicants had taken a loan to access subsidies.[7]

The middle-aged and senior farmers I spoke to in Ode praised Modi and were happy with the GGRC. The subsidies had allowed them to expand and professionalize their farming business, for instance, by taking more land under irrigated cultivation or by switching to higher value crops that often consume more water, for instance, from maize to capsicum, chili peppers or banana. In fact, the subsidies for drip have enabled farmers in Ode to engage in the commercial production of high value crops throughout the year. These observations – increase of total irrigated area, better net returns and additional income for farmers, especially for chili farmers – resonate with a recent government evaluation in which Gujarat was one of the 10 sampled states (GOI, 2014). Beneficiaries also praised the GGRC for the actual (physical) implementation of the scheme, the easy access to information, quick approval, the immediate release of the subsidy, prompt attention to farmers' complaints, strict monitoring and inspection, for the fact that no ceilings on area were imposed and that there was no corruption (see also Singh, 2010). The only complaint of the farmers was an occasional delay in the processing of documents by MI companies – not by the GGRC.

This account reveals that medium and large farmers are critical for the GGRC, because they represent the main group of beneficiaries. They enable the GGRC to present itself as a responsive government agency, efficiently disbursing subsidies and making MI implementation a success. Put differently, medium and large farmers in Gujarat enable the GGRC to claim effective implementation of drip and sprinkler irrigation, portraying its achievements as a success in comparison to other and less successful (earlier) central schemes for MI in India, including the National Mission on Micro Irrigation, which always had (and have) a clear objective to reach out to smallholders (as noted above).

For smallholders, the story is different, also in Gujarat. To reiterate: although the GGRC reports that marginal and small farmers are beneficiaries (40% of the total number of recipients of subsidies), it appears that most smallholders can only potentially benefit from subsidies of the GGRC through (additional) support programmes (of NGOs), especially when it is for drip irrigation. For instance, starting in 2010, AKRSP is assisting marginal and small farmers in Surendranagar, Rajkot and Botad districts in adopting standardized drip for the production of organic cotton, aiming to provide economic livelihood to them in an environmentally sustainable way. For this purpose, AKRSP was granted US$2 million by the Cotton Connect Foundation, a social enterprise that is funded by the C&A Foundation, Shell Foundation and Textile Exchange. AKRSP has used the money to create a pool to provide interest-free loans to farmers who invest in drip, to be paid back within 24 months. The rationale of the drip pool programme is to allow smallholders to plug in with the scheme of the GGRC. The AKRSP model offers smallholders an opportunity to cover an additional 33% of the initial capital costs through a soft loan, meaning that only 17% has to be paid upfront by the beneficiary.[8] Up to mid-2015, 1200 smallholders had benefitted from this scheme, and about 1400 ha had been put under drip irrigation. The C&A Foundation had recently asked the AKRSP to continue with the programme until 2019–2020, aiming to put a total area of 8000 ha of organic cotton under drip irrigation. A senior manager of AKRSP stated that significant impacts were achieved: drip irrigation enables farmers to use water efficiently (producing more crop per drop), they can irrigate larger areas and increase the production per hectare.

This account illustrates that subsidies of the GGRC, especially for standardized drip equipment, are difficult to access for smallholders without additional financial support. At the same time, it is possible that smallholders acquire and use non-standardized, low-cost drip technology through the non-subsidy market in Gujarat, but this I did not investigate. In other words, the observation that smallholders benefit less from GGRC subsidies for drip irrigation does not necessarily mean that they do not actually use drip equipment in Gujarat.

Focus on private sector actors

Another group of actors that features prominently in the positive attention for the GGRC, clearly gaining from the state subsidies on drip irrigation, are private sector actors (system suppliers, MI companies and their dealers). Around 2005 in Gujarat, there was hardly any MI market for standardized technology. A decade later, in 2015, a huge and yet unsaturated market for standardized (certified by the Bureau of Indian Standards) MI equipment was thriving in the state. It is not difficult to see that the generous price subsidies of the GGRC helped create this market, having injected about US$686 million in it (GGRC, 2016a). In 2005, there were only a few MI companies operating in Gujarat; in 2010, the GGRC was dealing with 25 registered MI companies (Singh, 2010); and in 2015, it was dealing with 65 companies. Many of these companies manufacture and supply their own brand, others sell equipment of established companies like Jain and Netafim, focusing on the design and distribution of MI systems. A senior manager of the GGRC estimated that 50% of these companies were operating exclusively in Gujarat. This development was paralleled by the expansion of dealers' networks of MI companies. Each MI company recruits local dealers to sell their products. These often are town or village-based shopkeepers of agricultural supplies. As noted above, by 2010, in some places, dealer networks had already reached a density of one dealer per 350 ha.

One consequence of the (price) subsidy scheme of the GGRC may be that it has distorted competition in the market and is pushing prices up. Three indirect observations support this hypothesis. First, competition in the form of non-subsidized, non-standardized MI equipment appears to be almost completely absent in Gujarat. As noted, I have not explicitly investigated the non-subsidy market for drip in Gujarat and neither did Singh (2010). He focused on the marketing of MI through the GGRC and sampled among owners of MI equipment, but I consider his survey to provide an apt illustration: it notes that only 3% of the MI systems of the 34 respondents consisted of non-standardized equipment (p. 194). Second, dealers of MI companies share in the growth of the market. They operate on a commissions' basis of 7% to 10%, thus taking their cut of the profits. The practice of giving commission to dealers is likely to increase the price of standardized MI equipment. And third, as Phansalkar and Verma (2008) so eloquently put it: the 'presence of [price] subsidies diverts attention of manufacturers from value engineering to offer a better cost-quality deal to customers, to jockeying for ensuring that their products are preferred tacitly or explicitly by the officers [the GGRC] approving the applications and to garnering a large share of the artificially propped market' (p. 82).

To illustrate the dynamic of prices: the GGRC sets detailed norms for unit costs of MI components, and revises them annually, based on market developments. With an expanding market, one would expect prices for MI systems to go down. However, in Gujarat, the unit costs of MI components, set by the GGRC, have gradually increased. This is reflected in the maximum amount of subsidy that is granted. This increased from INR 50,000 per ha in 2010 (or 50% of the capital costs, whichever is less) (Singh, 2010), to INR 60,000 in 2015 (GGRC, 2015b). These numbers suggest an increase of market prices of 3.7% per year in the period 2010–2015. To clarify, this increase is only visible when using the Indian rupee (INR) as the currency for calculation, which is justified because the manufacturing of standardized drip systems (for the Indian market) almost exclusively takes place on Indian soil. In US dollars, it shows a decrease of US\$1064 per ha in 2010 to US\$1000 per ha in 2015. The calculation is based on the assumption that it is unlikely that any increases in market price in Gujarat have been caused by increases in cost for machinery, assembly or labour, or by increases in price of raw material (oil). In regard to the latter, for the period 2010 to 2015, the international oil price remained stable and reached, in fact, an all-time low in the beginning of 2015 (www.opec.org, 24–4–2016). I consider an annual increase of 3.7% in unit costs a conservative estimate for Gujarat. A recent report of the Global Green Growth Institute (2015) on MI in Karnataka, for instance, reports a 'business-as-usual scenario of 7% annual increase in unit costs' (p. 6).

The GGRC thus provides MI companies with good opportunities to make profits, particularly in the early years when the number of MI companies registered with GGRC was still small. For instance, the annual turnover of Netafim in Gujarat jumped from US\$143 million in 2005–2006 to USD 289 million in 2006/07, and that of Parixit Industries, an Indian MI company, from USD 1,8 to USD 2,7 million (Singh, 2010: see p. 163 and p. 169). These figures reflect a growth of respectively 106% and 63%. In summary, as a branch manager of Kisan Irrigations, an Indian MI company in Ahmedabad commented: 'Modi [has been] good for business'.

The cooperation between the GGRC and MI companies thus works well for the promotion and selling of standardized drip and sprinkler equipment – a larger area in Gujarat is now irrigated under MI technologies. In regard to drip irrigation, it works especially well for those more well-to-do farmers who can mobilize the money and networks to engage in high-profit crops. In the GGRC model, MI companies and their dealers have little incentive to reach out to more marginal and small farmers. Hence, as a Netafim dealer in Himatnagar, Sabarkantha district (north Gujarat), reported in Singh (2010), explained: 'it is very difficult to penetrate below large and medium size category farmers, due to their low 'purchase capacity" (p. 184). In other words, dealing with smallholders, transaction costs (and risks) for MI companies and dealers will rise, returns decrease and profits dwindle. It is a formula bad for business.

Conclusions

My analysis has shown, through a case study of the GGRC in Gujarat, how the portrayal of drip irrigation technology as an important generator of a second green revolution – consisting of increasing productivities in water-conscious ways – provides the rationale of a novel subsidy programme that importantly relies on private sector actors. The positive imagery of drip is not new; it has been present in India from the 1970s when the technology was introduced in the country, to the present. Enduring low adoption rates among a majority of farmers, now and in the past, have done little to dampen the enthusiasm about (standardized) drip as a generator of green progress. The technology continues to be seen as technically good, being treated as a 'special case of hardware', to use the words of Phansalkar and Verma (2008: 7), a goodness that is attributed to its intrinsic engineering characteristics. As the technology itself is beyond question, the most common explanation for low adoption rates are the high initial capital costs of (standardized) drip systems. This is also the explanation that supports and justifies (new) state subsidy programmes: these have as their objective to allow farmers to purchase and install (certified) MI systems, something that is hoped will spark an intensification of agriculture. To date, the performance of the GGRC programme has been impressive in terms of subsidies disbursed, MI systems installed and areas brought under modern types of irrigation.

My analysis has shown, first of all, that the GGRC has raised much positive acclaim in India; it is seen and used as a model for other states. Its obvious success in efficiently disbursing enormous amounts of subsidies and disseminating large numbers of MI systems is importantly related to how the GGRC effectively targets larger farmers and mobilizes private entrepreneurs in distributing public goods. This resonates well with a larger development narrative that promotes private initiative and emphasizes the importance of business actors in bringing about growth. GGRC proudly aligns itself with this narrative, even though it officially is a public agency. For instance, it calls itself the 'first PPP [public private partnership] model in implementing [a] socio-economic scheme in the country' (GGRC, 2015a: 8) in India. In a circular fashion, the country-wide praise for GGRC is likely to also stem from how it provides further support to this narrative. Of course, the GGRC is and was also one clear testimony of how Modi lives up to his political promises. Overall, my analysis suggests that the success of the GGRC and the success of drip promotion in Gujarat primarily makes sense within a relatively closed network of government and private sector actors, and large- and medium-scale farmers, who align their interests.

In this regard, the positive image of the GGRC model is less connected to the promotion and actual use of drip and sprinkler technology than it may seem at first sight. For one, mere amounts of subsidies disbursed or irrigation systems adopted say little about who is actually making use of MI systems and how, which makes it difficult to establish if the subsidies are indeed bringing about the expected gold-green revolution. Based on the numbers provided by the GGRC, it is clear that the gold mainly accrues to a relatively small number of medium and large farmers and to private sector actors. Smallholders continue to be seen as a difficult group to reach, being perceived in state subsidy programmes as non-adopters of drip technology, real or imagined. As for the green: there is suspiciously little information on this. Yet, the few farmers that I interviewed mainly adopted drip irrigation to intensify agricultural production and increase their revenues. Expanding the area under irrigation, switching to higher value and more water consumptive crops, or growing a third season in the year may re-vitalize agriculture, but certainly do not reduce the overall amounts of water used for farming. Hence, whether the water preservation ideal of the GGRC is or can be met remains very much an open question.

Acknowledgments

The research described in this chapter was financially supported by the Netherlands Organisation for Scientific Research (NWO). The analysis benefitted greatly from the thoughtful and challenging comments of the editors and three anonymous reviewers.

Notes

1   Pepsee systems are low-cost water distribution lines. Pepsee is made of light density plastic. The plastic is used to fill ice candies and sold as Pepsee in the local markets. The plastic is available in rolls, which are used by farmers in place of drip tubes to irrigate cotton (Verma et al., 2004).
2   Standardized implies (a) that system components need to adhere to quality criteria imposed by the Bureau of Indian Standards and (b) that system implementation needs to adhere to stipulated installation norms.
3   In many states in India, the state government provides more than 10% subsidy, up to 40%. In such cases, the total subsidy for MI technologies sums up to 90%. In addition, there is a 20% subsidy scheme on MI under the National Horticultural Board (Singh, 2010: see p. 200).
4   In 2015, the National Mission on Micro Irrigation was subsumed in a new scheme, the National Mission for Sustainable Agriculture (GGGI, 2015: see p. 11).
5   Previously, this was only 5% management fee (Pullabhotla et al., 2012) but GGRC has recently changed its policy (pers. comm. senior manager of the GGRC, 4–11–2015).
6   A recent change in policy has changed this division in 60% central government and 40% state governments (pers. comm. senior manager of the GGRC, 4–11–2015).
7   GOI (2014) reports a higher percentage of applicants in Gujarat that had taken a loan, namely 14% (p. 28).
8   Interestingly, in AKRSP's experience, marginal and small farmers do not get 60% subsidy for drip irrigation, not even 50%, as the GGRC follows fixed unit prices for system components.

References

Berbel, J., Gutiérrez-Martín, C., Rodríguez-Díaz, J.A., Camacho, E., and Montesinos, P. (2015). Literature review on rebound effect of water saving measures and analysis of a Spanish case study. *Water Resources Management* 29: 663–678.
Global Green Growth Institute (GGGI). (2015). *Implementation roadmap for Karnataka Micro Irrigation Policy*. Seoul: GGGI.

Government of Gujarat (GOG). (2005). *Speech of Shri Narendra Modi, Chief Minister of Gujarat, fifty first meeting of the national development council*, New Delhi, June 27, 2005. New Delhi: GOG.

Government of India (GOI). (1990). *Status, potential and approach for adoption of drip and sprinkler irrigation systems*, National Committee on the Use of Plastics in Agriculture. Pune: GOI.

Government of India (GOI). (2006). *Micro irrigation (Sprinkler and drip irrigation): Guidelines*, Ministry of Agriculture, Department of Agriculture and Cooperation. New Delhi: GOI.

Government of India (GOI). (2010). *National mission on micro irrigation: Operational guidelines*, Ministry of Agriculture, Department of Agriculture and Cooperation, New Delhi: GOI.

Government of India (GOI). (2014). *Evaluation study on integrated scheme of micro irrigation.* Planning Commission, Programme Evaluation Organisation, PEO Report No. 222. New Delhi: GOI.

Gujarat Green Revolution Company (GGRC). (2015a). *GGRC – a role model for following best practices in implementation of micro-irrigation scheme in India.* MS PowerPoint presentation, 44 slides. Vadodara: GGRC.

Gujarat Green Revolution Company (GGRC). (2015b). *Yielding green gold, anytime, every time for lifetime*, promotional leaflet. Vadodara: GGRC.

Gujarat Green Revolution Company (GGRC). (2015c). *MISwiseTPA Summary*, information obtained from the website of the GGRC (15–11–2015). Vadodara: GGRC.

Gujarat Green Revolution Company (GGRC). (2016a). *Achievements*, information obtained from the website of the GGRC (20–7–2016). Vadodara: GGRC.

Gujarat Green Revolution Company (GGRC). (2016b). *MISwiseTPA Summary*, information obtained from the website of the GGRC (20–7–2016). Vadodara: GGRC.

Indian National Committee on Irrigation and Drainage (INCID). (1994). *Drip irrigation in India.* New Delhi: INCID.

Jalajakshi, C.K., Pal, R.C., and Jagadish, N. (2006). *A comparative analysis of drip versus flood irrigation techniques: An assessment of crop productivity and water conservation from India.* New Delhi: The Energy and Resources Institute.

Kumar, S.D., and Palanisami, K. (2011). Can drip irrigation technology be socially beneficial? Evidence from Southern India. *Water Policy* 13: 571–587.

Nabard Consultancy Services Private Limited (NABCONS). (2009). *Evaluation study of centrally sponsored scheme on micro irrigation: Executive summary*, National Committee on Plasticulture Applications in Horticulture (NCPAH), Ministry of Agriculture, Department of Agriculture and Cooperation. Mumbai: NABCONS.

Narayanamoorthy, A. (1997a). Economic viability of drip irrigation: An empirical analysis from Maharasthra. *Indian Journal of Agricultural Economics* 52(4): 728–739.

Narayanamoorthy, A. (1997b). Drip irrigation: A viable option for future irrigation development. *Productivity* 38(3): 504–511.

Narayanamoorthy, A. (2004). Drip irrigation in India: Can it solve water scarcity? *Water Policy* 6(2): 117–130.

Narayanamoorthy, A. (2005). *Potential for drip and sprinkler irrigation in India.* Unpublished paper, obtained through an internet search, November 22, 2015.

Palanisami, K., Mohan, K., Kakumanu, K.R., and Raman, S. (2011). Spread and economics of micro-irrigation in India: Evidence from nine states. *Economic and Political Weekly, Review of Agriculture* XLVI(26 and 27): 81–86.

Phansalkar, S., and Verma, S. (2008). *Silver bullets for the poor: Off the business mark?* Anand: IMMI-Tata Water Policy Programme.

Prakash, A. (2005). *The dark zone: Groundwater, irrigation, politics and social power in North Gujarat*, Wageningen University Water Resources Series. New Delhi: Orient Longman.

Pullabhotla, H.K., Kumar, C., and Verma, S. (2012). *Micro-irrigation subsidies in Gujarat and Andhra Pradesh: Implications for market dynamics and growth*, Water Policy Research Highlight. Anand: IWMI-TATA Water Policy Programme.

Shah, T., and Keller, J. (2002). *Micro irrigation and the poor: A marketing challenge in smallholder irrigation development*. Review for International Development Enterprise, India, unpublished paper, obtained through an internet search, November 22, 2015.

Singh, S. (2010). *Agricultural machinery industry in India: Growth, structure, marketing and buyer behavior*, Centre for Management in Agriculture, Indian Institute of Management, Ahmedabad. New Delhi: Allied Publishers.

Sivanappan, R.K. (1994). Prospects of micro-irrigation in India. *Irrigation and Drainage Systems* 8(1): 49–58.

Verma, S., Tsephal, S., and Jose, T. (2004). Pepsee systems: Grassroots innovation under groundwater stress. *Water Policy* 6: 303–318.

Postscript

A dialectic inquiry in the world of drip irrigation

Henk van den Belt

As a philosopher to have the privilege of writing a 'post-face' to this volume feels like being in the position of the owl of Minerva, which according to Hegel's well-known saying, 'only spreads its wings at the falling of dusk'. At the end of the day, when all has already been said and done, it is apparently up to someone not directly involved in the world of drip irrigation to wrap up the story and gauge its wider significance. I must confess that I am a complete stranger to 'waterland' or 'the water world', but I hope that my critical reflection on the contributions to this volume can offer a refreshing perspective on the truly impressive amount of admirable work that has been done. I should add, however, that my perspective is that of a sympathetic and charitable outsider, because I share many of the general interests and concerns that informed the research work reported in the chapters of this volume. As a philosopher, I have a special interest and some expertise in science and technology studies, global justice and social and political philosophy, but my primary object of study has been the modern life sciences (biotechnology, genomics, synthetic biology) – which offers a notable contrast to water engineering and drip irrigation. Most, if not all, of the authors in this volume appear thoroughly familiar with such approaches as actor-network theory (ANT), social construction of technology and the mutual shaping of technology and society. Many also explicitly deal with normative issues related to water justice and some have even dared to enter the arena of political philosophy by engaging, for example, in the debate on neo-liberalism. So there seems to be enough commonality to have a critical and fruitful discussion.

The seventeenth-century roots of modern narratives

The aim of this volume is to tell the 'untold stories' of drip irrigation and to critically engage with the discourses and narratives that have been spun around this subject, which are mainly about water savings and efficiency, modernization of agriculture, and poverty alleviation and development. For those acquainted with the history of Western philosophy, it is not difficult to recognize that these dominant narratives faithfully reflect the hopes and ambitions for humankind formulated already in the seventeenth century by such early-modern thinkers as Sir Francis Bacon and John Locke. According to Bacon, science and technology were destined to regain human dominion over nature ('knowledge is power') and give rise to practical applications that would alleviate the human predicament, promote health and well-being, and bring comfort and prosperity. Never short of hints and suggestions to put his program into action (advocating, among other things, a kind of biotechnology avant-la-lettre), he also on occasion discussed the judicial use of irrigation to improve agricultural yields. On his part, Locke argued that a social order based on private property would be ideally suited for the productive development of the world's natural resources. Although God had given the things of the earth to humankind in common, Locke held that He had nonetheless gifted humans with 'property in their persons' (self-ownership) and therefore entitled them to the work of their hands: 'As much land as a man tills, plants, improves, cultivates, and can use the product of, so much is his property. He by his labour does, as it were, enclose it from the common' (Locke, 1978 [1790]: 132). The object of property is (part of) 'the earth itself, as that which takes in and carries all the rest', not just the fruits and 'the beasts that subsist on it' (Locke, 1978 [1790]: 132), but presumably also the water resources that it carries. In a society without money, there would be a natural limit to the private appropriation of land, as people were not permitted by natural law to leave part of their harvest to waste and rot. Besides, according to the famous Lockean proviso, a person who carved out his own property from the commons had to leave 'enough and as good' to others. However, Locke argued that the introduction of money would relax these restrictions on appropriation and accumulation. In the end, he justified the ongoing 'enclosure of the commons' at home and the appropriation of 'uncultivated' overseas territories by European settlers, in the name of productive use and increased efficiency. By failing to properly cultivate the land and effectively letting it to 'waste', Locke opined, the native peoples of the Americas had actually forfeited their property rights. After all, it is human labour and industry, not 'unassisted Nature', that in his view accounted for the greater part of value creation.

Contemporary discussions on land grabbing and water grabbing, and the accompanying justifications invoking 'efficiency', 'waste' and 'scarcity', do not seem too far removed from the themes already broached in earlier debates during Locke's time. Marxist critics of Locke's 'possessive individualism' have later pointed out the subtle dialectical twist in his reasoning: starting out from the assumption that the right to property is grounded on labour, he ended up with legitimating the process of capital accumulation in which property ownership provides the opportunity to exploit other people's labour. Similarly, starting from the theological premise that God had given the earth to humankind in common, he nonetheless justified the privatization of the commons (for the sake of alleged greater efficiency) and the 'accumulation by dispossession' this entails. Those who would be left out in the cold (in apparent violation of the Locke's own proviso), could still improve their lot by seeking employment with the property owners who had been more alert in securing possessions. Indeed, Locke claimed, rather tendentiously, that an Amerindian chief 'feeds, lodges and is clad worse than a day labourer in England' (Locke, 1978 [1790]: 136). The implied suggestion was that the material situation of the English workers who had been deprived from their access to land through the enclosure of the commons had ultimately improved due to the increased efficiency in the exploitation of natural resources resulting from the enclosures. Needless to say that this is a highly contested claim. The consequences of European colonization for the native peoples of overseas continents may have been even worse.

The presumed historical mission (the 'white man's burden') to promote the productive and efficient use of the earth's natural resources also links up with the themes of modernization, development and technological progress. Writing more than half a century before Locke, in 1620, Bacon already claimed that it was technology ('the arts'), not soil, climate or race, that set the 'civilized' Europeans apart from the inhabitants of 'the wildest and most barbarous districts of New India' – creating a difference so big that it would justify the saying that 'man is a god to man' (Bacon, 1858 [1620]: paragraph CXXIX). Thus technology marks the distinction between 'barbarity' and 'civilization' and is therefore the true sign of modernity: 'Modernity is associated with rationality, empiricism, efficiency and change [. . .] Modern men [. . .] make use of sophisticated technology to remake their environment and change their social systems in ways intended to advance both their own careers and the development of their societies as a whole' (Adas, 1989: 413).

Drip irrigation as an elusive object

How to tell the 'untold story' of drip irrigation? How to escape from the enchantment of the dominant discourses and narratives that are all predicated on the great 'potential' of this 'promising' technology? As the literature in the sociology of expectations shows, a strong faith in the 'promises' of a new technology is generally a crucial factor in mobilizing and sustaining social support. Such beliefs would be inappropriate, however, for the analysts trying to understand the developmental path of the technology. Hence, the research strategy chosen by many contributions to this volume is to put the promises of drip irrigation within brackets and to focus instead on its actual use in a variety of practical contexts. This does not deny that if we want to understand the attractiveness of the technology for various actors, we need to delve deeper into the promises it offers them.

A characteristic and highly consequential tenet of most chapters in this volume is that drip irrigation is not taken as a stable material object with inherent and well-defined technical attributes, in contrast to the conventional engineering view. Instead, as Chapter 2 argues, the technology is to be considered as something that comes into being in and through day-to-day sociomaterial practices (including engineering practices). The implication is that drip irrigation will be different things in different places, given the variety of these practices in different parts of the world. It also means that the technology becomes a rather elusive 'fluid' object. This approach is reminiscent of the so-called translation model that Bruno Latour set out in his book *Science in Action* (Latour, 1988). Latour opposed what he called the diffusion model, according to which an artifact moves through social space as if driven by its own momentum. Instead, he proposed the 'translationist' view which emphasizes that the fate of an artifact is in the hands of the users, where each next actor along the road may in principle change and modify the artifact or fail to take it up at all. Yet, Latour also argued that engineers ('technoscientists') will often try to give their products the character of black boxes which may circulate through networks without alteration. So translation (i.e. displacement with modification) is the (default) rule, but there are also important exceptions to the rule.

The critique of the diffusion model is highly relevant for the debate on the success or failure of drip irrigation in sub-Saharan Africa. As the Israeli researchers Garb and Friedlander argue, the mostly failed uptake of the technology in this geographical area can be seen as a consequence of 'the transfer of static physical artifacts into new contexts lacking similar local systems [as in Israel] into which these could be absorbed and evolve (re-innovated)' (Garb and Friedlander, 2014: 13). By the same token, as Mohamed Naouri and his co-authors argue in Chapter 16, the successful adoption of drip irrigation in Algeria (in contrast to the failed take-up in sub-Saharan Africa) can be traced to the substantial adaptations and re-engineering of the original technical designs undertaken by a farmer-led innovation system. Interestingly, Garb and Friedlander also discuss two strategies that seem to follow Latour's black-boxing approach in that they attempt to make drip irrigation installations more self-contained and thereby to reduce their dependence on a supportive local context, but these strategies do not appear to be successful either. These same authors also point out that later generations of Israeli drip irrigation engineers lost the awareness that their irrigation systems were highly sensitive to the local innovation context, as the latter gradually came to be taken for granted and the technology came to be identified with the material system alone (Garb and Friedlander, 2014: 18). Perhaps such 'context forgetfulness' is characteristic for an engineering approach more generally.

Engineers also find it sometimes difficult to accept the widespread phenomenon of bricolage, whereby farmers and others reconfigure and tinker with the technical designs and standardized equipment for drip irrigation to make them work under local conditions and adapt them to special needs and purposes. Farmers do not shy away from opening up the black boxes that engineers had put together so carefully and with so much effort. As Chapter 15 explains, engineers often see bricolage as a synonym for amateurism, leading to systems that deviate from technical specifications and norms. They are nonetheless confronted with a dilemma: either allow for local adaptations so that the system will actually work (and run the risk of losing the respect of your fellow engineers), or stick to the ideal models of the official design procedures (thus, maintaining the respect of your peers), but at the price of accepting that the system will not work in practice. What may be operative here is a certain professional ideology of engineering that is especially strong in France and francophone countries.

Indeed, when back in the 1960s the French anthropologist Claude Lévi-Strauss introduced the famous distinction between the 'bricoleur' and the 'ingénieur' as representing two contrasting modes of thought (Lévi-Strauss, 1962), he was undoubtedly influenced by the strongly mathematical-rationalist conception of engineering that has prevailed in France since the Enlightenment (Alder, 1997). What this distinction obscures is that in practice the creative work of an engineer rarely if ever lives up to the professional ideal of rational design and actually contains many elements that can be more properly characterized as bricolage or tinkering. The same distinction has also been used to describe the differences between the do-it-yourself biology of 'biohackers' and the synthetic biology of academic and industrial researchers, but here too the contrast seems rather exaggerated (Keulartz and van den Belt, 2016). Officially, synthetic biology as the avowed attempt to make biology 'easy to engineer' through standardization subscribes to an ideology of rational design, but the discipline does not and cannot live up to this ideal:

> Synthetic biology's design processes always, so far, end up as iterative rounds of trial, error, and pragmatic solutions – sometimes referred to as 'debugging', 'tweaking', 'retrofitting', or 'parameter tuning' – to make systems behavior fit design specifications [. . .]. Rather than exemplifying rational, elegant, and efficient design, many devices work because they are kludges.
>
> (O'Malley, 2009: 382)

Something similar undoubtedly applies to the designs of irrigation engineers. One wonders what engineers would lose if they abandoned their professional ideology of rational design.

Efficiency and water savings

Drip irrigation is widely held to be the obvious and straightforward answer to the global problem of increasing water scarcity. As a technology that is deemed to be inherently efficient, it is supposed to have a huge potential for saving water. Contributions to this volume call this dominant narrative in the contemporary discourse on drip systems into question.

'Efficiency' is a tricky concept that must be handled with care. When the notion is used in engineering, the general idea is to get the most of a particular desired effect out of a given input or alternatively to realize the desired effect with a minimal expenditure of the input. The concept is especially used for purposes of comparison. The objective is to rank the merits of different methods or techniques by comparing their respective 'input-output ratios'. But many pitfalls are looming here. For one thing, focusing on a single output and a single input variable figuring in a proposed efficiency ratio might mean that other relevant variables are unduly ignored and neglected. 'To be efficient in one respect [. . .], often requires being inefficient in others [. . .]' (Tiles and Oberdiek, 1995: 52). Farmers, for example, may be concerned not only with achieving 'more crop for the drop' but also with minimizing labour input or extending the area of cultivation. Efficiency is thus dependent on the context and the perspectives of the actors. The concept must therefore be taken in a much less absolute sense than is normally done. As the American economist Sumner Schlichter already wrote in 1937 in an encyclopedic entry on this notion: 'There is no such thing as efficiency in general or efficiency as such – there are simply a multitude of particular kinds of efficiency. Actions and procedures which are efficient when measured with one measuring stick may be inefficient when measured with a different measuring stick' (Schlichter, 1937: 439; quoted in Mitcham, 1994: 338).

When discussing the 'efficiency' of various irrigation systems, it is also important to point out that the concept is also sensitive to the *scale* of the system under consideration. It matters whether we are talking about the efficiency of water use at the level of the plot, the farm, the river basin or the water shed (van der Kooij et al., 2013). Water 'wasted' or 'lost' by one farmer may be captured and reused by other farmers. One person's loss is another person's gain. This means that the common-sense and seemingly sound reasoning that if individual farmers use water-efficient irrigation technologies and thereby save water (for themselves), they will also *ipso facto* save water as a group in the aggregate is flawed. The reasoning error here is known as the fallacy of composition. This type of fallacy also plays a role in the well-known paradox of saving (or paradox of thrift) in economics. If each of the citizens of a country decides to increase his or her savings, the final result may well be that aggregate savings at the national level will remain constant and national income be reduced. Ironically, the widely held confident belief in the huge water-saving potential of drip irrigation as an answer to the global water crisis may be partly based on the fallacy of composition.

The authors of Chapter 4 argue that one of the key attractions of introducing modern irrigation systems with efficient technologies is precisely that it allows a multiplicity of actors to pretend as if they are playing a non-zero sum game. This seems to be happening in the Ain Bittit area in Northern Morocco. By treating the water 'savings' to be realized through efficient irrigation as if they were net additions of water to the system, they can be assigned to new users and new uses from the start, without anybody having to pay the price for this new arrangement – or so it seems. The whole process involves a de facto reallocation of water, but this reallocation is de-politicized through the use of technology. Farmers who lose access to water sources downstream refrain from complaining because any form of protest would inevitably stigmatize them as backward opponents of modernization (apart from the difficulty of definitively establishing a causal link between their loss of access to water and upstream events). It is also notable that the entire transition to modern irrigation systems depends on maintaining a tacit conspiracy of willful ignorance. As the authors write: 'A water meter on a tube-well would make the water savings tangible, leading to discussions about how the savings should be divided, *which is exactly something that is carefully avoided*' (my italics). The case thus represents an outstanding example of the social production of ignorance, a theme that is currently popular in the literature on 'agnotology' and the sociology of ignorance (Proctor and Schiebinger, 2008). This same theme of willful ignorance is further pursued in Chapter 5, which explores the reasons for the effective conspiracy of silence among so many parties resisting the almost obvious conclusion that the widespread use of drip irrigation systems exercises increasing pressure on groundwater resources.

In Chapter 6, Lisa Bossenbroek links the reallocation of water and the seizing of groundwater in Morocco's Saïss Region with the critical literature on land and water grabbing in the neo-liberal era. We are witnessing indeed a new round in the enclosure of the commons. Bossenbroek and her co-authors highlight the technological, material and discursive aspects of this process. In their retelling, the story literally starts with the enclosure or fencing off of a piece of land that has been bought by an outside investor. The land acquisition is immediately followed by the drilling of deep well-tubes affording access to remote reservoirs of groundwater. The next step is the installation of drip irrigation to serve the cultivation of commercial crops. Water flows are physically 'enclosed' in technology and kept hidden from view by disappearing underground. There is also a discursive enclosure. The authors point out that the dominant modernization narrative portrays existing uses of water resources (by farmers) as 'inefficient', 'underutilized' and 'below potential' – hence, it is allegedly no social loss when traditional users are forced to terminate their farming activities, leaving the field to purportedly more efficient modern 'entrepreneurs'. There is not that much difference with the justification John Locke offered at the end of the seventeenth century for the enclosure of the commons at home and the colonization of the overseas territories of native peoples abroad.

Drip irrigation and poverty alleviation

There is no better proof of the widespread and deeply held belief in the benign potential of drip irrigation than the enthusiastic and sustained attempt by international donors, NGOs, firms and other parties to develop and promote low-cost versions of this technology for smallholders in sub-Saharan Africa, Asia and Latin America. Given the high expecations for drip irrigation in general, it would presumably be a moral failure if we withheld this blessing from the poor farmers in the developing world. It must be admitted that commercial high-pressure drip systems are generally unaffordable and unsuitable to smallholders, but the low-cost low-pressure drip irrigation kits that have been developed might perhaps offer better opportunities. In fact, the latter have been invested with high hopes and expectations. Drip irrigation kits, it was surmised, could raise rural incomes and save water in times of drought, provide food security, lift people out of poverty and even support the empowerment of women. In short, they could become a veritable game changer. Over the last decades, several developmental NGOs backed by international donors have acted on these expectations by continuously promoting low-pressure drip irrigation among smallholders in developing countries. There is only one major drawback. This entire effort has largely been an exercise in futility. Evidence showing sustained adoption of drip irrigation kits by smallholders is lacking, yet developmental NGOs remain undeterred and continue to invest in their programs. So the task for the analyst is not only to explain why there has been so little adoption of low-cost drip irrigation, but also to explain, as Doug Merrey suggests in Chapter 14, why NGOs and international donors have been so 'persistent' in promoting this technology.

The colourful story of the failed take-up of low-pressure drip systems in Burkina Faso is told in Chapter 13. Jonas Wanvoeke *et al.* explain the apparent contradiction between sustained development enthusiasm and limited adoption and use by farmers from the interests of the development advocacy coalition, which defines its own metrics of success in terms of the number of systems disseminated or sold rather than actually used. As an ethnographer Wanvoeke visited the smallholder sites in Burkina Faso where drip irrigation was supposed to be in operation and found that women there were instead using buckets and hoses to irrigate their plots. The official experimental sites, by contrast, were a kind of *Potemkin village* where the success of the technology could be shown to international visitors ('a bill board to advertise the success of the projects'). Somewhat cynically, Wanvoeke and his co-authors argue that for the enthusiastic development coalition smallholder drip irrigation is indeed 'successful'. For Doug Merrey, such a conclusion seems a little too cynical, but nonetheless he seriously entertains what he calls the 'vested interest hypothesis' endorsed by Wanvoeke *et al.* With characteristic academic caution he writes: 'One wonders whether there is active resistance to assessing [the impacts of drip irrigation kits]. The absence of such impact assessments after over a decade of pilot testing and dissemination of free kits, combined with hints from colleagues working with and within donor programs, suggests this is a possibility'. So we may well be confronted here with another striking example of agnotology or the social production of ignorance.

It might be useful to put the story of the failed uptake of low-pressure drip irrigation in Burkina Faso and the rest of sub-Saharan Africa in a wider, comparative perspective, as it is by no means unique. Anthropologist Carol Hunsberger tells a similar story about the vicissitudes of the oilseed shrub *Jatropha curcas* in Kenya (Hunsberger, 2014). During the first decade of the new millennium, reflecting a general atmosphere of high expectations for biofuels as a climate-neutral and sustainable solution to our energy problems, NGOs were glad to be able to offer what looked like the poor man's (or poor woman's) version of this source of clean energy. As Hunsberger points out, NGO promotors of *Jatropha* reinforced an optimistic discourse, defended it against dissent and linked the oilseed shrub to global, national and local goals. While busily distributing *Jatropha* seeds among Kenyan smallholders, NGOs continued to see their promotional campaign as a success, but in the end no usable drop of oil was ever obtained from the shrub's berries (here too the metric of success was in terms of the number of seeds disseminated rather than the number of shrubs actually used as sources for oil). Hunsberger speaks of the 'economy of appearances' to explain the persistence of NGO's beliefs.

The wider issue here is the always-in-principle-problematic definition of success or failure and the sociology of expectations. In this regard one is reminded of what Thomas Kuhn said about new scientific paradigms: 'The success of a paradigm [. . .] is at the start largely a promise of success [. . .]' (Kuhn, 1970: 23). The same applies to technological innovations. So this inevitably introduces a large measure of ambiguity: how long do we have to wait before the promise of success will finally materialize or before we may legitimately conclude that it most probably won't pan out? The whole issue also tangentially touches the wider debate on the effectiveness of development aid (expressed in two catchy names: Jeffrey Sachs versus William Easterly) and the new fashion of 'effective altruism' in organized philanthropy with a much stronger emphasis on well-chosen metrics of success. In any case, it is no secret that developmental NGOs rarely if ever set up systematic and independent evaluations of their aid projects (Wenar, 2011).

The politics of water governance

Several contributions to this volume show that the implementation of drip irrigation in farmer's fields is closely associated with the social construction of identities. Farmers who install drip systems are considered 'modern' and 'entrepreneurial', while those who fail to do so are often branded 'traditional' and 'backward'. In Zambia, according to Gert Jan Veldwisch *et al.* in Chapter 12, the social enterprise iDE (International Development Enterprises) frames the target group to which it aims to bring low-cost drip kits as consisting of agricultural entrepreneurs looking for profit opportunities on commercial markets for whom access to water is a major bottleneck − a neo-liberal framing that fits only a small part of the poor rural population and does not sufficiently recognize the formidable structural obstacles to lifting oneself out of poverty through one's own efforts. Moreover, this framing also has definite gender effects, as it limits the adoption of drip systems by women.

Such constructions of identities are by no means innocuous and inconsequential, though on occasion they might be challenged and contested. Chapter 6 suggests rather ironically that the outside investors who acquire land and install deep tube-wells and irrigation systems in the Saïss region in Morocco, and who are so much celebrated and glorified as 'entrepreneurs' by the agricultural policymakers, may not be at all such undisputed paragons of the entrepreneurial virtues of efficiency, productivity and innovative creativity for which they are commonly held. In fact, their investments are channelled into the heavily subsidized routine production of fruit and grapes leading to saturation of domestic markets. Nonetheless, as the authors state, 'performing an entrepreneurial identity helps to access and secure water'. Water resources are thus diverted away from persons with the wrong identity ('backward peasants') toward persons with the right identity ('modern entrepreneurs'). The stakes are apparently set for opening up a new arena of identity politics.

Nowhere are the stakes so high as in various countries of Latin America where neo-liberalism appears to have become the grim political reality of daily life. In Chile, according to Daniela Henriquez *et al.* in Chapter 7, the state-driven implementation of drip irrigation even amounts to a 'depeasantization process'. In order to qualify for state support for drip irrigation (necessary to prove the 'efficiency' of their farming operations), farmers must effectively become small entrepreneurs.

In Chapter 8 Jeroen Vos and Anaïs Marshall cite Peruvian legislation, which gives preference in the granting of new water rights to 'users or operators of hydraulic infrastructure that generates surpluses of water resources and who have a certificate of efficiency' (apparently, surpluses are miraculously created *ex nihilo*). Yet the Chavimochic project in the dry North Coast of the country is not meant to save water, but to 'conquer the desert' or extend the agricultural frontier. Large tracts of land were auctioned off to big companies for growing luxury export crops like asparagus and avocado, after government investments had prepared the irrigation infrastructure of the entire area (thereby granting the companies a subsidy of 85%). The companies can earn their 'certificate of efficiency' by implementing drip irrigation. The small farmers in the older irrigation areas, by contrast, are stigmatized for using 'inefficient' methods. They are also blamed for the problems of waterlogging and salinization occurring in their areas as a consequence of more water use and deficient drainage in the entire system. The Chavimochic project may help sustain Peru's agricultural export boom, but its costs and benefits are unevenly distributed over the Peruvian population (and indeed between Peru and the developed countries where the multinational companies are headquartered). The authors also duly note such external costs of the project as water pollution by agrochemicals and the loss of biodiversity.

The experiences with drip irrigation in the Mexican state of Guanajuato as reported in Chapter 9 hardly make for more cheerful reading. Jaime Hoogesteger shows that a similar blaming game is going on here. While the state subsidy schemes for drip irrigation have disproportionately benefited the large capitalized, export-oriented elite farmers, who are also for the most part responsible for the overexploitation of groundwater, it is the small farmers (*ejidatarios*) who are blamed for being traditional and 'inefficient' groundwater users blocking the further development of agriculture. It is a 'debate in which commercial producers wash their hands under the claim that they use the [groundwater] resource efficiently because of the use of drip irrigation'.

Drip irrigation thus figures prominently in the politics of water management. As water is a finite resource, the way the steadily increasing demands for this resource are governed and regulated assumes ever greater importance. In more and more countries today, water governance is subjected to neo-liberal policies that claim to allocate this scarce resource in an economically efficient manner by extending the operation of the market. That does not mean that the role of the state is fading away, far from it! As Peru's water legislation shows so clearly, it is the state that regulates the distribution of water resources by selling new user rights to newly created 'surpluses' to the highest bidders, ostensibly without harming the interests of existing water users as the 'surpluses' are seemingly created *ex nihilo*. Just like in John Locke's seventeenth-century scenario, the granting of water rights is commonly justified by appealing to productive use and efficiency, but the famous proviso that he added, to leave 'enough and as good' to others, remains blatantly unfulfilled.

References

Adas, M. (1989). *Machines as the measure of men: Science, technology, and ideologies of Western Dominance*. Ithaca, NY: Cornell University Press.

Alder, K. (1997). *Engineering the revolution: Arms and enlightenment in France, 1763–1815*. Chicago: The University of Chicago Press.

Bacon, F. (1858 [1620]). *Instauratio Magna*. Trans. J. Spedding et al. *The Works of Francis Bacon* (Vol. IV). London: Longman and Co.

Garb, Y., and Friedlander, L. (2014). From transfer to translation: Using systemic understandings of technology to understand drip irrigation uptake. *Agricultural Systems* 128: 13–24.

Hunsberger, C. (2014). Jatropha as a biofuel crop and the economy of appearances. *Review of African Political Economy* 41: 216–231.

Keulartz, J., and Belt, H. van den. (2016). DIY-Bio – economic, epistemological and ethical implications and ambivalences. *Life Sciences, Society and Policy* 12(7): 1–19.

Kuhn, T. (1970). *The structure of scientific revolutions*. Chicago: The University of Chicago Press.

Latour, B. (1988). *Science in action: How to follow scientists and engineers through society*. Cambridge MA: Harvard University Press.

Lévi-Strauss, C. (1962). *La Pensée Sauvage*. Paris: Librairie Plon.

Locke, J. (1978 [1790]). *Two treatises of government*. London: Everyman's Library.

Mitcham, C. (1994). *Thinking through technology: The path between engineering and philosophy*. Chicago: The University of Chicago Press.

O'Malley, M. (2009). Making knowledge in synthetic biology: Design meets kludge. *Biological Theory* 4(4): 378–389.

Proctor, R.N., and Schiebinger, L. (2008). *Agnotology: The making and unmaking of ignorance*. Stanford, CA: Stanford University Press.

Schlichter, S.H. (1937). Efficiency. In E.R.A. Seligman and A. Johnson (Eds.), *Encyclopedia of the Social Sciences* (Vol. 3). New York: Macmillan, p. 439.

Tiles, M., and Oberdiek, H. (1995). *Living in a technological world: Human tools and human values*. London: Routledge.

Wenar, L. (2011). Poverty is no pond: Challenges for the affluent. In P. Illingworth, T. Pogge, and L. Wenar (Eds.), *Giving well: The ethics of philanthropy*. Oxford: Oxford University Press, pp. 104–132.

Index